Digital Signal Processing using Arm Cortex-M based Microcontrollers

Digital Signal Processing using Arm Cortex-M based Microcontrollers

Theory and Practice

arm Education Media

arm Education Media

Arm Education Media is an imprint of Arm Ltd. 110 Fullbourn Road, Cambridge, CBI 9NJ, UK

Notices
Knowledge and best practice in this field are constantly changing. As new research and experiences broaden our understanding, changes in research methods, professional practices, or medical treatment may become necessary.

Practitioners and researchers must always rely on their own experience and knowledge in evaluating and using any information, methods, compounds, or experiments described herein. In using such information or methods they should be mindful of their own safety and the safety of others, including parties for whom they have a professional responsibility.

To the fullest extent of the law, neither the Publisher nor the authors, contributors, or editors, assume any liability for any injury and/or damage to persons or property as a matter of product liability, negligence, or otherwise, or from any use or operation of any methods, products, instructions, or ideas contained in the material herein.

ISBN: 978-1-911531-16-6

British Library Cataloguing-in-Publication Data
A Catalogue record for this book is available from the British Library

Library of Congress Cataloging-in-Publication Data
A Catalog record for this book is available from the Library of Congress

Exercise solutions are available online via the Arm Connected Community: visit https://community.arm.com and search for the book title. You may need to create an account first.

For information on all Arm Education Media publications, visit www.arm.com/education

To report errors or send feedback, please email edumedia@arm.com.

To our families

Contents

Preface

We live in a digital world surrounded by signals. Therefore, understanding and processing these signals is one of the most important skills a graduating or postgraduate engineer can possess. Arm Cortex-M based microcontrollers provide a low-cost and powerful platform for this purpose. This book aims to introduce the basics of digital signal processing on these microcontrollers and the theory behind it and includes a set of labs that handle the practical side of digital signal processing. We believe that this book will help undergraduate and postgraduate engineering students to bridge the gap between the theoretical and practical aspects of digital signal processing, so they can grasp its concepts completely. As a result, they will be ready to apply digital signal processing methods to solve real-life problems in an effective manner. This will be a valuable asset in today's competitive job market.

About This Book

This book is not purely about the theory of digital signal processing, nor is it solely about the practical aspects of digital signal processing. We believe that it is not possible to implement a DSP algorithm without knowing its theoretical merits and limitations. Moreover, theory is not sufficient alone to implement a DSP algorithm. Therefore, we cover DSP theory in this book, and then we explore these theoretical concepts through practical applications and labs. The aim is to bridge the gap between theory and practice.

In this book, we assume that you have some basic knowledge of signal processing. Therefore, theory is only provided when necessary and we do not go into great detail. However, we do provide a further reading section at the end of each chapter, where we provide references that are relevant to the concepts covered in that chapter. These may help you to extend your knowledge further.

The topics covered in this book are as follows. Chapter 2 introduces the mathematical and practical basics that will be used throughout the book. Chapter 3 discusses the Z-transform used to analyze discrete-time signals and systems in the complex domain. It will become clearer that some analyses will be easier to perform in this domain. In relation to this, Chapter 4 introduces frequency domain analysis in discrete time. In this chapter, you will learn how to analyze a discrete-time signal and system in the frequency domain. This will also uncover a method for designing and implementing filters in the frequency domain. Chapter 5 covers analog–digital conversions. These concepts are vital in understanding the relationship between analog and digital signals. Chapter 6 introduces methods of processing analog signals in a digital system. This chapter also focuses on methods of representing an analog system in digital form. Chapter 7 covers ways of representing a discrete-time system in different structural forms. Chapter 8 covers filter design, which is an important topic in digital signal processing. Chapter 9 focuses on adaptive signal processing concepts. This is one of the strengths of digital systems because their parameters can be easily changed on the fly. Chapter 10 explores fixed-point number representation issues. In this chapter, we observe that these issues have a direct effect on hardware usage, computation load, and the obtained result. Chapter 11 covers real-time digital signal processing concepts, which are extremely important for real-life applications with timing constraints.

A Note about Online Resources

This book includes a number of online resources that are accessible to readers of both the print and ebook versions, including answers to the end-of-chapter exercises and lab tasks, as well as code and other materials useful for the lab tasks. To access these resources, please visit the Arm Connected Community, https://community.arm.com/, and search for the book title. You may need to create an account first.

Acknowledgments

We would like to thank Arm Education for encouraging us to develop this book.

Author Biographies

Cem Ünsalan
Marmara University

Dr. Cem Ünsalan has worked on signal and image processing for 18 years. After receiving a Ph.D. degree from The Ohio State University, USA in 2003, he began working at Yeditepe University, Turkey. He now works at Marmara University, Turkey. He has been teaching microprocessor and digital signal processing courses for 10 years. He has published 20 articles in refereed journals. He has published five international books and holds one patent.

M. Erkin Yücel
Yeditepe University

M. Erkin Yücel received his B.Sc. and M.Sc. degrees from Yeditepe University. He is pursuing a Ph.D. degree on embedded systems at the same university. He has guided microprocessor and digital signal processing laboratory sessions for three years. Currently, he is working in research and development in industry.

H. Deniz Gürhan
Yeditepe University

H. Deniz Gürhan received his B.Sc. degree from Yeditepe University. He is pursuing a Ph.D. degree on embedded systems at the same university. For six years, he has been guiding microprocessor and digital signal processing laboratory sessions. He has published one international book on microcontrollers.

Digital Signal Processing Basics

Contents

1.1 Introduction

The aim of this book is to introduce you to the concepts of digital signal processing (DSP). To do so, the first step is explaining what a signal is and how to process it using a system. This chapter briefly introduces these concepts and explains why digital signal processing is important in today's world.

1.2 Definition of a Signal and Its Types

We perceive the world around us through our sensory organs. When an event occurs, a physical effect is generated. A sensor in our body receives this effect, transforms it into a specific form, and sends the result to our brain. The brain processes these data and commands the body to act accordingly. An electronic system works in a similar manner. There are various sensors (such as microphones, cameras, and pressure sensors) that transform a physical quantity into electronic form. If a sensor output changes with time (or another variable), we call it a *signal.* More generally, we can define a signal as data that change in relation to a dependent variable. A signal can be processed by a *system* to either obtain information from it or to modify it.

1.2.1 Continuous-Time Signals

If an acquired signal is in analog form, then we call it a continuous-time signal. A system that processes a continuous-time signal is called a continuous-time system. Assume that we form an analog microphone circuitry to display audio signals on an ***oscilloscope*** screen. If we say the word "HELLO," we will see a continuous-time signal on the screen as shown in Figure 1.1. This signal can be further processed by a continuous-time system composed of analog devices.

> **oscilloscope**
> *A laboratory instrument commonly used to display and analyze the waveform of electronic signals.*

1.2.2 Discrete-Time Signals

Although processing a continuous-time signal with a continuous-time system may seem reasonable, this is not the case for most applications. An alternative method is to sample the signal discretely. This corresponds to a discrete-time signal. Let us take a section of a continuous-time signal as in Figure 1.2(a). Taking 44,100 samples per second from it, we

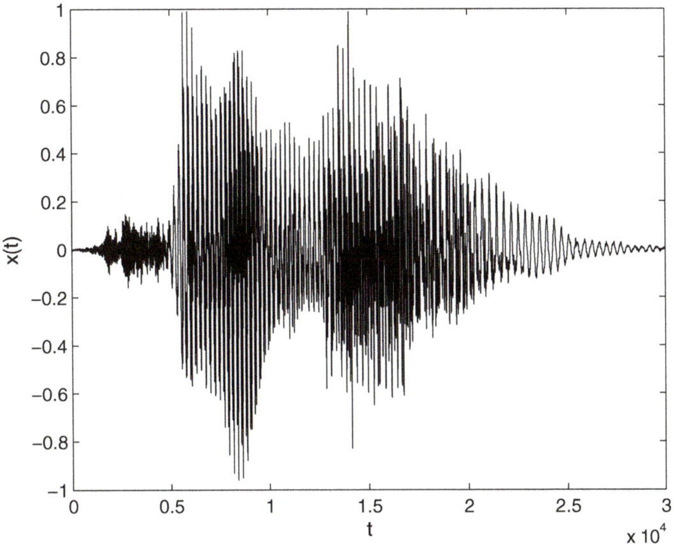

Figure 1.1 The word "HELLO" displayed as a continuous-time signal.

can obtain the corresponding discrete-time signal shown in Figure 1.2(b). In this case, the ***amplitude*** of each sample is real-valued.

> **amplitude**
> *The maximum deviation of a quantity from a reference level.*

1.2.3 Digital Signals

In a digital system (such as a microcontroller), a discrete-time signal can be represented as an array, where each sample in the signal is an array entry. This array cannot hold real values and can only store values with limited range and resolution. Think of an `integer` array in the C language. Each array entry can only be represented by a certain number of bits. This will also be the case if the array is composed of ***float*** or `double` values. Therefore, the amplitude of the discrete-time signal should be ***quantized***. The resulting signal is called a digital signal. Figure 1.3 shows the digital signal for the same section of signal we used in Figure 1.2. In this case, the signal is stored in a `float` array.

> **quantization**
> *The process of mapping sampled analog data into non-overlapping discrete subranges.*

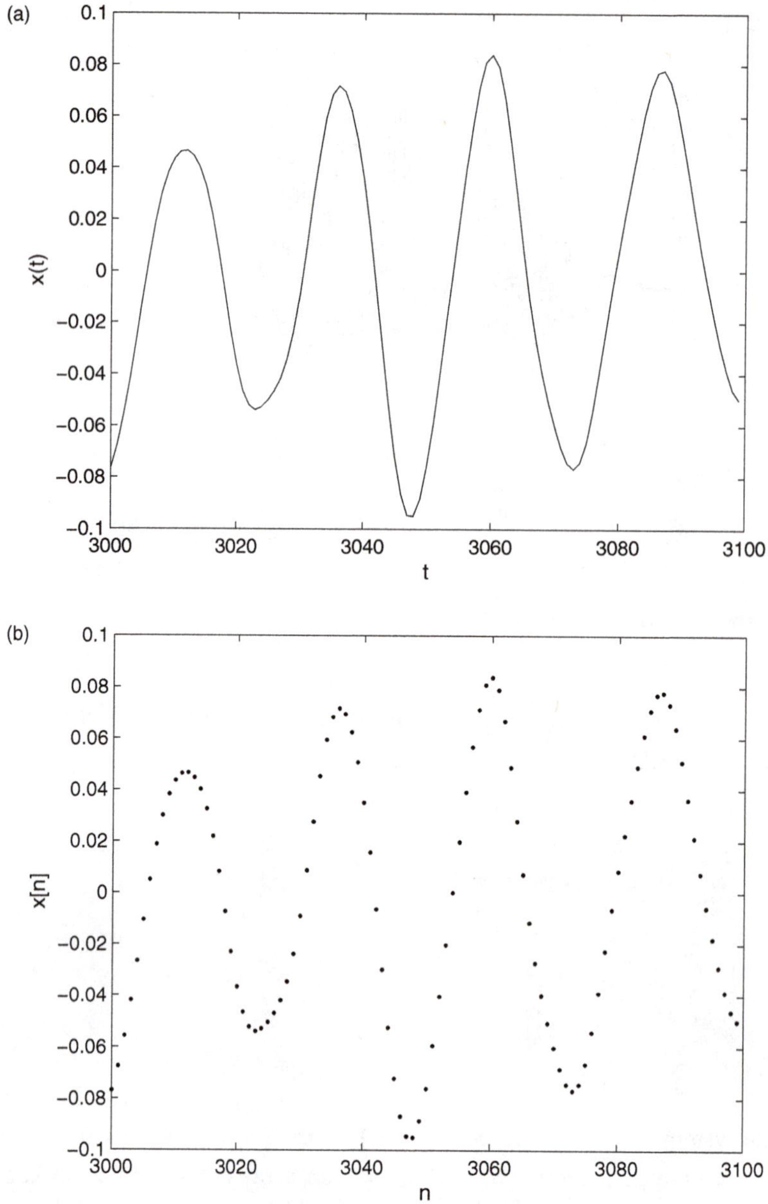

Figure 1.2 Part of the continuous-time signal and its sampled discrete-time version. (a) Continuous-time signal; (b) Discrete-time signal.

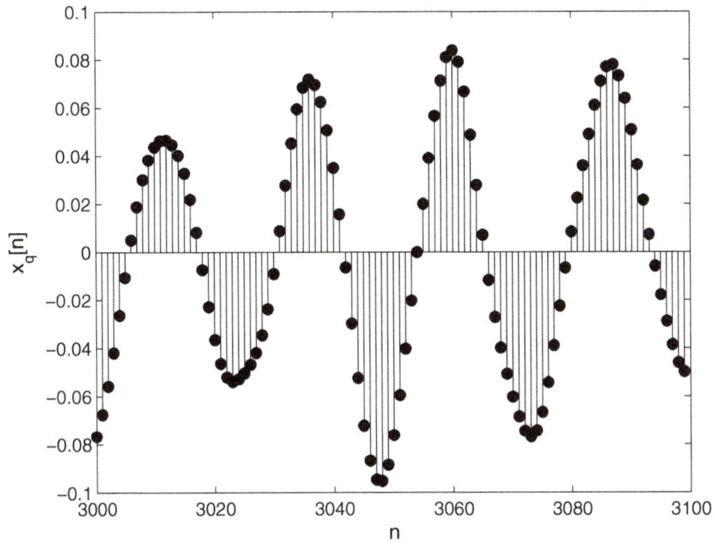

Figure 1.3 The digital signal.

float
The keyword for a floating-point data type with single precision.

Discrete-time signals and systems are preferred because they simplify the mathematical operations in theoretical derivations. On the contrary, digital signals and systems are necessary for implementation. This means that books focusing on the theoretical aspects of signal processing use discrete-time signal and system notations. Books on signal processing hardware and software use digital signal processing notations. Because this book aims to bridge the gap between theory and practice, we use the terms discrete time and digital interchangeably. To be more precise, we call a signal (or system) discrete-time when we explain its theoretical aspects, and we call it digital when we implement it.

1.3 Digital Signal Processing

In the past, analog systems were the only option for processing continuous-time signals. With the introduction of powerful hardware, digital signal processing has become more popular. Arm's Cortex-M based microcontrollers provide a low-cost and powerful platform for using DSP in practical applications. This book focuses on the theoretical and practical aspects of DSP concepts on Arm Cortex-M based microcontrollers.

The beauty of DSP is in its implementation because fixed analog hardware is generally not required. Instead, a suitable microcontroller is often sufficient for implementation. In fact, the DSP system will simply be a code fragment, which has several advantages. First, if the system does not satisfy design specifications, it can be redesigned by simply changing the relevant code block, and there is no need to change the hardware. Second, system parameters can be changed on the fly, which is extremely important for adaptive and learning systems. Moreover, if the microcontroller has communication capabilities, these parameters can be modified from a remote location. The abovementioned advantages make DSP more advantageous than continuous-time signal processing via analog equipment.

1.4 Lab 1

1.4.1 Introduction

The STM32F4 Discovery kit is the main hardware platform used in this book. It has a microcontroller, which is based on the Arm Cortex-M4 architecture. This lab introduces the properties of the kit and the microcontroller. While doing this, standard software packages will be used to simplify operations. Later on, the "AUP audio card"[1] will be introduced for use in audio processing. Being familiar with the properties of the discovery kit and the audio card will help in implementing and running the codes given.

1.4.2 Properties of the STM32F4 Discovery Kit

The STM32F4 Discovery kit has an STM32F407VGT6 microcontroller on it. In this section, we briefly introduce the microcontroller and the architecture of the kit.

The STM32F407VGT6 Microcontroller

The STM32F407VGT6 microcontroller is based on the 32-bit Arm Cortex-M4 architecture, which has ***Reduced Instruction Set Computing (RISC)*** structure. The microcontroller uses a decoupled three-stage pipeline to separately ***fetch***, ***decode***, and ***execute*** instructions. To be more specific, this operation can be summarized as follows. While the first instruction is being executed, the second one is decoded, and the third one is fetched. This way, most instructions are executed in a single CPU cycle. Related to this, with the help of pipelined RISC structure, Cortex-M4 CPU cores can reach a processing

1. Audio card from the Arm University Program

power of 1.27 and 1.25 *DMIPS*/MHz with and without the floating-point unit (FPU), respectively.

> **Reduced Instruction Set Computing (RISC)**
> *A microprocessor design strategy based on the simplified instruction set.*

> **fetch**
> *The first step of a CPU operation cycle. The next instruction is fetched from the memory address that is currently stored in the program counter (PC) and stored in the instruction register (IR).*

> **decode**
> *The second step of a CPU operation cycle, in which the instruction inside the instruction register is decoded.*

> **execute**
> *The last step of a CPU operation cycle in which the CPU carries out the decoded information.*

> **DMIPS (Dhrystone Million Instructions per Second)**
> *A measure of a computer's processor speed.*

The Cortex-M4 CPU core uses a ***multiply and accumulate (MAC) unit*** for fixed-point calculations. This unit can execute most instructions in a single cycle with 32-bit data multiplication and hardware divide support. The hardware FPU can carry out floating-point calculations in a few cycles, compared to hundreds of cycles without the FPU. The FPU supports 32-bit instructions for single-precision (C float type) data-processing operations. It has hardware support for conversion, addition, subtraction, multiplication with optional accumulate, division, and square root operations. The Cortex-M4 CPU core has an additional set of instructions to perform parallel arithmetic operations in a single processor cycle for DSP-specific applications.

> **multiply and accumulate unit**
> *A microprocessor circuit that carries a multiplication operation followed by accumulation.*

The Cortex-M4 CPU core also has a ***nested vectored interrupt controller (NVIC)*** unit, which can handle 240 interrupt sources. The NVIC, which is closely integrated with the processor core, provides low latency interrupt handling. It takes 12 CPU cycles to reach the first line of the interrupt service routine code. Additionally, the Cortex-M4 structure has a ***memory protection unit*** (MPU), a ***wake-up interrupt controller*** (WIC) for ultra-low power sleep, instruction trace (ETM), data trace (DWT), instrumentation trace (ITM), serial/parallel debug interfaces for low-cost debug and trace operations, an ***advanced high-performance bus*** (AHB), and an ***advanced peripheral bus*** (APB) interface for high-speed operations.

> **interrupt**
> *A signal to the processor emitted by hardware or software to indicate an event that needs immediate attention.*

> **nested vectored interrupt controller (NVIC)**
> *ARM-based interrupt handler hardware.*

> **memory protection unit**
> *The hardware in a computer that controls memory access rights.*

> **wake-up interrupt controller**
> *A peripheral that can detect an interrupt and wake a processor from deep sleep mode.*

> **advanced high-performance bus**
> *An on-chip bus specification to connect and manage high clock frequency system modules in embedded systems.*

> **advanced peripheral bus**
> *An on-chip bus specification with reduced power and interface complexity to connect and manage high clock frequency system modules in embedded systems.*

The STM32F407VGT6 microcontroller has an Arm Cortex-M4 core with a 168-MHz clock frequency, 1 MB flash memory, 192 KB ***static random access memory*** (SRAM), an extensive range of enhanced I/Os and peripherals connected to two APB buses, three AHB buses, and a 32-bit multi-AHB bus matrix. The STM32F407VGT6 microcontroller has three 12-bit ADCs, two *digital to analog converters* (DACs), a low-power real-time clock (RTC), two advanced-control timers, eight general-purpose timers, two basic timers, two *watchdog* timers, a SysTick timer, a true random number generator (RNG), three I^2C modules, three full-duplex *serial peripheral interface* (SPI) modules with I^2S support, four *universal synchronous/asynchronous receiver/transmitter* (USART) modules, two *universal asynchronous receiver/transmitter* (UART) modules, a high-speed and a full-speed USB *on-the-go* (OTG) module, two *controller area network* (CAN) modules, two *direct memory access* (DMA) modules, a *secure digital input output* (SDIO)/*multimedia card* (MMC) interface, an ethernet interface, and a camera interface.

> **static random access memory**
> *Static, as opposed to dynamic, RAM retains its data for as long as its power supply is maintained.*

> **digital to analog converter**
> *A device that converts a digital value to its corresponding analog value (e.g., voltage).*

> **watchdog**
> *A timer in an embedded system that is used to detect and recover from malfunctions.*

> **serial peripheral interface**
> *A serial communication bus used to send data, with high speed, between microcontrollers and small peripherals.*

> **universal synchronous/asynchronous receiver/transmitter**
> *A serial communication bus commonly used to send data, both synchronously and asynchronously, between microcontrollers and small peripherals.*

> **universal asynchronous receiver/transmitter**
> *A serial communication bus commonly used to send data, asynchronously, between microcontrollers and small peripherals.*

on-the-go
A USB specification that allows a USB device to act as a host, allowing other USB devices to connect to themselves. Also called USB on-the-go.

controller area network
A vehicle bus standard designed to allow microcontrollers and devices to communicate with each other, in applications, without a host computer.

direct memory access
A mechanism whereby data may be transferred from one memory location to another (including memory-mapped peripheral interfaces) without loading, or independently of, the CPU.

secure digital input output
A circuit that allows the sending of data to external devices using Secure Digital (SD) specification.

multimedia card
A memory card standard used for solid-state storage.

The STM32F4 Discovery Kit

The STM32F4 Discovery kit has an ST-Link/V2 in-circuit debugger and programmer, a USB mini connector for power and debug operations, a USB Micro-AB connector for USB OTG operations, a reset button, a user push button, and four LEDs available to the user. It also has an 8-MHz main oscillator crystal, ST-***MEMS*** three-axis ***accelerometer***, ST-MEMS audio sensor, and a CS43L22 audio DAC with an integrated class-D speaker driver. The kit supports both 3.3 V and 5 V external power supply levels.

MEMS (Micro-Electro-Mechanical Systems)
Microscopic mechanical or electro-mechanical devices.

accelerometer
A device that measures change in velocity over time.

1.4.3 STM32Cube Embedded Software Package

STM32Cube is the software package released by STMicroelectronics for the STM32 family of Arm Cortex-M based microcontrollers. It contains a low-level hardware abstraction layer (HAL) and board support package (BSP) drivers for on-board hardware components. Additionally, a set of examples and middleware components are included in the STM32Cube package.

Including STM32Cube in a Project

STM32Cube is available in Keil μVision. In order to use it in a project, STM32Cube should be enabled from the **Manage Run-Time Environment** window by selecting **Classic** under **STM32Cube Framework (API)** as shown in Figure 1.4. Desired HAL drivers can then be selected from the **STM32Cube HAL** list.

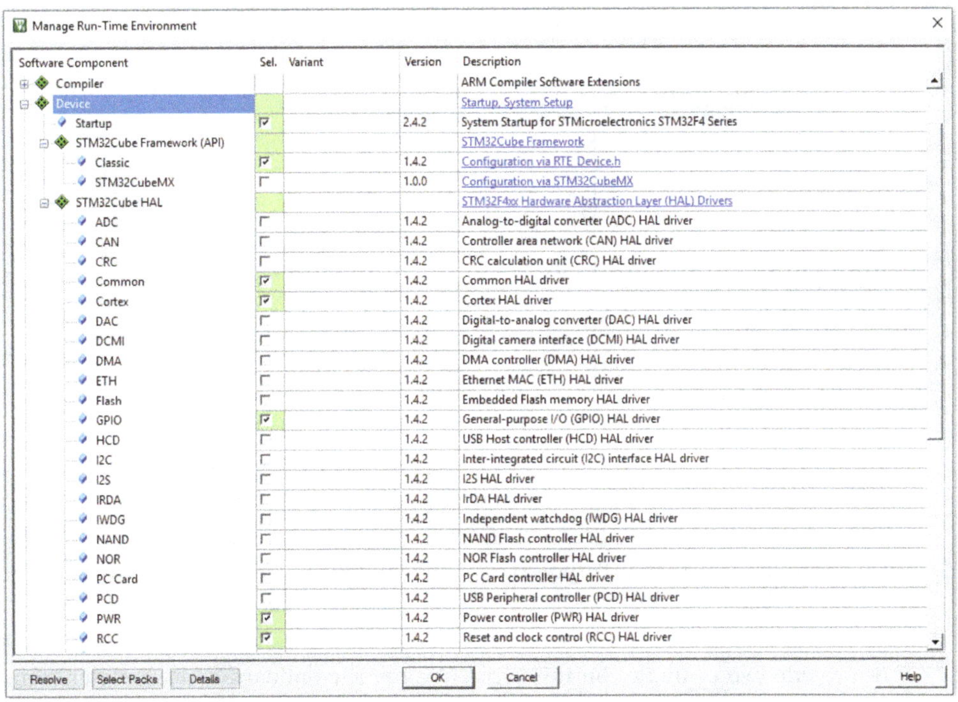

Figure 1.4 STM32Cube configuration by managing the run-time environment.

For our applications, **Common, Cortex, PWR, RCC,** and **GPIO** must be added before adding other HAL drivers. Some HAL drivers are connected, so they should be added together for them to work. Finally, the `stm32f4xx_hal.h` header file should be added at the top of the project's `main.c` file. Then, HAL drivers can be used in the source code.

Using BSP Drivers in the Project

To use BSP drivers in the project, first we need to specify the include folder for BSP drivers. To do this, click on the **Options for Target** button located in the toolbar, or select it from the **Project** tab. Click the **...** button next to the *Include Paths* text area from the **C/C++** tab. Then, add the full path of source files for BSP drivers under Keil's installation directory (`...\Arm\Pack\Keil\STM32F4xx_DFP\DFP_version\Drivers\BSP\`). Finally, click **OK** twice.

Application-specific header files must also be included in the source code. To add these specific files to the project, right click on the **Source Group 1** folder (in the **Project** window) and select **"Manage Project Items..."** as shown in Figure 1.5. From the pop-up window, click the **Add Files** button and navigate to the folders that contain the desired files.

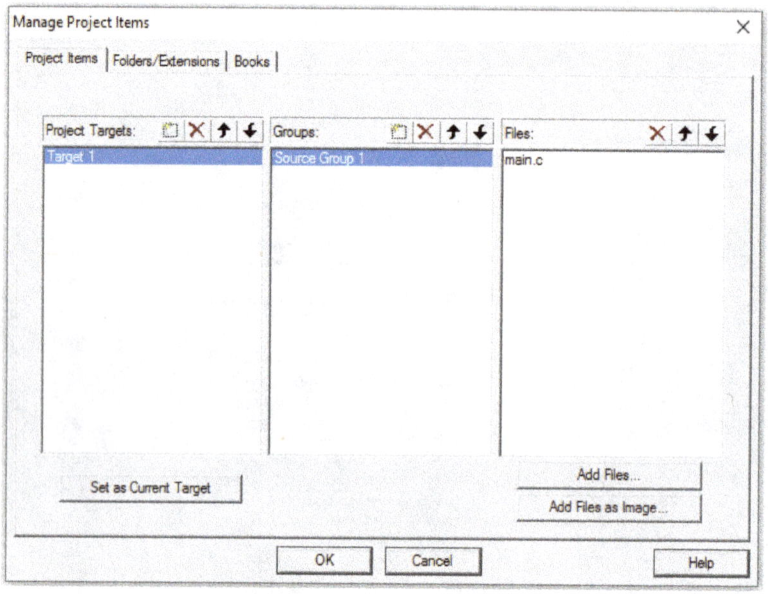

Figure 1.5 Adding BSP source files to the project.

Let us provide two examples on this topic. We use the onboard accelerometer from Section 1.4.6. To use it, you should add the `lis3dsh.c` and `lis302dl.c` files from the `...\Arm\Pack\Keil\STM32F4xx_DFP\DFP_version\Drivers\BSP\Components\lis3dsh` and `...\Arm\Pack\Keil\STM32F4xx_DFP\DFP_version\Drivers\BSP\Components\lis302dl` folders. You should also add the `stm32f4_discovery.c` and `stm32f4_discovery_accelerometer.c` files, along with their associated `.h` files, from the `...\Arm\Pack\Keil\STM32F4xx_DFP\DFP_version\Drivers\BSP\STM32F4-Discovery` folder. Similarly, if you want to use the onboard audio codec, then you should add the lis302dl.c file from the ...\Arm\Pack\Keil\STM32F4xx_DFP\DFP_version\Drivers\BSP\Components\lis302dl folder. The stm32f4_discovery.c and stm32f4_discovery_audio.c files should also be added from the ...\Arm\Pack\Keil\STM32F4xx_DFP\DFP_version\Drivers\BSP\STM32F4-Discovery folder.

> **codec (coder-decoder)**
> *A device or computer program that allows the encoding or decoding of a digital data stream or signal.*

1.4.4 STM32F407VGT6 Microcontroller Peripheral Usage

We use the STM32F407VGT6 microcontroller peripherals through HAL drivers. The HAL driver library contains a set of generic and extension APIs, which can be used by peripheral drivers. While generic APIs are applicable to all STM32 devices, extension APIs can be used by a specific family or part number. The HAL library should be initialized first in order to use peripheral-specific HAL drivers in a project. We provide the general header file hal_config.h for peripheral configuration using HAL drivers in Online_Student_Resources\Lab1.

Although the STM32F407VGT6 microcontroller has a wide variety of peripherals, we will only mention the ones used in this book. We now focus on the peripherals we use.

Power, Reset, and Clock Control

HAL power control functions can be used to adjust the power configuration of the STM32F407VGT6 microcontroller. To do so, the APB interface should be enabled first after reset. Furthermore, low-power modes can be configured through these functions.

As mentioned in the previous section, we have the general header file hal_config.h for peripheral configuration. After initializing the HAL library, if this header file is included in the project, the SystemClock_Config function can be used to set the maximum clock frequency. You can make necessary changes to the power, reset, and clock configuration through this function.

General-Purpose Input and Output

The STM32F407VGT6 microcontroller has six physical general-purpose input and output (GPIO) ports, namely Port A, B, C, D, E, and H. These ports support up to 82 programmable ***input/output*** pins. These pins are called PA0-15, PB0-15, PC0-15, PD0-15, PE0-15, and PH0-1. All programmable input/output pins are accessible from the STM32F4 Discovery kit.

input/output
A circuit in an embedded system that connects the system to the external world.

The STM32F4 Discovery kit has one push button and four LEDs. The connections for these components are provided in Table 1.1. The push button is connected as weak pull-down, so the internal ***pull-up/down resistors*** must be used for it.

resistor
A passive electrical component designed to implement electrical resistance as a circuit element.

pull-down resistor
A pull-down resistor (to a negative power supply voltage) ensures that a signal conductor adopts a valid (electrical) logic level (low) in the absence of any other connection.

pull-up resistor
A pull-up resistor (to a positive power supply voltage) ensures that a signal conductor adopts a valid (electrical) logic level (high) in the absence of any other connection.

Basic HAL GPIO functions can be used to configure GPIO pins. In these, GPIO pins are configured as ***floating input***. During and after reset, all alternate functions and external interrupts are disabled by default. Here, the AHB clock used for the GPIO port must be enabled first using the `__GPIOx_CLK_ENABLE()` function. Then, the GPIO pins should be initialized using the `HAL_GPIO_Init()` function. Here, each GPIO pin can be configured as a digital input, digital output, or peripheral specific pin. Optionally, output drive strength, pull-up/down resistors, speed, and ***open-drain*** options for a GPIO pin can also be configured.

Table 1.1 Push button and LED connections.

GPIO Pin	Component
PD13	LD3 LED (Orange)
PD12	LD4 LED (Green)
PD14	LD5 LED (Red)
PD15	LD6 LED (Blue)
PA0	Push button 1 (B1)

> **floating input**
> *An input pin with no signal source or termination connected.*

> **open-drain**
> *An output pin driven by a transistor, which pulls the pin to only one voltage.*

We provide a sample project on the usage of GPIO functions in `Online_Student_Resources\Lab1\GPIO_Example`. This project uses the `hal_config.h` header file for peripheral configuration. If another configuration setting is required, such as using more than one port as input or output, then it should be performed manually. Here, pins 12, 13, 14, and 15 of the GPIO D port are defined as output. These are connected to the onboard LEDs of the STM32F4 Discovery kit. Pin 0 of the GPIO A port is defined as input, which is connected to the onboard push button of the STM32F4 Discovery kit. The sample project controls the status of LEDs with this push button.

There is also a counter in the code. Pressing the push button increases this counter. When the counter is at zero, all LEDs are turned off. When the counter is at one, the orange LED turns on. When the counter is at two, the green LED turns on. When the counter is at three, the red LED turns on. Finally, when the counter is at four, the blue LED turns on. Furthermore, there is a software debouncer to prevent any glitches due to a button press.

Interrupts

A total of 82 maskable interrupt channels and 16 interrupt lines in the STM32F407VGT6 microcontroller can be prioritized as 16 main and 16 sub-levels. Interrupts are controlled by the NVIC.

A brief summary of how the NVIC works is as follows. While the CPU is executing a non-interrupt code, it is said to be in thread mode. When an ***interrupt flag*** is raised, the NVIC will cause the CPU to jump to the appropriate interrupt service routine (ISR) and execute the code. At this stage, the CPU is said to be in handler mode. While switching from thread to handler mode, the NVIC performs two operations in parallel. First, it fetches the exception vector, which holds the address of the related ISR. Second, it pushes necessary key register content to the ***stack***. Whenever the CPU is in thread mode, this process takes exactly 12 clock cycles (independent of the code executed).

> **interrupt flag**
> *A register bit used for indicating related interrupt status.*

> **stack**
> *A data structure in which items are removed in the reverse order from that in which they are added, so that the most recently added item is the first one removed.*

Exception vectors are stored in an interrupt vector table. This table is located at the start of the address space, which is predefined as part of the startup code. A label for each ISR is stored at each interrupt vector location. To create an ISR function, a void C function must be declared using the same name as the interrupt vector label. When the program counter reaches the end of the ISR function, the NVIC forces the CPU to return from handler mode to the point in thread mode that it left. At this point, key register data are also retained from the stack.

The interrupt flag should be cleared at the beginning of handler mode in order to not miss any other generated interrupts. If the interrupt flag is cleared at the beginning of the ISR function and a new interrupt flag is raised while in handler mode, then the latter interrupt is nested. It waits until the former ISR function is executed. If two interrupt flags are raised at the same time, the CPU fetches the highest priority one first and nests the lower priority one. Moreover, if the interrupt flag is not cleared before returning from handler mode, the NVIC causes the CPU to incorrectly jump back to handler mode.

The HAL library has special functions to handle interrupts. These functions enable and disable interrupts, clear pending interrupt requests, and set the priority of interrupts.

The ISR function must be linked to the related interrupt vector before an interrupt is used. Normally, all unused interrupt vectors are linked to `Weak` interrupt functions in the `startup_stm32f407xx.s` file. When an interrupt is used, the related ISR function should be redefined. All ports have external interrupt/event capability. Each of the 16 external interrupt lines are connected to the multiplexed output of six GPIO ports of the microcontroller. For example, pin 0 of each port is connected to Line0, pin 1 of each port is connected to Line1, and so on. The port must be configured as the input mode to use external interrupt lines.

We provide a sample project on the usage of interrupt and GPIO functions in `Online_Student_Resources\Lab1\Interrupts_Example`. This project does the same job as the one given in `GPIO_Example`. The only difference here is that the program execution is controlled by interrupts. Again, pins 12, 13, 14, and 15 of the GPIO D port are defined as outputs, which are connected to the onboard LEDs of the STM32F4 Discovery kit. Pin 0 of the GPIO A port is defined as an external interrupt source, which is connected to the onboard push button of the STM32F4 Discovery kit. This pin is set to generate an interrupt when the button is pressed.

Timers

There are 17 timer modules in the STM32F407VGT6 microcontroller. These are two advanced-control timers, 10 general-purpose timers, two basic timers, two watchdog timers, and one SysTick timer. These are briefly described below.

Advanced-control timers (TIM1, TIM8): 16-bit three-phase *pulse width modulation* (PWM) generators multiplexed on six channels with full modulation. Can be used as a general-purpose timer.

> **pulse width modulation**
> *A modulation technique that generates variable-width pulses to represent the amplitude of an analog input signal.*

General-purpose timers (TIM3 and TIM4): Full-featured general purpose 16-bit up, down, and up/down auto-reload counter. Four independent channels for input capture/output compare, and PWM or one-pulse mode output.

General-purpose timers (TIM2 and TIM5): Full-featured general-purpose 32-bit up, down, and up/down auto-reload counter. Four independent channels for input capture/output compare, and PWM or one-pulse mode output.

General-purpose timers (TIM10, TIM11, TIM13, and TIM14): 16-bit auto-reload up counter and a 16-bit prescaler. One independent channel for input capture/output compare, and PWM or one-pulse mode output.

General-purpose timers (TIM9 and TIM12): 16-bit auto-reload up counter and a 16-bit prescaler. Two independent channels for input capture/output compare, and PWM or one-pulse mode output.

Basic timers (TIM6 and TIM7): 16-bit auto-reload up counter and a 16-bit prescaler. Mainly used for DAC triggering and waveform generation.

Independent watchdog timer: 12-bit down counter and 8-bit prescaler. Clocked from an independent 32 kHz internal RC oscillator. Operates independently of the main clock. Can be used as a free-running timer.

Window watchdog timer: 7-bit down counter. Clocked from the main clock. Can be used as a free-running timer.

SysTick timer: 24-bit down counter. Integrated into Cortex M4 core. Mainly used for generating a timer-based interrupt for use by an operating system. Whenever the predefined function `HAL_Init()` is added to the project, it is set to interrupt every millisecond. The `HAL_Delay()` function uses the SysTick to generate delay.

We will only use timers to generate time bases in this book. We will not consider other timer modes, such as capture, compare, PWM, one-shot, SysTick, and watchdog.

The HAL library provides generic and extended functions to control timers. These functions are used for enabling and disabling timers, setting time bases, configuring clock sources, and starting and stopping timers. In order to use a timer module, its clock source, clock divider, count mode, frequency, and period must first be configured and initialized. Then, the AHB clock and interrupt for the timer module in use should be enabled. Interrupt priorities should also be set in the specific package initialize function. Finally, the timer module should be started in blocking (polling), non-blocking (interrupt), or DMA mode. The time duration generated by the timer module in seconds can be calculated as

$$Duration = \frac{(TimerPrescaler + 1)(TimerPeriod + 1)}{TimerClock} \tag{1.1}$$

We provide a sample project on the usage of timer functions and interrupts in `Online_Student_Resources\Lab1\Timers_Example`. This project uses the `hal_config.h` file to configure the timer module. Here, pins 12, 13, 14, and 15 of the GPIO D port are defined as outputs, which are connected to the onboard LEDs of the STM32F4 Discovery kit. The timer frequency is set to 10 kHz, and the timer period is set to 10000. This means an interrupt is generated every 10000 clock cycles. In other words, the interrupt period is 1 s. In the timer ISR, onboard LEDs are turned on one by one in a similar way to the `GPIO_Example`.

Analog to Digital Converter

There are three identical 12-bit ***analog to digital converter*** (ADC) modules in the STM32F407VGT6 microcontroller. These are called ADC0, ADC1, and ADC2. Each ADC module has 16 external inputs, two internal inputs, a V_{BAT} input, analog to digital converter block, data register blocks, interrupt control block, and trigger blocks. ADC modules share 16 analog input channels. The channels and pins related to them are listed in Table 1.2.

> **analog to digital converter**
> *A device that samples and quantizes an analog input signal to form a corresponding/representative digital signal.*

ADC modules in the STM32F407VGT6 microcontroller are based on the successive approximation register method. These can run in independent or dual/triple conversion modes. In the independent conversion mode, the ADC module can conduct single conversion from single-channel, single conversion from multichannel (scan), continuous conversion from single-channel, continuous conversion from multichannel (scan), and injected conversion. In dual or triple conversion modes, the ***sampling*** rate of the ADC module can be increased with configurable interleaved delays. Here, sampling rate is the number of digital samples obtained per second (sps). The speed of the analog to digital converter block defines the sampling rate of the ADC module in single mode. The clock for the ADC block (ADCCLK) is generated from the APB2 clock divided by a programmable prescaler. This allows the ADC module to work with clock speeds of

Table 1.2 STM32F407VGT6 microcontroller analog input channels.

Pin Number	Pin Name	Channel Name
23	PA0	ADC{1, 2, 3}_IN0
24	PA1	ADC{1, 2, 3}_IN1
25	PA2	ADC{1, 2, 3}_IN2
26	PA3	ADC{1, 2, 3}_IN3
29	PA4	ADC{1, 2}_IN4
30	PA5	ADC{1, 2}_IN5
31	PA6	ADC{1, 2}_IN6
32	PA7	ADC{1, 2}_IN7
35	PB0	ADC{1, 2}_IN8
36	PB1	ADC{1, 2}_IN9
15	PC0	ADC{1, 2, 3}_IN10
16	PC1	ADC{1, 2, 3}_IN11
17	PC2	ADC{1, 2, 3}_IN12
18	PC3	ADC{1, 2, 3}_IN13
33	PC4	ADC{1, 2}_IN14
34	PC5	ADC{1, 2}_IN15

up to 42 MHz. The total conversion time is the sampling time plus 12 clock cycles. This allows a sample and conversion rate of up to 2.8 Msps.

> **sampling**
> *The process of obtaining values at specific time intervals.*

The analog to digital conversion operation is initiated by a trigger in the ADC module. The source for this can be software, internal hardware, or external hardware in the STM32F407VGT6 microcontroller. A software trigger is generated by the `HAL_ADC_Start` function. The internal hardware trigger can be generated by timer events. The external hardware trigger can be generated by EXTI_11 or EXTI_15 pins.

In this book, we only use the ADC modules in single-channel and single conversion mode.

The HAL library provides generic and extended functions to control ADC modules. In order to use the ADC module, its clock divider, conversion mode, and resolution must first be configured and initialized. Then, the AHB clock and the interrupt used for the ADC module should be enabled. Next, the GPIO pin used in operation should be configured as analog input. Interrupt priorities should also be set in the specific package initialize function. Then, the sampling time and channel should be configured. The ADC module should be started in blocking (polling), non-blocking (interrupt), or DMA mode. When the conversion is complete, the digital value should be read from the ADC register.

We provide three sample projects on the usage of ADC modules in `Online_Student_Resources\Lab1\ADC_Examples`. The first project uses the internal temperature sensor. Here, ADC1 is configured for 12-bit single conversion from Channel 16 (internal temperature sensor). It is set to be triggered with software. ADCCLK is set to 42 MHz. Sampling time is set to 84 cycles, which leads to approximately 437.5 Ksps. When the converted data are read, they are scaled to yield a temperature output in degrees Celsius. The second project uses the software trigger in the ADC module. Here, the PA1 pin is used as the ADC channel. The ADC1 module is configured for 12-bit single conversion from Channel 1. ADCCLK is set to 42 MHz. Sampling time is set to three cycles, which leads to approximately 2.8 Msps. The third project uses the ADC module with a timer trigger. Here, the PA1 pin is used for the ADC channel. The ADC1 module is configured for 12-bit single conversion from Channel 1. The timer module triggers the ADC every 1/10000 s. Moreover, the ADC interrupt is enabled so that an interrupt is generated when the conversion ends.

Digital to Analog Converter

There is one module to handle the digital to analog conversion operation in the STM32F407VGT6 microcontroller. There are two independent 12-bit DACs within this module. Each DAC has a buffered independent monotonic voltage output channel. Each DAC has one output, one reference input, a digital to analog converter block, a data register block, a control block, and a trigger block. Each DAC has a separate output channel. These channels and their related pins are listed in Table 1.3.

DACs can run in single or dual conversion modes. In the single conversion mode, each DAC can be configured, triggered, and controlled independently. In dual mode, DACs can be configured in 11 possible conversion modes. The DAC module is clocked directly from the APB1 clock and can give an output with or without a trigger. There can be no DAC triggers, or they can be set as internal hardware, external hardware, or software. If no hardware trigger is selected, data stored in the DAC data holding register, (`DAC_DHRx`), are automatically transferred to the DAC output register, (`DAC_DORx`), after one APB1 clock cycle. When a hardware trigger is selected and a trigger occurs, the transfer is performed three APB1 clock cycles later. The analog output voltage becomes available after a settling time that depends on the power supply voltage and the analog

Table 1.3 The STM32F4 microcontroller analog output channels.

Pin Number	Pin Name	Channel Name
29	PA4	DAC_CHANNEL1
30	PA5	DAC_CHANNEL2

output load. The DAC output register (DOR) value is converted to output voltage on a linear scale between zero and the reference voltage. This value can be calculated as

$$DAC_{output} = V_{REF} \times \frac{DOR}{4095} \tag{1.2}$$

DACs have output *buffers* that can be used to reduce the output impedance and to drive external loads directly without having to add an external operational amplifier. These buffers can be enabled or disabled through software. In this book, we will only use the single conversion operation mode in DAC.

> **buffer**
> *A circuit in which data are stored while it is being processed or transferred.*

The HAL library provides generic and extended functions to control the DAC module. The DAC module must be initialized before it is used. Then, the AHB clock should be enabled. The GPIO output pin should be configured as analog output in the specific package initialize function. Then, the output buffer and DAC channel trigger should be configured. Next, the DAC module should be started in blocking (polling), non-blocking (interrupt), or DMA mode. Finally, the digital data should be transferred to the output register to set the analog output value.

We provide three sample projects on the usage of DAC functions in `Online_Student_Resources\Lab1\DAC_Examples`. In the first project, the DAC module is configured for 12-bit single conversion to output from Channel 1 (PA4 pin) without triggering. In the second project, the DAC module is used with a software trigger. Here, the PA4 pin is used as the DAC channel output and DAC is configured for 12-bit conversion every second using delays. In the third project, the DAC module is used with a timer trigger. Here, the PA4 pin is used as the DAC channel output and the DAC module is configured for 12-bit conversion every second via the basic timer TIM6. TIM6 is configured to update TRGO output once the counter reaches its maximum value. DAC is triggered with this TIM6 TRGO output.

Direct Memory Access

DMA is a special module of the microcontroller. Through it, data transfer can be conducted in the background without using CPU resources, which means the CPU can conduct other tasks at the same time.

The STM32F407VGT6 microcontroller has two DMA modules with a total of 16 streams (8 per unit). Each stream has up to eight selectable channels. Each channel is dedicated to a specific peripheral and can be selected by software. Each channel has a *First In First Out* (FIFO) buffer with a length of four words (4×32 bits). These buffers can be used in FIFO mode to temporarily store the incoming data. Then, data transfer can be initiated when the selected threshold value is reached. Threshold values can be 1/4, 2/4, 3/4, or full. FIFO buffers can also be used in direct mode to transfer data immediately. To transfer data, either normal or circular mode can be selected. The second mode will be especially useful while handling circular buffers, which we will consider in Chapter 11.

> **First In First Out**
> *FIFO is a method for organizing and manipulating a data buffer, where the oldest (first) entry, or "head" of the queue, is processed first.*

DMA modules in the STM32F407VGT6 microcontroller can initiate data transfer from memory to memory, from memory to peripheral, and from peripheral to memory. When the memory to memory transfer mode is selected, only the second DMA module can be used. Here, circular and direct modes are not allowed. The data width for the source and destination can be set as one byte (8 bits), half-word (16 bits), or word (32 bits). The DMA module can be configured for incrementing source and destination addresses automatically after each data transfer.

There are four different stream priorities for handling multiple DMA streams. If multiple DMA streams have the same priority, hardware priority is used (e.g., stream 0 has priority over stream 1). Burst transfer can be used when FIFO mode is selected. Here, the burst size can be selected as $4\times$, $8\times$, or $16\times$ a data unit. FIFO buffers can also be configured as double buffer to support the ping–pong data transfer structure, which we will consider in Chapter 11.

We provide two sample projects on the usage of the DMA module in `Online_Student_Resources\Lab1\DMA_Examples`. In the first project, the content of a buffer is copied to another buffer in direct mode. In the second project, data are read from the ADC module and written into a buffer. The ADC module is triggered with software as explained in Section 5.3.

1.4.5 Measuring Execution Time by Setting Core Clock Frequency

We will use the DWT unit introduced in Section A.5 to measure the execution time of a code block. We will demonstrate its usage when the core clock is set to a certain frequency. As in Section A.5, the `time.h` header file should be added to the project. In order to set the core clock frequency, we will use the function `SystemClock_Config` in the `hal_config.h` header file.

Measuring Execution Time without the SysTick Timer

The first method of measuring the execution time does not use the SysTick timer. This timer is used as a time base. When the `HAL_Init` function is called in the code, it is configured to create an interrupt every millisecond. However, the SysTick timer is not used when the DWT unit is used for measuring the execution time. If used by mistake, its interrupt generates additional clock cycles.

We provide a sample project on measuring the execution time of a code block in `Online_Student_Resources\Lab1\Measure_Execution_Time_I`. Here, the core clock is set to 168 MHz. There are some HAL functions used in this code. Therefore, you should arrange the project as explained in Section 1.4.3. Then, you can add the code block to be measured in the indicated place.

There are some differences between this code and the code mentioned in Section A.5. The first difference is in writing the core clock frequency. Here, the line `CoreClock = SystemCoreClock` is used instead of writing the core clock frequency manually to the `CoreClock` variable. This feature can only be used when the STM32Cube is enabled in the project. The second difference is in the `HAL_Init` and `SystemClock_Config` functions. They are used for setting the core clock frequency. The third difference is in the `HAL_SuspendTick` function. This function is used for disabling the SysTick timer interrupt.

In order to test the method introduced in this section, use the example given in Section A.5. Add a `for` loop with 2000000 iterations in the indicated place in the example code and run it. Observe the number of clock cycles and execution time. You can also use the State and Sec registers described in Section A.5 to measure the execution time.

Measuring Execution Time with the SysTick Timer

The second method of measuring the execution time is by using the SysTick timer. You should be aware that this method gives a time resolution at the millisecond level. We provide a sample project on measuring the execution time in `Online_Student_Resources\Lab1\Measure_Execution_Time_II`. Here, the core clock is set to 168 MHz. There are some HAL functions used in this code. Therefore, you should arrange the project as explained in Section 1.4.3. Then, you can add the code block to be measured in the indicated place.

If there is some additional code between the clock configuration and the code block to be measured, the SysTick timer must be disabled after configuring the clock. Then, using the `HAL_ResumeTick` function, the SysTick timer should be enabled before the code block to be executed. Execution time can be obtained in milliseconds using the `HAL_GetTick` function after the code block to be measured. Test this method by adding a `for` loop with 2000000 iterations in the indicated place in the code. Observe the execution time.

1.4.6 STM32F4 Discovery Kit Onboard Accelerometer

There are two accelerometer types available depending on the STM32F4 Discovery kit *printed circuit board* (PCB) revision. If the kit's PCB revision is MB997B (revision B), then the *LIS302DL* accelerometer is available on the board. If the kit's PCB revision is MB997C (revision C), then the *LIS3DSH* accelerometer is available on the board. Both devices are controlled by the SPI interface.

> **printed circuit board**
> *A board made of fiberglass, composite epoxy, or other laminate material and used for connecting different electrical components via etched or printed pathways.*

Basic BSP accelerometer functions can be found in `Keil_v5\Arm\Pack\Keil\STM32F4xx_DFP\2.8.0\Drivers\BSP\STM32F4-Discovery\stm32f4_discovery_accelerometer.c`. In order to use these functions, the BSP must be included in the project as explained in Section 1.4.3. A sample project on the usage of accelerometer functions is given in `Online_Student_Resources\Lab1\Accelerometer_Usage_I`. Here, LEDs are turned on or off based on the board's motion. Four LEDs form a cross form between the two switches on the ST Discovery board. Each LED is closer to one edge of the board than the others. We process the accelerometer data so that the LED closest to the edge, facing downward, turns on.

1.4.7 The AUP Audio Card

The AUP audio card is a multipurpose audio expansion board. It is based on the TLV320AIC23B chip, which is a stereo audio codec with a headphone amplifier. The board is shown in Figure 1.6. We will use the AUP audio card to acquire audio signals. Therefore, we discuss its features in detail below.

- Analog line-level input and output via 3.5 mm line in and out sockets.
- High quality headphone output and microphone input via 3.5 mm headphone and microphone sockets.
- 8 to 96 kHz sampling-frequency support.
- *Inter-IC sound* (I2S)-compatible interface for audio stream.
- I2C-compatible interface for configuration.

> **Inter-IC sound**
> *A serial communication bus interface designed to connect digital audio devices.*

Connecting the AUP Audio Card

The AUP audio card is compatible with the STM32 Discovery kit. It can be connected as shown in Figure 1.6.

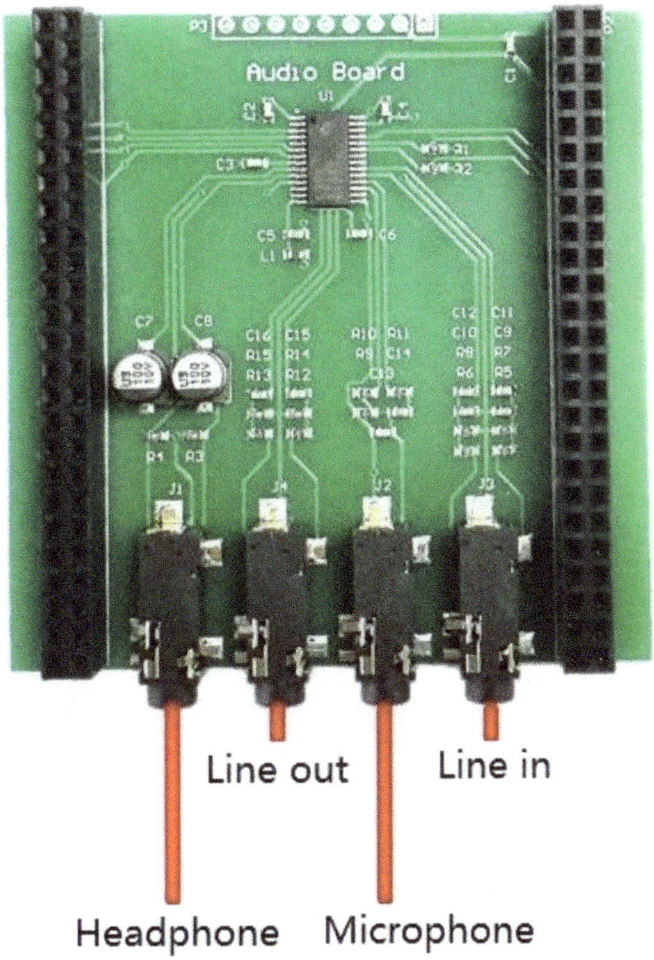

Figure 1.6 The AUP audio card. Photo by Arm Education.

Using the AUP Audio Card

In order to use the AUP audio card with the STM32F4 Discovery kit, the card must first be configured. The function to configure the AUP audio card is `AudioInit`. While calling the configuration function, the ***sampling frequency***, input source, and mode should be set. You can set the input source to digital microphone, analog microphone, or line-in using one of the code lines in Listing 1.1. You can set the sampling rate using one of the code lines in Listing 1.2 and set the mode using one of the code lines in Listing 1.3. We will only use the `IO_METHOD_INTR` mode, which is an interrupt-driven mode.

> sampling frequency
> *The reciprocal of the time delay between successive samples in a discrete-time signal.*

We provided the "predefined constants" for the "input source," "sampling rate," and "acquisition mode" of the audio card used in Listing 1.1, 1.2, and 1.3. In fact, all these definitions have been made in the `hal_config.h` file. The user can use these within the `AudioInit` function. For example, within the `LTEK_Example_Keil` project, they are used as `AudioInit(FS_32000_HZ,AUDIO_INPUT_MIC,IO_METHOD_INTR)`.

```
AUDIO_INPUT_LINE
AUDIO_INPUT_MIC
```

Listing 1.1 Setting the input source to digital microphone.

```
FS_8000_HZ
FS_16000_HZ
FS_22050_HZ
FS_32000_HZ
FS_44100_HZ
FS_48000_HZ
FS_96000_HZ
```

Listing 1.2 Setting the sampling rate.

```
IO_METHOD_INTR
IO_METHOD_DMA
IO_METHOD_POLL
```

Listing 1.3 Setting the mode.

```
void ProcessData(I2S_Data_TypeDef* I2S_Data)
/* Process data of the left channel. */
if (IS_LEFT_CH_SELECT(I2S_Data->mask))
I2S_Data->output_l=I2S_Data->input_l;

/* Process data of the left channel. */
if (IS_RIGHT_CH_SELECT(I2S_Data->mask))
I2S_Data->output_r=I2S_Data->input_r;

}
```

Listing 1.4 The ProcessData function.

The function for transmitting and receiving data from the AUP audio card is `ProcessData`. You should provide this function in the main code as given in Listing 1.4. The left and right audio channels can be selected using the `IS_LEFT_CH_SELECT` and `IS_RIGHT_CH_SELECT` macros. The `I2S_Data` structure contains the mask, input, and output data.

We provide a sample project on the usage of the TLV320AIC23B codec functions in `Online_Student_Resources\Lab1\L-Tek_Example`. The code in the project simply forms a loop between ADC input and DAC output on the AUP audio card. Data coming from microphones are directly fed to headphone output in the `ProcessData` subroutine.

1.4.8 Acquiring Sensor Data as a Digital Signal

We will use sensors on the STM32F4 Discovery kit and the AUP audio card to form digital signals. We will begin with the accelerometer.

Task 1.1

We provide the sample project to acquire accelerometer data in `Online_Student_Resources\Lab1\Accelerometer_Usage_II`. Use this project to acquire the sensor data for 1000 samples. Data are stored in the `buffer` array. While acquiring the data, shake the STM32F4 Discovery kit along the x-axis. To note here, the direction of this movement should be along the widest side of the ST Discovery kit with the mini USB connector facing forward. Observe the acquired data through the watch window as explained in Chapter 0.

Task 1.2

Now, we will use the internal temperature sensor. Use the ADC functions to acquire the temperature data for 1000 samples. To obtain data from the temperature sensor, make necessary adjustments as explained in Section 1.4.4. Observe the acquired data through the watch window as explained in Chapter 0.

Task 1.3

Finally, we will use the AUP audio card. Use the AUP audio card to acquire the audio signal when the word “HELLO” is spelled. Set the sampling rate to 32 kHz and buffer size to 32000. Store 1 s of audio signal in an array. Please refer to Section 1.4.7 to use the AUP audio card effectively. Observe the acquired data through the watch window as explained in Chapter 0.

Here, you should observe what actual sensor data look like. To note here, we will not process these data in real time until Chapter 11. We will only acquire the sensor data and process them offline. We follow this strategy so we do not have to deal with real-time signal processing concepts while introducing the basics of digital signal processing.

2

Discrete-Time Signal Processing Basics

Contents

2.1 Introduction

Throughout this book, we will be considering discrete-time signals and systems. In this chapter, we introduce the basics of discrete-time signal processing. We will begin with discrete-time signals and operations on them. Then, we will focus on discrete-time systems. We will specifically handle linear and time-invariant (LTI) systems as they will be the dominant system type in the proceeding chapters. In relation to LTI systems, we will review constant-coefficient ***difference equations***.

> **difference equation**
> *A difference equation characterizes a discrete-time system in terms of its input and output sequences.*

2.2 Discrete-Time Signals

A discrete-time signal can originate from various sources [1]. In this book, we will represent a discrete-time signal as $x[n]$, where n is an integer index. This allows us to represent the discrete-time signal as an array. As mentioned previously, a discrete-time signal can take any real value. Hence, the array representing it will be real-valued. In order to obtain the corresponding digital signal, we must quantize these values. As a result, the array representing the digital signal can be represented directly in any programming language. The entries of the array can be `integer`, `float`, or any other type. Because digital signal values are stored in an array, the length of the digital signal must also be limited. These constraints are not valid in discrete-time signal representation, but they are important because they affect the way we process signals.

2.2.1 Basic Discrete-Time Signals

One can define many discrete-time signals, but five fundamental signals are sufficient for our purposes. These are discrete unit impulse, unit step, unit ramp, ***exponential***, and sinusoidal. They are defined as follows.

> **exponential**
> *A function that raises a given constant to the power of its argument.*

The discrete unit impulse signal has a value of one when its index is zero and a value of zero otherwise. It is formally defined as

$$\delta[n] = \begin{cases} 1 & n = 0 \\ 0 & \text{otherwise} \end{cases} \tag{2.1}$$

Discrete unit impulse is a discrete representation of the continuous-time unit impulse signal

$$\delta(t) = \lim_{\epsilon \to 0} \frac{1}{\epsilon}\left(u(t) - u(t-\epsilon)\right) \tag{2.2}$$

The discrete unit impulse signal is well-defined compared to its continuous-time counterpart as there is no limit operation in its definition. We plot the discrete unit impulse signal in Figure 2.1.

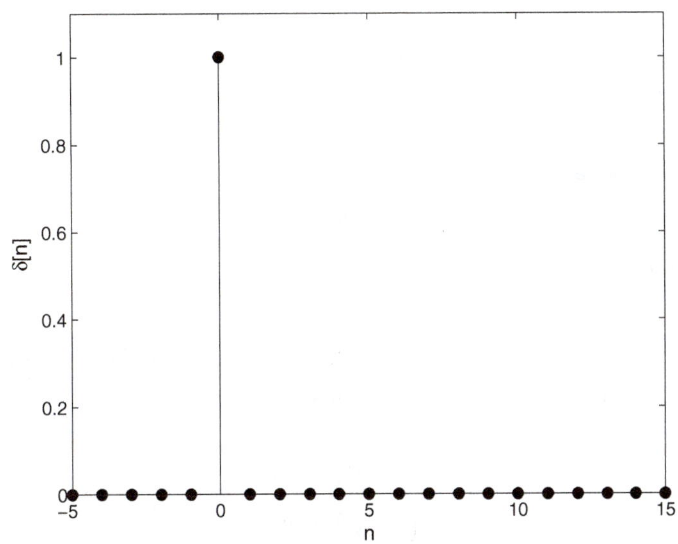

Figure 2.1 Discrete unit impulse signal.

The discrete-time unit step signal is equal to one when the index of the signal is greater than or equal to zero, and equal to zero otherwise. This signal is formally defined as

$$u[n] = \begin{cases} 1 & n = 0, 1, 2, \cdots \\ 0 & \text{otherwise} \end{cases} \tag{2.3}$$

This signal is a discrete representation of the continuous-time unit step signal

$$u(t) = \begin{cases} 1 & t > 0 \\ 0 & t < 0 \end{cases} \tag{2.4}$$

The discrete-time unit step signal is also well-defined as there is no discontinuity in its definition. To note here, we will refer to the discrete-time unit step signal simply as the unit step signal as we will not use the continuous and discrete-time representations together. We plot the unit step signal in Figure 2.2.

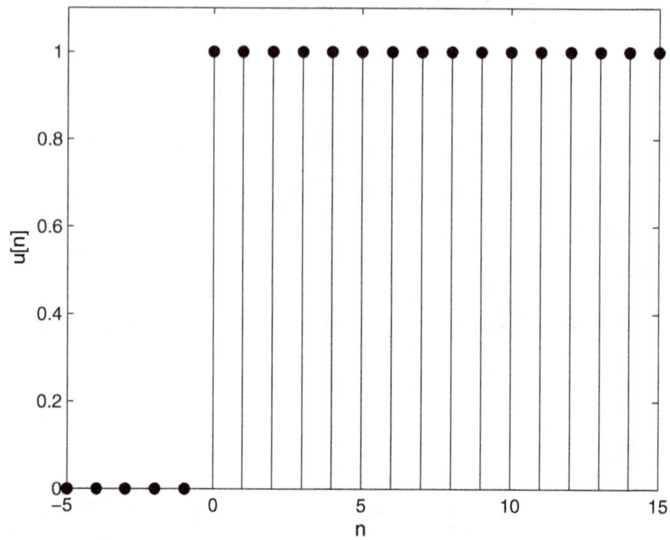

Figure 2.2 Unit step signal.

The unit ramp signal's value increases as its index increases. Assuming that the unit ramp signal starts at index value zero, it can be represented as

$$r[n] = \begin{cases} n & n = 0, 1, 2, \cdots \\ 0 & \text{otherwise} \end{cases} \tag{2.5}$$

We plot the unit ramp signal in Figure 2.3.

The discrete-time exponential signal can be defined as

$$e[n] = a^n \quad n = \cdots, -2, -1, 0, 1, 2, \cdots \tag{2.6}$$

where a is a positive real number. We plot the exponential signal (with $a = 0.8$) for positive n values in Figure 2.4.

We can represent a discrete-time sinusoidal signal as

$$s[n] = \sin(\omega_0 n) \quad n = \cdots, -2, -1, 0, 1, 2, \cdots \tag{2.7}$$

where ω_0 is the ***normalized angular frequency***.

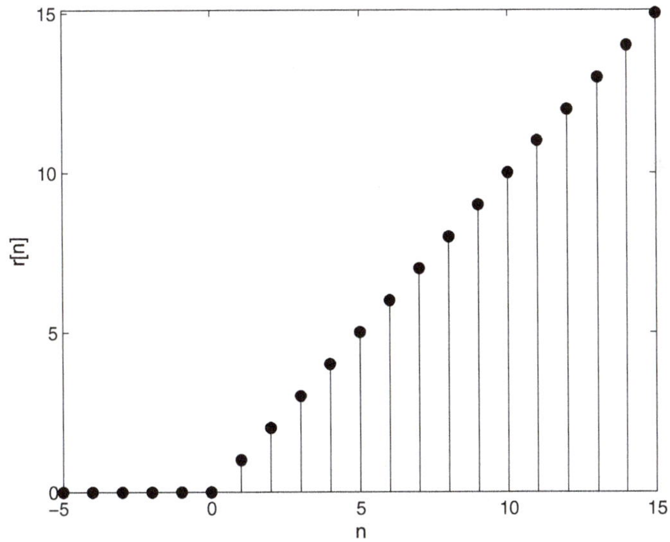

Figure 2.3 Unit ramp signal.

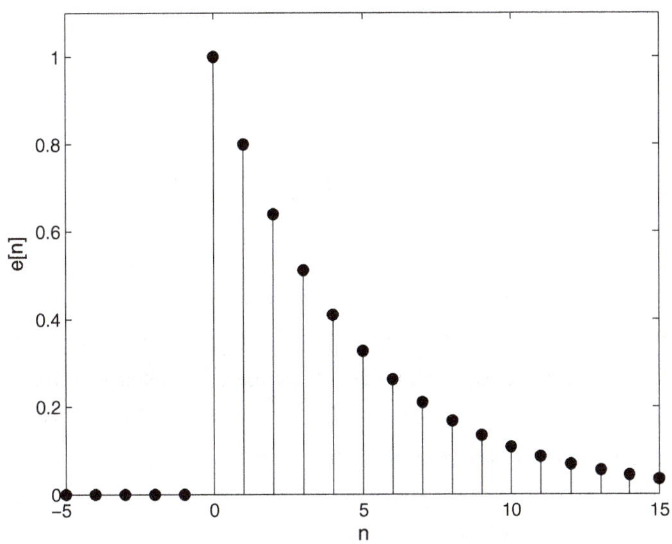

Figure 2.4 An exponential signal.

normalized angular frequency
f/f_s, where f is the angular frequency (rad/sec), which is the phase change of a sinusoidal signal per unit of time, and f_s is the sampling frequency (samples/sec). It has the unit of rad/sample.

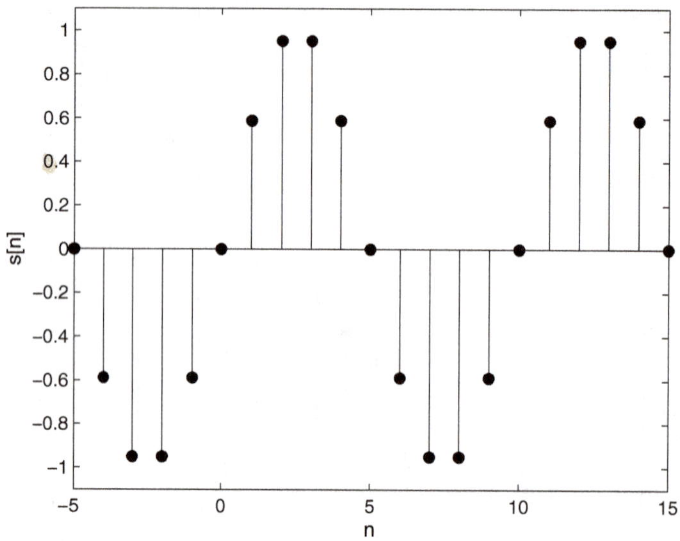

Figure 2.5 A sinusoidal signal.

These signals are constructed mathematically for demonstration purposes. In practice, most digital signals are acquired by sensors. We provide such signals in Lab 2.9. These will demonstrate what an actual signal looks like.

2.2.2 Operations on Discrete-Time Signals

We may need to modify a discrete-time signal. This can be done by shifting and arithmetic operations. We can shift a discrete-time signal by changing its index value. We can add a constant to a discrete-time signal and multiply it by a constant. We can also apply arithmetic operations between two or more discrete-time signals.

Let us start with shifting a discrete-time signal. The values of the discrete-time signal will not change as a result this operation, but the index representing them will change. Therefore, the signal will move to either the left or right. We can demonstrate the shifting operation using the discrete unit impulse signal.

Example 2.1 Shifting the discrete unit impulse signal

Assume $\delta[n]$ is the discrete unit impulse defined earlier. If we define a new signal as $x_1[n] = \delta[n-3]$, both signals will have the same shape, but the signal $x_1[n]$ will be shifted to the right by three increments. In other words, it will be delayed by three samples. The new signal will be

$$x_1[n] = \begin{cases} 1 & n = 3 \\ 0 & \text{otherwise} \end{cases} \tag{2.8}$$

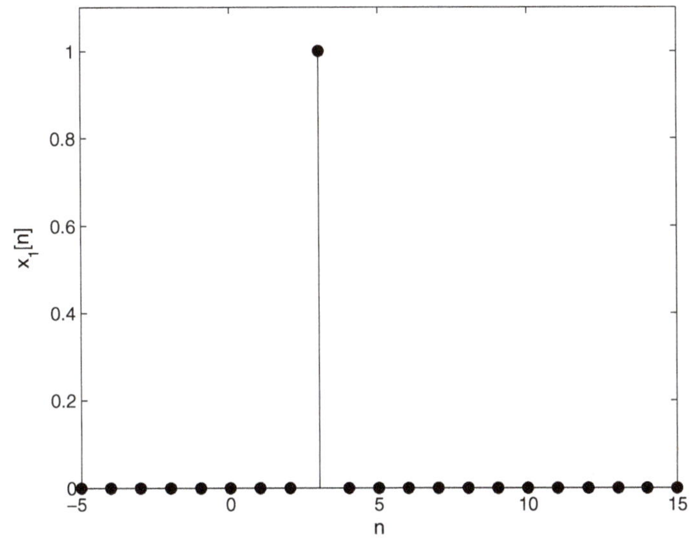

Figure 2.6 Shifted discrete unit impulse signal $x_1[n]$.

We plot the shifted discrete unit impulse signal $x_1[n]$ in Figure 2.6.

We can add a constant to (or subtract a constant from) a discrete-time signal. This constant will change the value of each entry of the discrete-time signal without shifting it in time. We can demonstrate this operation using a sinusoidal signal.

Example 2.2 Adding a constant to a sinusoidal signal

Assume that we have a sinusoidal signal $s[n] = \sin(\omega_0 n)$ with $\omega_0 = \pi/5$ rad/sample. If we define a new signal $x_2[n] = \sin(\omega_0 n) + 1$, we will have an elevated sinusoidal signal. Within the new signal, each entry will be elevated by one. We plot the elevated sinusoidal signal $x_2[n]$ in Figure 2.7.

We can multiply (or divide) a discrete-time signal by a constant. This means the value of each entry in the discrete-time signal will be multiplied (or divided) by that constant. Again, the signal will not be shifted in time by this operation. We can demonstrate this operation using the unit step signal.

Example 2.3 Multiplying the unit step signal by a constant

Assume that we have a unit step signal $u[n]$ as defined earlier. If we multiply this signal by -1, we obtain $x_3[n] = -u[n]$. This signal can be represented as

$$x_3[n] = \begin{cases} -1 & n = 0, 1, 2, \cdots \\ 0 & \text{otherwise} \end{cases} \tag{2.9}$$

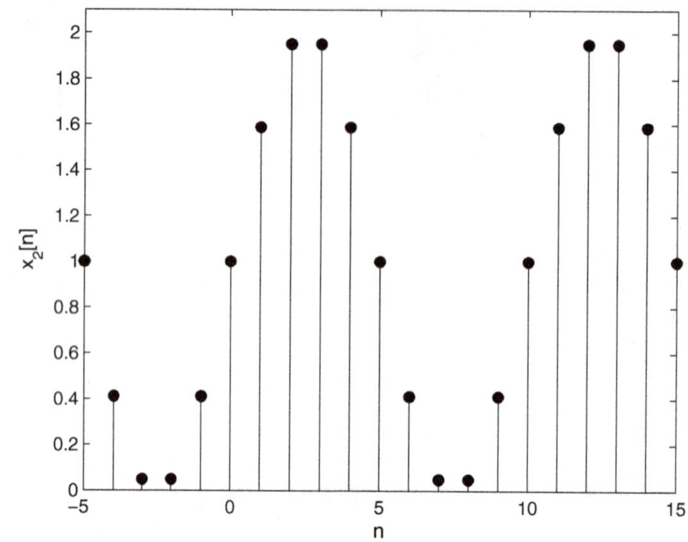

Figure 2.7 Sinusoidal signal with a constant added to it.

We plot the signal $x_3[n]$ in Figure 2.8.

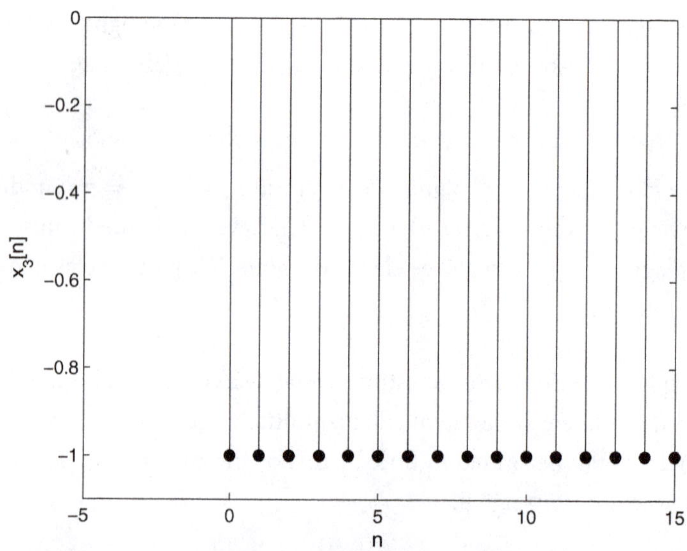

Figure 2.8 Unit step signal multiplied by -1, $x_3[n]$.

We can also apply any combination of the above arithmetic operations. Let us take the sinusoidal signal as an example and apply shifting, add a constant value to it, and multiply it by a constant.

Example 2.4 Shifting, adding a constant, and multiplying by a constant all at once

Assume that we have a sinusoidal signal $s[n] = \sin(\omega_0 n)$ with $\omega_0 = \pi/5$ rad/sample. If we shift it by -2, subtract 2 from it, and multiply it by 3, we obtain a new signal $x_4[n] = 3\sin(\omega_0(n-2)) - 2$. We plot the signal $x_4[n]$ in Figure 2.9.

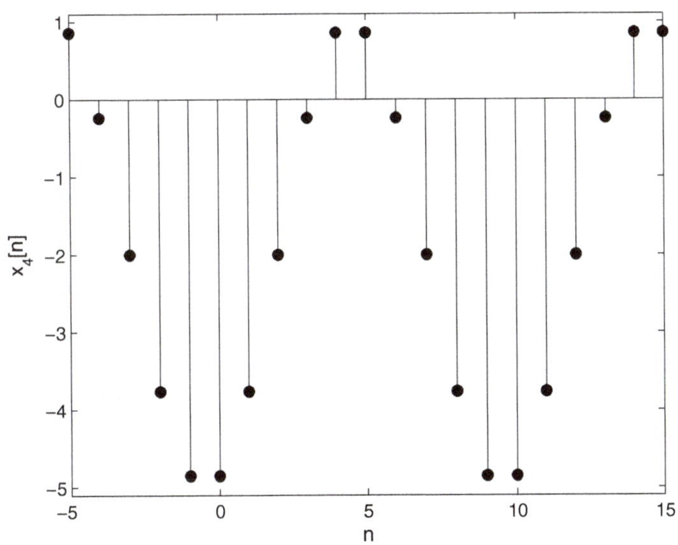

Figure 2.9 Applying all operations to a sinusoidal signal.

The above examples demonstrate how a discrete-time signal can be modified by basic arithmetic and shifting operations. This will come in useful later on.

Arithmetic operations can also be applied between two or more signals. Here, it is important to remember that all operations are to be applied to each individual discrete-time signal entry. Let us start with the subtraction operation. We can subtract one signal from another to obtain a new signal. Let us demonstrate this on two unit step signals.

Example 2.5 Subtracting one unit step signal from another

Assume that we have unit step signals $u[n]$ and $u[n-4]$ as defined earlier. If we define a new signal $x_5[n] = u[n] - u[n-4]$, we obtain

$$x_5[n] = \begin{cases} 1 & n = 0, 1, 2, 3 \\ 0 & \text{otherwise} \end{cases} \tag{2.10}$$

This signal takes the value one for a finite number of index values. This means it belongs to the general class of signals called window signals. We plot the signal $x_5[n]$ in Figure 2.10.

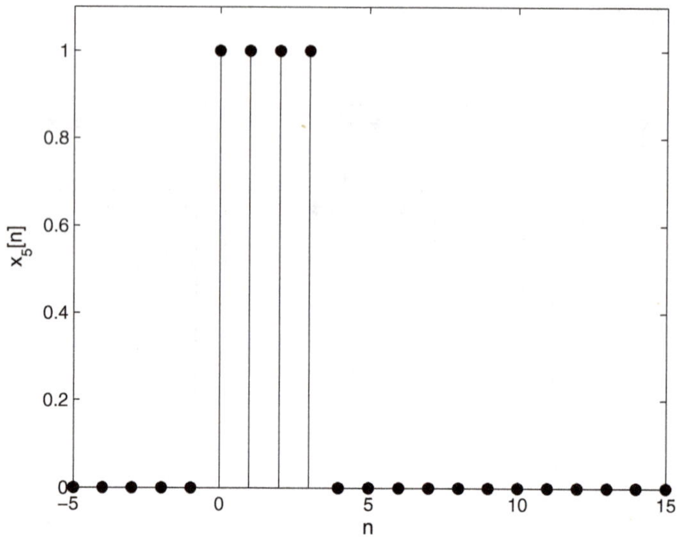

Figure 2.10 Subtraction between two unit step signals.

We can multiply two discrete-time signals together. Again, the multiplication operation is performed on each entry separately. Let us give an example of this operation by multiplying the unit step and exponential signals together. To note here, multiplying a discrete-time signal by the unit-step is typically done to limit its index range.

Example 2.6 Multiplying the unit step and exponential signals

Assume that we have the two signals $u[n]$ (unit step) and $e[n] = a^n$ (exponential) with $a = 0.8$. Multiplying these two signals together yields $x_6[n] = a^n u[n]$. We plot the signal $x_6[n]$ in Figure 2.11.

Another example of multiplying two signals together can be given using a sinusoidal signal and an exponential signal. This results in a damped sinusoidal signal.

Example 2.7 Damped sinusoidal signal

We can multiply $x_6[n] = a^n u[n]$ and $\sin(\omega_0 n)$ together to obtain a damped sinusoidal signal. The resulting signal is $x_7[n] = a^n \sin(\omega_0 n) u[n]$. As in previous examples, let us take $a = 0.8$ and $\omega_0 = \pi/5$ rad/sample. We plot the signal $x_7[n]$ in Figure 2.12.

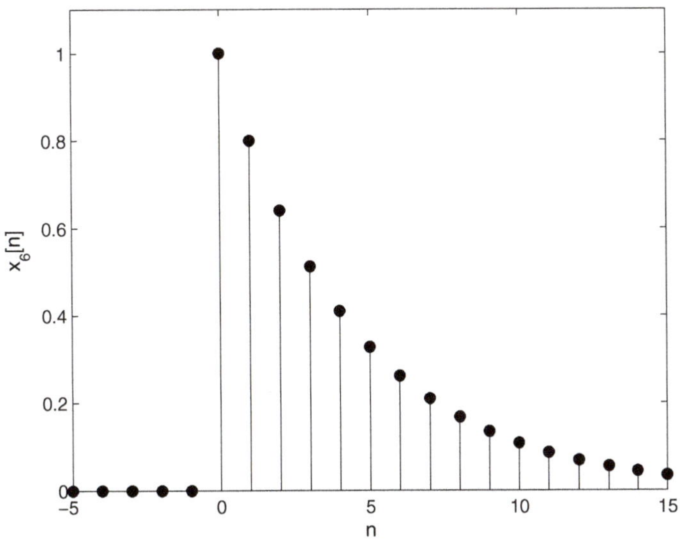

Figure 2.11 Multiplying the unit step and exponential signals together.

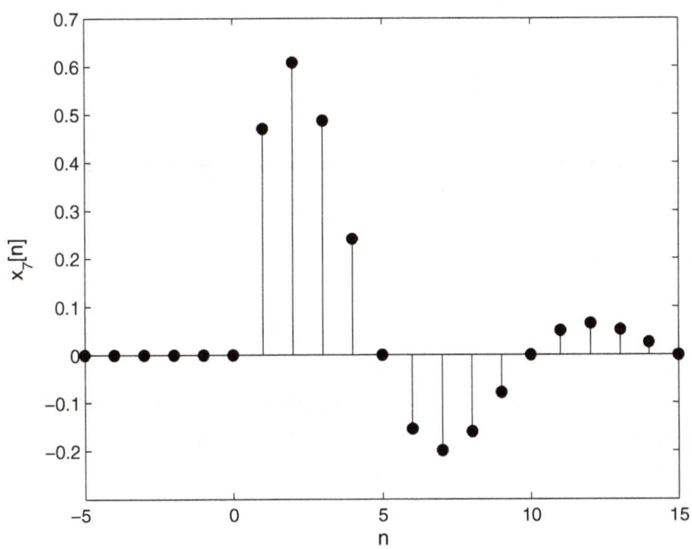

Figure 2.12 Damped sinusoidal signal.

The final example involves multiplying a signal and a window signal together. A window signal has a value of one for some index values and a value of zero otherwise. Therefore, it can be used to pass certain signal values, while suppressing the rest. Let us give an example on the usage of the window signal.

Example 2.8 Multiplying a signal by a window

The signal $x_5[n]$ from Eqn. 2.10 can be used as a window. We can multiply it with $x_6[n] = a^n u[n]$ to obtain

$$x_8[n] = \begin{cases} a^n & n = 0, 1, 2, 3 \\ 0 & \text{otherwise} \end{cases} \tag{2.11}$$

We plot the signal $x_8[n]$ in Figure 2.13.

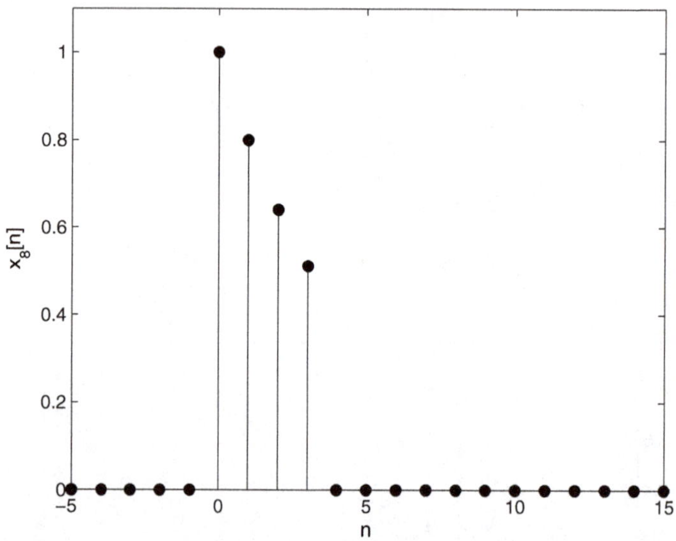

Figure 2.13 Multiplying a signal by a window.

Operations on digital signals are used extensively on real-life signals. Applying a windowing operation to a sensor signal is a good example. We provide other similar examples in Lab 2.9. These will provide further insight.

2.2.3 Relationships in Discrete-Time Signals

There are two well-known relationships in discrete-time signals. The first is between the discrete unit impulse signal and the unit step signal. The discrete unit impulse signal can be constructed from two unit step signals as

$$\delta[n] = u[n] - u[n-1] \tag{2.12}$$

In a similar manner, the unit step signal can be constructed from discrete unit impulse signals as

$$u[n] = \sum_{k=-\infty}^{n} \delta[k] \tag{2.13}$$

The second relationship in discrete-time signals is based on *Euler's formula*. This formula can be used to represent the exponential signal in terms of sinusoidal signals as

$$e^{j\omega_0 n} = \cos(\omega_0 n) + j\sin(\omega_0 n) \tag{2.14}$$

where $j = \sqrt{-1}$. Eqn. 2.14 leads to

$$\cos(\omega_0 n) = \frac{1}{2}\left(e^{j\omega_0 n} + e^{-j\omega_0 n}\right) \tag{2.15}$$

and

$$\sin(\omega_0 n) = \frac{1}{2j}\left(e^{j\omega_0 n} - e^{-j\omega_0 n}\right) \tag{2.16}$$

Eqns. 2.15 and 2.16 will be helpful in the frequency domain representation of discrete-time signals.

> **Euler's formula**
> *Also called Euler's expansion formula.* $e^{jx} = \cos(x) + j\sin(x)$, *where e is the base of the natural logarithm and j is the imaginary unit.*

2.2.4 Periodicity of Discrete-Time Signals

A periodic signal repeats itself an infinite number of times. We can represent a continuous-time periodic signal as

$$x(t) = x(t+T) \quad \text{for all } t \tag{2.17}$$

where T is the period of the signal. The smallest value of T that satisfies Eqn. 2.17 is called the fundamental period.

The definition of periodicity in discrete-time signals is slightly different than in continuous-time signals. The difference is in the value of the period. In discrete-time signals, the period must be an integer. Otherwise, the signal is said to be aperiodic. The reason for this is as follows. If a signal is periodic, it should repeat itself after a given period. However, this period must be an integer as it has to correspond to an index value for the discrete-time signal. Otherwise, the signal will never repeat itself. Therefore, it will be aperiodic. Based on this definition, we can define the periodicity condition for discrete-time signals as

$$x[n] = x[n+N] \quad \text{for all } n \tag{2.18}$$

where $N \in \mathbb{I}$ is the period of the signal. The smallest value of N that satisfies Eqn. 2.18 is called the fundamental period.

The integer period constraint can be relaxed to a rational number for discrete-time signals. The reason for this is as follows. If a signal is periodic with period N, then it is also periodic with integer multiples of N. Thus, if N is a rational number then we can multiply it by its denominator to obtain the actual integer period. Let us present an example to explain the periodicity of discrete-time signals.

Example 2.9 Periodicity of discrete-time signals

Assume that we have the sinusoidal signal $s[n] = \cos(\omega_0 n)$. We can represent the normalized angular frequency as $\omega_0 = 2\pi/N$ rad/sample, where N is the candidate discrete-time period. If $\omega_0 = \pi/10$ rad/sample, the period of the discrete-time signal will be $N = 20$. If $\omega_0 = 3/10$ rad/sample, then the candidate period will be $N = 20\pi/3$. This is an irrational number and there is no way to represent it as an integer. Therefore, the corresponding discrete-time signal will be aperiodic. You should be extremely cautious with this issue. Although the signal is sinusoidal, it is aperiodic in discrete time.

We plot the signals with $\omega_0 = \pi/10$ rad/sample and $\omega_0 = 3/10$ rad/sample in Figures 2.14 (a) and (b), respectively. Although both signals seem similar, we know

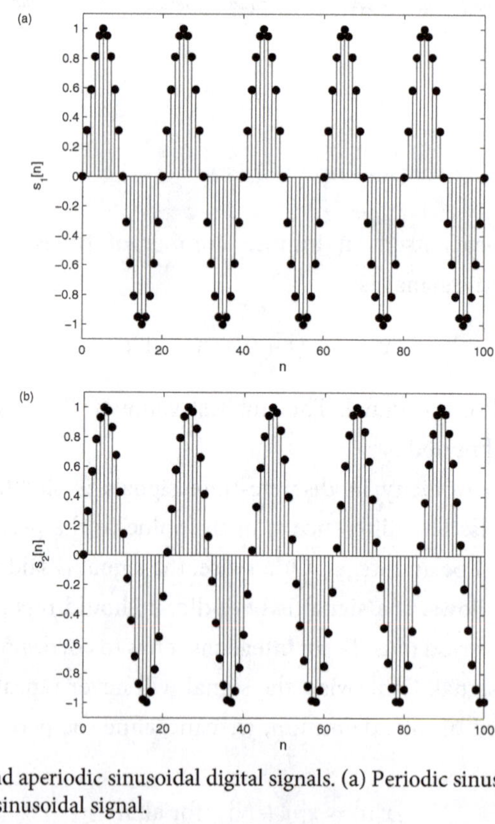

Figure 2.14 Periodic and aperiodic sinusoidal digital signals. (a) Periodic sinusoidal signal; (b) Aperiodic sinusoidal signal.

that $s_1[n]$ in Figure 2.14 (a) is periodic and repeats itself every 20 samples. $s_2[n]$ in Figure 2.14 (a) also repeats itself, but the repetition period is $N = 20\pi/3$. Because this value is not an integer, $s_2[n]$ can never satisfy Eqn. 2.18; so it is aperiodic.

We provide other periodic discrete-time signal examples in Lab 2.9. Specifically, we introduce square, triangle, and sawtooth signals. These periodic signals will also be used in later chapters of this book.

2.3 Discrete-Time Systems

A discrete-time system processes a given input signal and produces a new output signal. We can represent such a discrete-time system as in Figure 2.15. Here, $x[n]$ is the input signal fed to the system and $y[n]$ is the output signal obtained from the system. Throughout this book, we will use this notation for discrete-time systems.

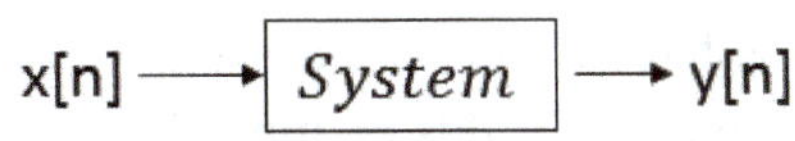

Figure 2.15 Representation of a discrete-time system.

The relationship between the input signal and the output signal should be identified in advance of analyzing the system. Systems are categorized based on this relationship. This relationship is useful in implementing a discrete-time system.

In digital signal processing literature, a system is usually called a filter. We will use these two terms interchangeably in this book.

2.3.1 Sample Discrete-Time Systems

In the following examples, we demonstrate several discrete-time systems (with sample input signals). We assume that the systems are initially at rest. Please consult Lab 2.9 for the full implementation of these systems. We expect these examples will allow you to grasp how a discrete-time system behaves. The properties of these systems are analyzed in Section 2.3.2.

Example 2.10 Multiplier (first system)

The first discrete-time system is $y[n] = \beta_0 x[n]$. This system multiplies the current input value by the constant β_0 and feeds it to the output. In this example, the input signal is

$x[n] = u[n]$. We provide the output signal obtained from this system (with $\beta_0 = 2.2$) in Figure 2.16.

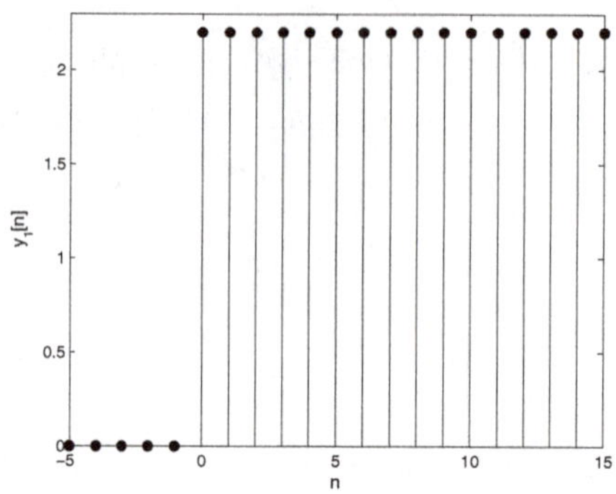

Figure 2.16 Output signal obtained from the first system.

Example 2.11 Adder accumulator (second system)

The second discrete-time system is $y[n] = x_1[n] + x_2[n]$. This system feeds the sum of two different signals to the output. In this example, the input signals are $x_1[n] = u[n]$ and $x_2[n] = \sin(\pi n/5)u[n]$. We provide the output signal obtained from this system in Figure 2.17.

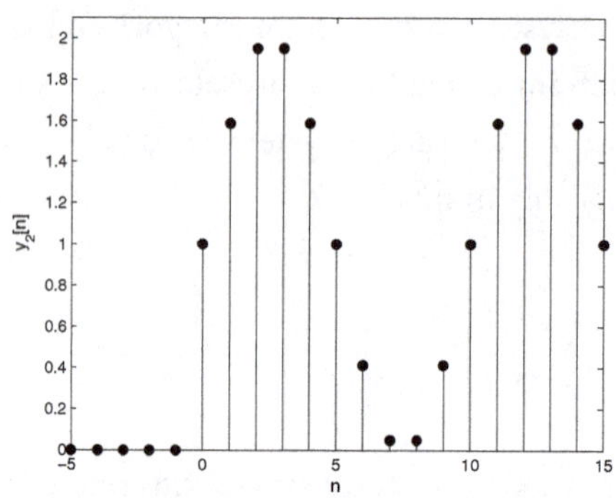

Figure 2.17 Output signal obtained from the second system.

Example 2.12 Squaring device (third system)

The third discrete-time system is $y[n] = x^2[n]$. This system takes the square of the input signal value and feeds it to the output. In this example, the input signal is $x[n] = \sin(\pi n/5)u[n]$. We provide the output signal obtained from this system in Figure 2.18.

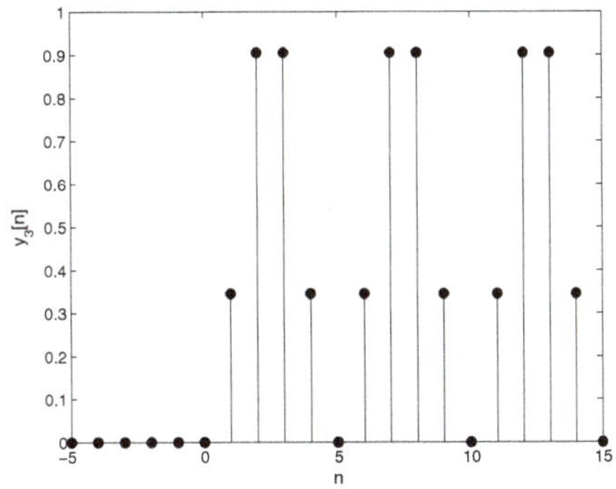

Figure 2.18 Output signal obtained from the third system.

Example 2.13 Multiplier and accumulator (fourth system)

The fourth system is $y[n] = \beta_0 x[n] + \beta_1 x[n-1]$. This system multiplies the previous and current input signals by constants β_0 and β_1, respectively, and feeds the sum to the output. In this example, the input signal is $x[n] = \sin(\pi n/5)u[n]$. We provide the output signal obtained from the fourth system (with $\beta_0 = 2.2$ and $\beta_1 = -1.1$) in Figure 2.19.

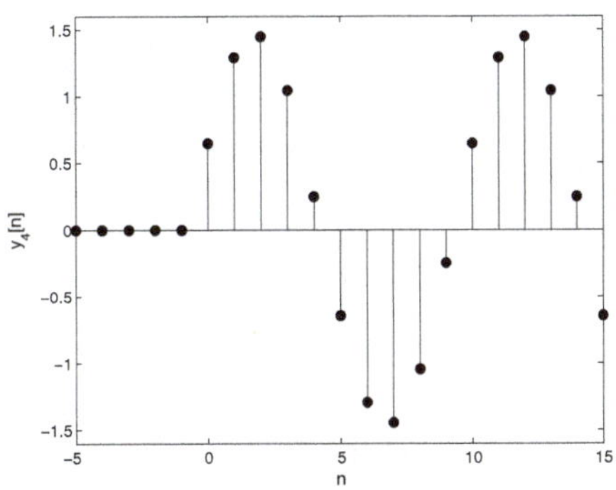

Figure 2.19 Output signal obtained from the fourth system.

Example 2.14 Multiplier and accumulator with feedback (fifth system)

The fifth discrete-time system is $y[n] = \beta_0 x[n] + \beta_1 x[n-1] + \alpha_1 y[n-1]$. This system multiplies the current input, previous input, and previous output by constants β_0, β_1, and α_1, respectively, and feeds the sum to the output. In this example, the input signal is $x[n] = \sin(\pi n/5)u[n]$. We provide the output signal obtained from the fifth system (with $\beta_0 = 2.2$, $\beta_1 = -1.1$, and $\alpha_1 = 0.7$) in Figure 2.20.

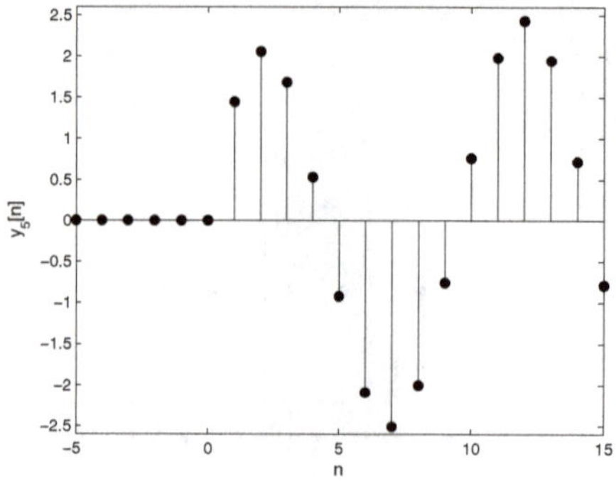

Figure 2.20 Output signal obtained from the fifth system.

Example 2.15 Multiplier and accumulator with future input (sixth system)

The sixth discrete-time system is $y[n] = \beta_0 x[n+1] + \beta_1 x[n]$. This system multiplies the future and the current input values by constants β_0 and β_1, respectively, and feeds the sum to the output. In this example, the input signal is $x[n] = u[n]$. We provide the output signal obtained from the sixth system (with $\beta_0 = 2.2$ and $\beta_1 = -1.1$) in Figure 2.21.

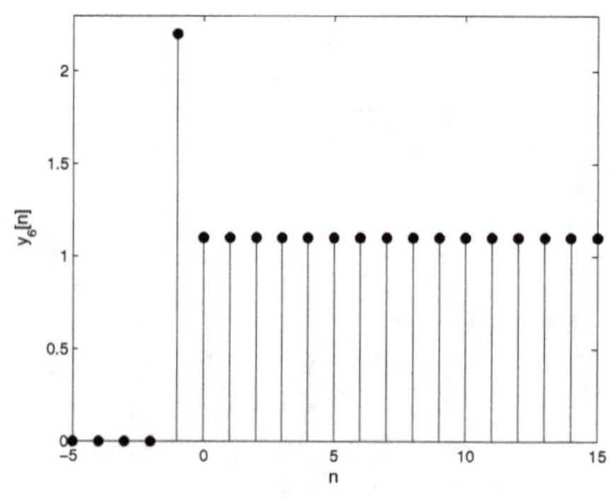

Figure 2.21 Output signal obtained from the sixth system.

Example 2.16 Multiplier and accumulator with unbounded output (seventh system)

The seventh discrete-time system is $y[n] = \beta_0 x[n] + \alpha_1 y[n-1]$. This system multiplies the current input and previous output values by constants β_0 and α_1, respectively, and feeds the sum to the output. In this example, the input signal is $x[n] = \delta[n]$. We provide the output signal obtained from the seventh system (with $\beta_0 = 1$ and $\alpha_1 = 2$) in Figure 2.22.

Example 2.17 Multiplier with a time-based coefficient (eighth system)

The eighth discrete-time system is $y[n] = nx[n]$. This system multiplies the current input value with the current index value and feeds the result to the output. In this example,

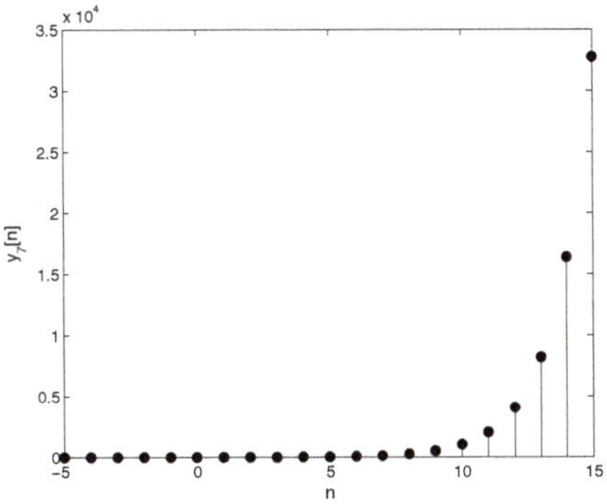

Figure 2.22 Output signal obtained from the seventh system.

the input signal is $x[n] = \sin(\pi n/5)u[n]$. We provide the output signal obtained from the eighth system in Figure 2.23.

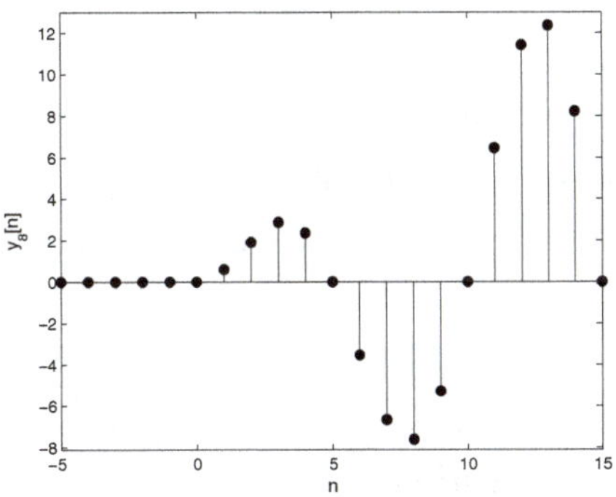

Figure 2.23 Output signal obtained from the eighth system.

The aforementioned discrete-time system examples are not an exhaustive list. However, they present a general idea of the types of systems you will encounter. The aforementioned example systems have different properties, and we will analyze these in detail next.

2.3.2 Properties of Discrete-Time Systems

Discrete-time systems have several properties. In this section, we explore the ones that are most important for our purposes.

Memory

A discrete-time system is said to be memoryless if it requires only the current input value to operate. A system with memory requires the previous (or future) input or output samples to calculate the current output value. In the following examples, we discuss systems with and without memory.

Example 2.18 A system without memory

The first discrete-time system, $y[n] = \beta_0 x[n]$, is memoryless. It multiplies the current input value with a constant and feeds the result to the output. This system does not require any previous (or future) input or output values.

Example 2.19 A system with memory

The fifth discrete-time system, $y[n] = \beta_0 x[n] + \beta_1 x[n-1] + \alpha_1 y[n-1]$, has memory. It requires the previous input and output values to calculate the current output value.

Causality

A discrete-time system with the causality property does not anticipate future input values. We now provide examples of causal and noncausal systems.

Example 2.20 A causal system

The fifth discrete-time system, $y[n] = \beta_0 x[n] + \beta_1 x[n-1] + \alpha_1 y[n-1]$, is causal. It does not require any future input or output values to calculate the current output value.

Example 2.21 A noncausal system

The sixth discrete-time system, $y[n] = \beta_0 x[n+1] + \beta_1 x[n]$, is noncausal. It requires the next input value to calculate the current output.

Stability

There are several different definitions of the stability property [2]. Here, we only consider the bounded input bounded output (BIBO) stability. BIBO stability means that if a system's input signal is bounded, then the system's output signal is also bounded. We now provide examples of stable and unstable systems.

Example 2.22 A stable system

The second discrete-time system, $y[n] = \beta_0 x[n] + \beta_1 x[n-1]$, is stable. In this system, the output is calculated by adding two input samples. If the input signal is bounded, then the sum of any two samples must also be bounded.

Example 2.23 An unstable system

The seventh discrete-time system, $y[n] = \beta_0 x[n] + \alpha_1 y[n-1]$, is unstable for $|\alpha_1| > 1$. Assume that we have $x[n] = \delta[n]$, $\beta_0 = 1$, and $\alpha_1 = 2$. The output from this input signal will tend to infinity as the index n tends to infinity. Thus, this system is unstable.

Stability is an important system property. However, determining the stability of a system is not as simple as just looking at its input–output relationship as we have done here.

Linearity

Linearity is an important system property. Let us say the input–output relationship of a system is represented by $y[n] = H\{x[n]\}$. A system will be linear if it satisfies the following two conditions.

$$H\{\gamma_1 x[n]\} = \gamma_1 H\{x[n]\} \tag{2.19}$$

and

$$H\{\gamma_1 x_1[n] + \gamma_2 x_2[n]\} = \gamma_1 H\{x_1[n]\} + \gamma_2 H\{x_2[n]\} \tag{2.20}$$

where γ_1 and γ_2 are constants. If the system does not satisfy these constraints, it is nonlinear. Next, we provide examples of linear and nonlinear systems.

Example 2.24 A linear system

The sixth discrete-time system, $y[n] = \beta_0 x[n+1] + \beta_1 x[n]$, is linear. It satisfies the constraints given in Eqns. 2.19 and 2.20.

Example 2.25 A nonlinear system

The third discrete-time system, $y[n] = x^2[n]$, is nonlinear. It does not satisfy the constraints given in Eqns. 2.19 and 2.20.

Time-Invariance

Time-invariance is also a very important system property. It states that the system characteristics do not change with time. To test it, we should shift the input signal and observe the effects. If the output signal shifts in the same way as the input signal, the system is said to be time-invariant. Otherwise, it is time-varying. Next, we provide examples of time-invariant and time-varying systems.

Example 2.26 A time-invariant system

The sixth discrete-time system, $y[n] = \beta_0 x[n+1] + \beta_1 x[n]$, is time-invariant. The characteristics of this system do not change with time.

Example 2.27 A time-varying system

The eight discrete-time system, $y[n] = nx[n]$, is time-varying. The multiplier of the input signal is directly related to the index value. Hence, the system characteristics change with time.

Not all systems need to be time-invariant. Adaptive filters are good examples of time-varying systems. They are deliberately designed to be time-varying. Hence, their characteristics can change with time based on a design criterion. We will discuss adaptive filters in detail in Chapter 9.

2.4 Linear and Time-Invariant Systems

Discrete-time systems with both linearity and time-invariance properties are called linear and time-invariant (LTI) systems. LTI systems deserve special consideration because their input–output relationship can be represented in a standard form called a convolution sum.

2.4.1 Convolution Sum

Before deriving the convolution sum, we should first define the impulse response of an LTI system. As the name implies, the impulse response of a system is the output obtained when its input is the discrete unit impulse signal, $\delta[n]$. An LTI system can be completely defined by its impulse response. Throughout this book, we will use $h[n]$ to represent the impulse response of a system. We will cover the derivation of impulse response in Chapter 3. For now, we assume that it is already known.

The impulse response of an LTI system can be used to derive the convolution sum formula. There are three main steps to this.

First, we multiply a signal by $\delta[n]$. This step indicates that every discrete-time signal can be decomposed as the sum of shifted discrete unit impulse signals. Let us assume that the input signal to the LTI system is $x[n]$. By multiplying it by $\delta[n]$, we obtain $x[0]\delta[n]$. Here, the zeroth value of the input signal becomes the coefficient of $\delta[n]$. If we shift $\delta[n]$ to $\delta[n-n_0]$ and multiply it by $x[n]$, we obtain $x[n_0]\delta[n-n_0]$. Now, the n_0th value of the input signal becomes the coefficient of $\delta[n-n_0]$. We can generalize this operation to obtain

$$x[n] = \sum_{k=-\infty}^{\infty} x[k]\delta[n-k] \tag{2.21}$$

Second, we use the time-invariance property. We know that when the input signal is $\delta[n]$, the output will be $h[n]$. Because we have a time-invariant system, if we shift the input signal by n_0 samples, the output should be shifted by the same amount. Therefore, when the input signal is $\delta[n-n_0]$, the output signal will be $h[n-n_0]$.

Third, we benefit from the linearity property. Because we have a linear system, if we have the input signal $x[0]\delta[n]+x[n_0]\delta[n-n_0]$, the output becomes $x[0]h[n]+x[n_0]h[n-n_0]$. Note here that $x[0]$ and $x[n_0]$ are constant values.

These three steps lead to the following general relationship between the input and the output of an LTI system.

$$y[n] = \sum_{k=-\infty}^{\infty} x[k]h[n-k] \tag{2.22}$$

This operation is called the convolution sum. In the above equation, it is represented in the form $y[n] = x[n] * h[n]$. Through a change of variables, the convolution sum can also be represented as

$$y[n] = h[n] * x[n] = \sum_{k=-\infty}^{\infty} h[k]x[n-k] \tag{2.23}$$

Eqn. 2.22 states that as long as we know $h[n]$, we can find the output of an LTI system from any input signal. This general form makes LTI systems favorable to others.

2.4.2 Infinite and Finite Impulse Response Filters

Based on the length of their impulse response, $h[n]$, discrete-time LTI systems can be divided into two categories. If the impulse response of an LTI system is defined for infinite index values, then it is called an Infinite Impulse Response (IIR) filter. If the impulse response of an LTI system is defined only for a finite number of index values, then it is called a Finite Impulse Response (FIR) filter.

In practice, almost all continuous-time filters are IIR filters. A discrete-time filter can be an FIR or IIR filter. You should keep this in mind while reading the remainder of this book.

Theoretically, the convolution sum is the same for IIR and FIR filters. However, practical calculations will yield different results. Assume that $h[n]$ is defined for index values $n \in [0, \infty]$. We require an infinite sum to calculate each output value, which is not possible in practice. Therefore, the convolution sum cannot be used in practice for an IIR filter. Instead, we use the constant coefficient difference equation representation, which we introduce in Section 2.5. On the contrary, the practical version of the convolution sum for an FIR filter is

$$y[n] = \sum_{k=0}^{K-1} h[k]x[n-k] \tag{2.24}$$

where $h[k]$ is defined for $k = 0, \cdots, K-1$.

2.4.3 Causality of LTI Systems

Causality of an LTI system can be observed by looking at its impulse response. If the impulse response is zero for $n < 0$, then the LTI system will be causal. We can justify this property as follows. Let us start with Eqn. 2.23. Assume that we want to calculate the output at index value zero. We can rewrite Eqn. 2.23 as

$$y[0] = \sum_{k=-\infty}^{-1} h[k]x[0-k] + \sum_{k=0}^{\infty} h[k]x[0-k] \tag{2.25}$$

The first sum in Eqn. 2.25 depends on future input values. The second sum depends on current and past input values. For an LTI system to be causal, future input values should not be used in calculating the current output value. This can only be achieved by taking $h[n] = 0$ for $n < 0$. Therefore, an LTI system will be causal if and only if $h[n] = 0$ for $n < 0$. Based on this constraint, the convolution sum formula for a causal LTI system becomes

$$y[n] = \sum_{k=0}^{\infty} h[k]x[n-k] \tag{2.26}$$

2.5 Constant Coefficient Difference Equations

The input–output relationship of a discrete-time LTI system can be represented by a constant coefficient difference equation. The difference equation has the general form

$$y[n] + \alpha_1 y[n-1] + \cdots + \alpha_{L-1} y[n-L+1] = \beta_0 x[n] + \cdots + \beta_{K-1} x[n-K+1] \tag{2.27}$$

We can rewrite Eqn. 2.27 in closed form as

$$y[n] = \sum_{k=0}^{K-1} \beta_k x[n-k] - \sum_{l=1}^{L-1} \alpha_l y[n-l] \tag{2.28}$$

There are two sum terms in Eqn. 2.28. The first one only takes current and past input signal values. This sum is called the *feedforward* term. The second sum only takes past output signal values. This sum is called the feedback term.

> feedforward
> *A non-recursive path from the input to the output of a system.*

The feedback term does not exist in an FIR filter. This can be observed by modifying Eqn. 2.24 with impulse response $h[k] = \beta_k$ for $k = 0, \cdots, K-1$. This also shows the relationship between the convolution sum and difference equation representations for FIR filters. The feedback term will exist in IIR filters. We will demonstrate why this is the case in Chapter 3. The feedback term is the fundamental difference between FIR and IIR filters.

As discussed in the previous section, FIR filters can be implemented by the convolution sum. Because this is not possible for IIR filters, we require the difference equation between the input and output signals to implement them. Before going further, let us present examples of the difference equation representation of FIR and IIR filters.

Example 2.28 FIR filter

Consider the fourth discrete-time system from Section 2.3.1. The input–output relationship of this system is $y[n] = \beta_0 x[n] + \beta_1 x[n-1]$. This is a good example of a constant coefficient difference equation representation of the corresponding FIR filter.

Example 2.29 IIR filter

Consider the fifth discrete-time system from Section 2.3.1. The input–output relationship of the system is $y[n] = \beta_0 x[n] + \beta_1 x[n-1] + \alpha_1 y[n-1]$. This is an example of the constant-coefficient difference equation representation of an IIR filter.

In this chapter, we assumed that the difference equation representing the input–output relationship of a discrete-time system is known. In the next chapter, we will use a Z-transform to obtain the difference equation from the impulse response of an LTI system. We will apply the same procedure to obtain the impulse response of an LTI system represented by a constant coefficient difference equation.

2.6 Further Reading

The theoretical concepts considered in this chapter are explored further in a number of texts [1, 3, 4, 5, 6, 7, 8, 9].

2.7 Exercises

2.7.1 A discrete-time system is defined as $x[n] = u[n+1] - u[n-5]$. The following operations are applied to this signal. Plot $x[n]$ and the generated signals for $n = 0, \cdots, 10$. Comment on the results.

(a) $x[n-2]$.

(b) $x[-4n]$.

(c) $x[2n]$.

(d) $x[n]u[2-n]$.

(e) $x[n-1]\delta[n-3]$.

2.7.2 A discrete-time signal is defined as $x[n] = \sin(3\pi n/5)(u[n] - u[n-5])$. The following operations are applied to this signal. Plot $x[n]$ and the generated signals. Comment on the results.

(a) $3x[n-5]$.

(b) $x[n] - 3x[n-5]$.

2.7.3 Three discrete-time signals are defined below. Plot these signals for $n = -20, \cdots, 20$. Comment on their periodicity.

(a) $x_1[n] = \sin(\pi n/5) + \cos(n/5)$.

(b) $x_2[n] = \cos(\pi n/5) - \cos(n/5)$.

(c) $x_3[n] = x_1[n] + x_2[n]$.

2.7.4 A discrete-time signal is defined as $x[n] = \cos(\pi n/4) + \sin(\pi n/3)$. Plot this signal for $n = -20, \cdots, 20$. Comment on the periodicity of the signal. If this signal is periodic, what is the fundamental period?

2.7.5 Comment on the memory, causality, linearity, and time-invariance properties of the following systems.

(a) $y[n] = 5x[n] + 3$.

(b) $y[n] = \cos(x[n+1])$.

2.7.6 Comment on the linearity and time-invariance properties of the system $y[n] = \lfloor x[n] \rfloor$, where $\lfloor \cdot \rfloor$ is the floor operation.

2.7.7 A discrete-time system is represented as $y[n] = y[n-1] + \cos(x[n+1])$. The system is initially at rest. What will be the output of this system to the below input signals? Plot the obtained output signals for $n = 0, \cdots, 20$.

(a) $x[n] = \delta[n]$.

(b) $x[n] = u[n]$.

2.7.8 An LTI system has impulse response $h[n] = 0.3^n u[n]$. The below input signals are applied to this system. Plot the obtained output signals for $n = 0, \cdots, 20$.

(a) $x[n] = u[n-5]$.

(b) $x[n] = (-0.3)^n u[n]$.

2.7.9 Find $y[n] = x[n] * h[n]$, where $x[n] = (-1/3)^n (u[n] - u[n-20])$ and $h[n] = \sin(\pi/3n)(u[n] - u[n-20])$. Plot $y[n]$ for $n = 0, \cdots, 40$.

2.7.10 An input–output relation of a discrete-time system is given by $y[n] = \frac{n}{n+1} y[n-1] + x[n]$.

(a) Can this system be time invariant?

(b) Find and plot the output of this system for $n = 0, \cdots, 40$ when $x[n] = \delta[n]$. Assume the system is initially at rest.

2.7.11 Determine and plot the impulse response of the LTI system defined by the difference equation $y[n] = -2x[n] + 4x[n-1] - 2x[n-2]$.

2.7.12 The input–output relation of a discrete-time system is given by $y[n] - 2y[n-1] + 3y[n-2] = -4x[n-1]$.

(a) Implement this system. Assume that the system is initially at rest.

(b) Find and plot the output of this system for $n = 0, \cdots, 40$ when $x[n] = \delta[n+1]$.

2.8 References

[1] Smith, S.W. (1997) *The Scientist and Engineer's Guide to Digital Signal Processing*, California Technical Pub.
[2] Ogata, K. (1987) *Discrete-Time Control Systems*, Prentice Hall.
[3] Oppenheim, A.W. and Schafer, R.W. (2009) *Discrete-Time Signal Processing*, Prentice Hall, Third edn.
[4] Mitra, S.K. (2010) *Digital Signal Processing*, McGraw-Hill, Fourth edn.
[5] Proakis, J.G. and Manolakis, D.K. (1995) *Digital Signal Processing: Principles, Algorithms and Applications*, Prentice Hall, Third edn.
[6] Orfanidis, S. (1995) *Introduction to Signal Processing*, Prentice-Hall.
[7] McClellan, J.H., Schafer, R.W., and Yoder, M.A. (2003) *Signal Processing First*, Prentice-Hall.
[8] Manolakis, D.G. and Ingle, V.K. (2011) *Applied Digital Signal Processing*, Cambridge University Press.
[9] Oppenheim, A.W., Willsky, A.S., and Nawab, S.H. (1997) *Signals and Systems*, Prentice Hall, Second edn.

2.9 Lab 2

2.9.1 Introduction

This lab focuses on the basics of discrete-time signals and systems from a practical point of view. We assume that you are familiar with the tools explained in Chapter 0. Hence, a new project has been created and basic adjustments have been made for the STM32F4 microcontroller.

2.9.2 Digital Signals

We will start with digital signals. There are two aims here. First, to make you familiar with basic digital signals. Second, for you to observe real-life data acquired by a sensor.

Basic discrete-time signals are given in Section 2.2.1. The C code for samples of these signals is given below in Listing 2.1. The complete project related to this code can be found in `Online_Student_Resources\Lab2\Basic_Signals`.

```
#include "math.h"
#include "arm_math.h"

#define N 10
int n;
int d1[N], u1[N], r[N];
float e1[N], s[N];
float a=0.8;
float w0=PI/5;

int main(void)
{
```

```
        //unit pulse signal
        for(n=0;n<N;n++)
        if (n==0) d1[n]=1;
        else d1[n]=0;

        //unit step signal
        for(n=0;n<N;n++)
        u1[n]=1;

        //unit ramp signal
        for(n=0;n<N;n++)
        r[n]=n;

        //exponential signal
        for(n=0;n<N;n++)
        e1[n]=pow(a,(float)n);

        //sinusoidal signal
        for(n=0;n<N;n++)
        s[n]=sin(w0*(float)n);

        return(0);
}
```

Listing 2.1 Basic digital signals.

Discrete unit impulse, unit step, and unit ramp functions can be represented as `integer` arrays as shown in Listing 2.1. However, the exponential and sinusoidal signals must be represented as `float` arrays. Therefore, we had to apply casting to the variable `n` in operation.

Task 2.1

Run the code in Listing 2.1 and plot the digital signals you obtain. This step is necessary to become familiar with basic digital signals.

Task 2.2

We will extensively use the sum of sinusoidal signals in the following exercises. Therefore, we will generate two sinusoidal signals with normalized angular frequencies of $w = \pi/128$ and $w = \pi/4$ rad/sample, respectively. Write a C code to construct these two signals for 512 samples. Plot the obtained signals. Comment on the effect of normalized angular frequency on the sinusoidal signal.

Sinusoidal signals are extensively used in digital signal processing. You can understand how they are generated in the C code with this application. To note here, these two generated sinusoidal signals will be used in examples throughout this book, so they should be generated correctly in this lab.

Task 2.3

We acquired actual digital signals from the accelerometer sensor and AUP audio card in Chapter 1. Plot these acquired signals, and comment on the complexity of these signals compared to the ones given in Listing 2.1.

Here, you should observe the differences between digital signals created by a code line and ones obtained from the real world. This should highlight the complexity of digital signals in real-life applications.

2.9.3 Operations on Digital Signals

The aim here is to emphasize how arithmetic and shifting operations affect a digital signal. We considered operations on digital signals in Section 2.2.2. The C code for the examples given there are provided in Listing 2.2. The complete project related to this code can be found in `Online_Student_Resources\Lab2\Operations_on_Signals`.

```
#include "math.h"
#include "arm_math.h"


#define N 10

int n;
int d[N], u[N];
float e[N], s[N];
float a=0.8;
float w0=PI/5;

int x1[N], x3[N], x5[N];
float x2[N], x4[N], x6[N], x7[N], x8[N];

int main(void){

      //unit pulse signal
      for(n=0;n<N;n++)
      if (n==0) d[n]=1;
      else d[n]=0;

      //unit step signal
      for(n=0;n<N;n++)
      u[n]=1;

      //exponential signal
      for(n=0;n<N;n++)
      e[n]=pow(a,(float)n);

      //sinusoidal signal
      for(n=0;n<N;n++)
      s[n]=sin(w0*(float)n);

      //shifted unit pulse signal
      for(n=0;n<N;n++)
      if (n<3) x1[n]=0;
      else x1[n]=d[n-3];

      //elevated sinusoidal signal
```

```
        for(n=0;n<N;n++)
        x2[n]=s[n]+1;

        //negated unit step signal
        for(n=0;n<N;n++)
        x3[n]=-u[n];

        //applying all operations on the sinusoidal signal
        for(n=0;n<N;n++)
        if (n<2) x4[n]=0.0;
        else x4[n]=3.0*s[n-2]-2.0;

        //subtracting two unit step signals
        for(n=0;n<N;n++)
        if (n<4) x5[n]=u[n];
        else x5[n]=u[n]-u[n-4];

        //multiplying the exponential signal with the unit step signal
        for(n=0;n<N;n++)
        x6[n]=e[n]*u[n];

        //multiplying the exponential signal with the sinusoidal signal
        for(n=0;n<N;n++)
        x7[n]=e[n]*s[n];

        //multiplying the exponential signal with the window signal
        for(n=0;n<N;n++)
        x8[n]=e[n]*x5[n];

        return(0);
}
```

Listing 2.2 Digital signals obtained by shifting and arithmetic operations.

Task 2.4

Run the code in Listing 2.2 and plot the obtained digital signals. This step is necessary for you to become familiar with operations on digital signals.

Task 2.5

Now, we provide more examples related to operations on digital signals. Apply the following operations and plot the obtained signals for 10 samples.

1. $x_1[n] = 0.6r[n+4]$
2. $x_2[n] = u[n-3] - u[n-8]$
3. $x_3[n] = u[n] - u[n-3] + u[n-8]$
4. $x_4[n] = x_2[n]s[n] + d[n]$
5. $x_5[n] = -2.4e[n]s[n]$

You should observe that digital signals can be modified using different arithmetic operations with this application. Moreover, new digital signals can be obtained by combining different digital signals.

Task 2.6

Finally, we focus on the sinusoidal signals generated in Task 2.2. Write a C code to sum these two sinusoidal signals. Comment on the obtained signal.

Here, you should observe that a signal can be constructed using mathematical operations. Further, note that adding two sinusoidal signals will actually generate a new signal.

2.9.4 Periodic Digital Signals

In this section, we deal with periodic digital signals. The aim here is for you to become familiar with the concept of periodicity in digital signals. We will also introduce different types of periodic digital signals.

We introduced periodic signals in Section 2.2.4. There, we used Example 2.9 to discuss the periodicity of discrete-time signals. We provide the C code for this example in Listing 2.3. The complete project related to this code can be found in `Online_Student_Resources\Lab2\Periodic_Signals`.

```
#include "math.h"
#include "arm_math.h"

#define N 100

int n;

float w1=PI/10;
float w2=3.0/10.0;

float sin1[N], sin2[N];

int main(void){

        //sinusoidal signals
        for(n=0;n<N;n++)
        {
                sin1[n]=cos(w1*(float)n);
                sin2[n]=cos(w2*(float)n);
        }

        return(0);
}
```

Listing 2.3 Sinusoidal periodic and aperiodic digital signals.

Task 2.7

Run the code in Listing 2.3 and plot the obtained digital signals. This step is necessary to become familiar with periodic digital signals.

A periodic signal need not be sinusoidal only. We can generate periodic square and triangle signals using the C code in Listing 2.4. The complete project related to this code can be found in `Online_Student_Resources\Lab2\Periodic_Square_Triangle_Signals`. Here, the periodic square signal has amplitude 2.4 and a period of 50 samples. The periodic triangle signal has amplitude 1.5 and a period of 40 samples. We generate 100 samples of both signals in Listing 2.4.

```
#define N 100

int n;

float Asq=2.4; // Amplitude of the square signal
float Atr=1.5; // Amplitude of the triangle signal
int Psq=50; // Period of the square signal
int Ptr=40; // Period of the triangle signal

float square[N], triangle[N];

int main(void){

        //One period of the square signal
        for(n=0;n<Psq/2;n++)
        if(n<(Psq/2)) square[n]=Asq;
        else square[n]=-Asq;

        // One period of the triangle signal
        for(n=0;n<Ptr;n++)
        if(n<(Ptr/2))triangle[n]=(2*Atr/(Ptr/2))*n - Atr;
        else triangle[n] = -(2*Atr/(Ptr/2))*(n-Ptr/2) + Atr;

        // Generating the square signal
        for(n=Psq;n<N;n++)
        square[n]=square[n%Psq];

        // Generating the triangle signal
        for(n=Ptr;n<N;n++)
        triangle[n]=triangle[n%Ptr];

        return(0);
}
```

Listing 2.4 The C code for obtaining periodic square and triangle signals.

Task 2.8

Run the code in Listing 2.4, and plot the obtained digital signals.

In Listing 2.4, we first generate one period of the periodic signal (square or triangle). Then, we extend it by copying its content. This method should emphasize the periodicity concept clearly.

Task 2.9

Modify the C code in Listing 2.4 to generate 100 samples of a periodic sawtooth signal. Let the amplitude of the signal be 0.75 and its period be 20 samples. Run your code and plot the obtained periodic digital signal.

We will use the periodic signals generated here in Chapter 4.

2.9.5 Implementing Digital Systems

One of the most important properties of a digital system is that it can be implemented in code form. If a suitable digital platform can execute this code block, then the system can be realized. The beauty of this framework is that if we want to change the system characteristics, only the code block corresponding to the system needs to be changed.

There are two ways to implement a digital system. The first is to process each input signal sample separately. This is called sample-based processing. The second is to process a block of input signal samples all at once. This is called frame-based (buffer-based) processing. Each implementation method has certain advantages and disadvantages. We will explore these in detail in Chapter 11.

We introduced eight digital systems with different properties in Section 2.3.1. We used the discrete unit impulse, unit step, and sinusoidal digital signals as inputs for these systems. We provide the sample-based implementation of these eight digital systems in Listing 2.5. The complete project related to this code can be found in `Online_Student_Resources\Lab2\Sample_based_Systems`.

```
#include "math.h"
#include "arm_math.h"

float digital_system1(float b, float input){
      return b*input;}

float digital_system2(float input1, float input2){
      return input1+input2;}

float digital_system3(float input){
      return pow(input,2);}

float digital_system4(float b0, float b1, float input0, float input1){
      return b0*input0+b1*input1;}

float digital_system5(float b0, float b1, float a1, float input0,
float input1, float output1){
      return b0*input0+b1*input1+a1*output1;}

float digital_system6(float b0, float b1, float input0, float input1){
      return b0*input0+b1*input1;}

float digital_system7(float b0, float a1, float input0, float output1){
      return b0*input0+a1*output1;}

float digital_system8(int n, float input0){
      return (float)n*input0;}

#define N 10

int n;
float d[N], u[N], s[N];
float w0=PI/5;
float y1[N],y2[N],y3[N],y4[N],y5[N],y6[N],y7[N],y8[N];

int main(void)
```

```
{

    //unit pulse signal
    for(n=0;n<N;n++)
    if (n==0) d[n]=1;
    else d[n]=0;

    //unit step signal
    for(n=0;n<N;n++)
    u[n]=1;

    //sinusoidal signal
    for(n=0;n<N;n++)
    s[n]=sin(w0*(float)n);

    //y[n] = b x[n]
    for(n=0;n<N;n++)
    y1[n]=digital_system1(2.2, u[n]);

    //y[n] = x1[n] + x2[n]
    for(n=0;n<N;n++)
    y2[n]=digital_system2(u[n], s[n]);

    //y[n] = x^2[n]
    for(n=0;n<N;n++)
    y3[n]=digital_system3(s[n]);

    //y[n] = b0 x[n] + b1 x[n-1]
    for(n=0;n<N;n++)
    if (n==0)
    y4[n]=digital_system4(2.2,-1.1,s[n],0);
    else
    y4[n]=digital_system4(2.2,-1.1,s[n],s[n-1]);

    //y[n] = b0 x[n] + b1 x[n-1] + a1 y[n-1]
    for(n=0;n<N;n++)
    if (n==0)
    y5[n]=digital_system5(2.2,-1.1,0.7,s[n],0,0);
    else
    y5[n]=digital_system5(2.2,-1.1,0.7,s[n],s[n-1],y5[n-1]);

    //y[n] = b0 x[n+1] + b1 x[n]
    for(n=0;n<N;n++)
    if (n<N)
    y6[n]=digital_system6(2.2,-1.1,u[n+1],u[n]);

    //y[n] = b0 x[n] + a1 y[n-1]
    for(n=0;n<N;n++)
    if (n==0)
    y7[n]=digital_system7(1.0,2.0,d[n],0);
    else
    y7[n]=digital_system7(1.0,2.0,d[n],y7[n-1]);

    //y[n] = n x[n]
    for(n=0;n<N;n++)
    y8[n]=digital_system8(n,s[n]);

    return(0);
}
```

Listing 2.5 Implementation of sample-based digital systems in code form.

Task 2.10

Run the code in Listing 2.5, and plot the obtained digital signals. This step is necessary to become familiar with sample-based implementation of digital systems. Please refer to the initial conditions for digital systems in Listing 2.5, where it can be seen that the initial conditions of a digital system should be handled in a specific way.

We provide the frame-based implementation of the eight digital systems in Listing 2.6. The complete project related to this code can be found in `Online_Student_Resources\Lab2\Frame_based_Systems`. As in Listing 2.5, we use the discrete unit impulse, unit step, and sinusoidal digital signals as inputs to these systems. Here, the input signal to be processed is stored in a buffer and is fed to the system all at once. The output of the system is also obtained through a buffer.

```
#include "math.h"
#include "arm_math.h"

void digital_system1(float b, float input[], float output[], int size){
        int n;
        for(n=0;n<size;n++)
        output[n]=b*input[n];
}

void digital_system2(float input1[], float input2[], float output[],int size){
        int n;
        for(n=0;n<size;n++)
        output[n]=input1[n]+input2[n];
}

void digital_system3(float input[], float output[], int size){
        int n;
        for(n=0;n<size;n++)
        output[n]=pow(input[n],2);
}

void digital_system4(float b[], float input[], float output[], int size){
        int n;
        for(n=0;n<size;n++)
        if (n==0)
        output[n]=b[0]*input[n];
        else
        output[n]=b[0]*input[n]+b[1]*input[n-1];
}

void digital_system5(float b[], float a, float input[], float output[], int size){
        int n;
        for(n=0;n<size;n++)
        if (n==0)
        output[n]=b[0]*input[n];
        else
        output[n]=b[0]*input[n]+b[1]*input[n-1]+a*output[n-1];
}

void digital_system6(float b[], float input[], float output[], int size){
        int n;
        for(n=0;n<size;n++)
        if (n<size)
        output[n]=b[0]*input[n+1]+b[1]*input[n];
}

void digital_system7(float b, float a, float input[], float output[],int size){
        int n;
        for(n=0;n<size;n++)
```

```
        if (n==0)
        output[n]=b*input[n];
        else
        output[n]=b*input[n]+a*output[n-1];
}

void digital_system8(float input[], float output[], int size){
        int n;
        for(n=0;n<size;n++)
        output[n]=(float)n*input[n];
}

#define N 10

int n;
float d[N], u[N], s[N];
float w0=PI/5;
float b[2];
float a;
float y1[N],y2[N],y3[N],y4[N],y5[N],y6[N],y7[N],y8[N];


int main (void){

        //unit pulse signal
        for(n=0;n<N;n++)
        if (n==0) d[n]=1;
        else d[n]=0;

        //unit step signal
        for(n=0;n<N;n++)
        u[n]=1;

        //sinusoidal signal
        for(n=0;n<N;n++)
        s[n]=sin(w0*(float)n);

        //y[n] = b x[n]
        b[0]=2.2;
        digital_system1(b[0],u,y1,N);

        //y[n] = x1[n] + x2[n]
        digital_system2(u,s,y2,N);

        //y[n] = x^2[n]
        digital_system3(s,y3,N);

        //y[n] = b0 x[n] + b1 x[n-1]
        b[0]=2.2;
        b[1]=-1.1;
        digital_system4(b,s,y4,N);

        //y[n] = b0 x[n] + b1 x[n-1] + a1 y[n-1]
        b[0]=2.2;
        b[1]=-1.1;
        a=0.7;
        digital_system5(b,a,s,y5,N);

        //y[n] = b0 x[n+1] + b1 x[n]
        b[0]=2.2;
        b[1]=-1.1;
        digital_system6(b,u,y6,N);

        //y[n] = b0 x[n] + a1 y[n-1]
        b[0]=1.0;
        a=2.0;
        digital_system7(b[0],a,d,y7,N);

        //y[n] = n x[n]
        digital_system8(s,y8,N);
```

```

        return(0);
}
```

Listing 2.6 Implementation of frame-based digital systems in code form.

Task 2.11

Run the code in Listing 2.6, and plot the obtained digital signals. This step is necessary to become familiar with frame-based implementation of digital systems.

Task 2.12

Now, we provide more examples on sample-based and frame-based implementation of digital systems. Implement the following systems in sample-based form and plot the obtained output signals.

1. $y_1[n] = x_1[n] + x_2[n]$, where $x_1[n] = r[n]$ and $x_2[n] = e[n]$
2. $y_2[n] = x_3[n]$, where $x_3[n] = r^2[n]$
3. $y_3[n] = 2.2y_1[n] - 1.1y_1[n-1] + 0.7y_3[n-1]$
4. $y_4[n] = 2.2y_2[n+1] - 1.1y_2[n]$

Task 2.13

Implement the systems in Task 2.12 in frame-based form. Plot the obtained signals. Is there a difference between the sample-based and frame-based implementations in terms of output?

The exercises in this section should clarify digital system concepts and how they can be implemented. We will extensively use sample-based and frame-based processing methods in the following chapters, so you should become familiar with them.

2.9.6 LTI System Implementation

This section covers the implementation of LTI systems. The first option here is direct implementation, where all necessary functions are constructed by the user. The second option is to use predefined functions in implementation. Arm offers the CMSIS-DSP library for implementing DSP algorithms. This library has many built-in functions for digital signal processing and its related mathematical operations. We will extensively use them in implementing DSP algorithms throughout this book. More information on the CMSIS-DSP library can be found at `http://www.arm.com/products/processors/cortex-m/cortex-microcontroller-software-interface-standard.php`. We start with FIR filter implementation.

FIR Filter Implementation

The FIR filter is easy to implement and is always stable, so we start this section with the implementation of FIR filters. We use the sum of sinusoidal signals obtained in Task 2.6 as the input for all examples in this section. We start with the direct implementation of FIR filters.

Direct Implementation

In this section, we will focus on the direct implementation of FIR filtering in terms of the convolution sum. The aim here is to help you understand how FIR filtering is implemented in C code. The C code for direct implementation is given in Listing 2.7. The complete project related to this code can be found in `Online_Student_Resources\Lab2\Direct_FIR_Filtering`. There, we also provide the filter coefficients to be used in the `FIR_lpf_coefficients.h` header file. We will explain how these filter coefficients are obtained in Chapter 6.

```
#include "arm_math.h"
#include "FIR_lpf_coefficients.h"

void convolution_sum (float x[], float h[], float y[], int
size1 , int size2){
          int n,k;
          float sum;
          for(n=0;n<size1;n++)
          {
                    sum=0.0;
                    for(k=0;k<size2;k++)
                    if(k<(n+1)) sum+=h[k]*x[n-k];
                    y[n]=sum;
          }
}

#define N 512

int n;
float x[N],y[N];

int main (void){

          for(n=0;n<N;n++)
          x[n] = arm_sin_f32(PI*n/128)+arm_sin_f32(PI*n/4);

          convolution_sum (x,h,y,N,K);

          return(0);
}
```

Listing 2.7 FIR filtering application with direct implementation.

The `convolution_sum` function has the basic operations of multiplication and addition. These are called multiply and accumulate (MAC) operations. Microcontroller and DSP hardware take this structure into account to speed up the convolution sum calculations.

```
void arm_fir_init_f32(arm_fir_instance_f32 *S, uint16_t numTaps, float32_t *pCoeffs,
float32_t *pState, uint32_t blockSize)
/*
S: the instance of FIR filter.
numTaps: the number of filter coefficients.
pCoeffs: the pointer for the filter coefficients buffer.
pState: the pointer for the state buffer.
blockSize: number of samples that are processed per call.
*/
```

Listing 2.8 FIR initialization function.

Task 2.14

Run the code in Listing 2.7, and plot the output. What do you observe?

Task 2.15

Replace the header file in Listing 2.7 with the file `FIR_hpf_coefficients.h` given in `Online_Student_Resources\Lab2\Direct_FIR_Filtering`. We will explain how these filter coefficients are obtained in Chapter 6. Run the code in Listing 2.7 again, and plot the output. What do you observe this time?

Task 2.16

Measure the memory usage of the filtering operations in this section. You will benefit from the method given in Section A.6. Then, rearrange the code as explained in Section A.5, and measure the execution time of the `convolution_sum` function. Comment on the results.

CMSIS-DSP Library-based Implementation

Because the convolution sum is extensively used in FIR filtering, the CMSIS-DSP library contains predefined functions for this. These take the Arm architecture into account. Therefore, they provide the result faster than direct implementation.

The CMSIS-DSP library has two function sets for FIR filtering. The first set is based on predefined FIR functions. The second set is based on the convolution sum operation. The two sets of functions provide the same result, but their usage is slightly different.

Let us start with the predefined FIR-based functions. These are the functions `arm_fir_init_f32` and `arm_fir_f32`. The first function is for initialization and is summarized in Listing 2.8.

Once initialization is completed, `arm_fir_f32` can be used for floating-point FIR filter implementation. This function is summarized in Listing 2.9.

We provide the CMSIS-DSP library-based FIR filtering function implementation in Listing 2.10. The complete project related to this code can be found in `Online_`

```
void arm_fir_f32(const arm_fir_instance_f32 *S, float32_t *pSrc, float32_t *pDst,
uint32_t blockSize)
/*
S: the instance of FIR filter.
pSrc: the pointer for the input data buffer.
pDst: the pointer for the output data buffer.
blockSize: the number of samples that are processed per call.
*/
```

Listing 2.9 FIR implementation function.

`Student_Resources\Lab2\CMSIS_FIR_Filtering`. There, we also provide the filter coefficients to be used in the `FIR_lpf_coefficients.h` header file.

```
#include "arm_math.h"
#include "FIR_lpf_coefficients.h"

#define N 512

float32_t x[N];
float32_t y[N];
float32_t firStateF32[N + K - 1];
int16_t n,k;
float32_t dummy;

arm_fir_instance_f32 S;

int main (void){

        for(n=0;n<N;n++)
        x[n] = arm_sin_f32(PI*n/128)+arm_sin_f32(PI*n/4);

        for(k=0;k<K/2;k++){
            dummy = h[k];
            h[k] = h[K-k-1];
            h[K-k-1] = dummy;
        }

        arm_fir_init_f32(&S, K, h, firStateF32, N);
        arm_fir_f32(&S, x, y, N);

        return(0);
}
```

Listing 2.10 FIR filtering with the CMSIS-DSP library FIR function.

Task 2.17

Run the code in Listing 2.10, and plot the output. What do you observe?

Task 2.18

Now, replace the header file in Listing 2.10 with the file `FIR_hpf_coefficients.h` given in `Online_Student_Resources\Lab2\CMSIS_FIR_Filtering`. Run the code in Listing 2.10 again, and plot the output. What do you observe?

```
void arm_conv_f32(float32_t * pSrcA, uint32_t srcALen, float32_t * pSrcB,
uint32_t srcBLen, float32_t * pDst)
/*
pSrcA: the pointer for the first input array.
srcALen: the length of the first input array.
pSrcB: the pointer for the second input array.
srcBLen: the length of the second input array.
pDst: the pointer for the output data buffer. Length srcALen+srcBLen-1.
*/

void arm_conv_partial_f32(float32_t *pSrcA, uint32_t srcALen, float32_t *pSrcB,
uint32_t srcBLen,
float32_t *pDst, uint32_t firstIndex, uint32_t numPoints)
/*
pSrcA: the pointer for the first input array.
srcALen: the length of the first input array.
pSrcB: the pointer for the second input array.
srcBLen: the length of the second input array.
pDst: the pointer for the output data buffer.
firstIndex: the first output sample to start with.
numPoints: the number of output points to be computed.
*/
```

Listing 2.11 Convolution functions.

Task 2.19

Measure the memory usage of the filtering operations in this section. You will benefit from the method given in Section A.6. Then, rearrange the code as described in Section A.5, and measure the execution time of the FIR filtering function. Comment on the results. Compare the results obtained here with the ones obtained in Task 2.16.

The second set of functions for FIR filtering are based on the convolution sum. The CMSIS-DSP library has two different convolution sum functions to implement an FIR filter. These are `arm_conv_f32` and `arm_conv_partial_f32`. The first function does not allow any parameters to be set. The second function allows the user to decide the number of output samples to compute. We demonstrate the usage of both convolution functions in Listing 2.11.

We provide the CMSIS-DSP library-based convolution function implementation in Listing 2.12. The complete project related to this code can be found in `Online_Student_Resources\Lab2\CMSIS_Convolution`. There, we also provide the filter coefficients to be used in the `FIR_lpf_coefficients.h` header file.

Task 2.20

Run the code in Listing 2.12, and plot the output. What do you observe?

Task 2.21

Replace the header file in Listing 2.12 with the file `FIR_hpf_coefficients.h` given in `Online_Student_Resources\Lab2\CMSIS_Convolution`. Run the code in Listing 2.12 again, and plot the output. What do you observe this time?

```
#include "arm_math.h"
#include "FIR_lpf_coefficients"

#define N 512

float32_t x[N];
float32_t y[N];
int16_t n;

int main (void){

        for(n=0;n<N;n++)
        x[n] = arm_sin_f32(PI*n/128)+arm_sin_f32(PI*n/4);

        arm_conv_f32(x, N, h, K, y);

        return(0);
}
```

Listing 2.12 FIR filtering with the CMSIS-DSP library convolution function.

Task 2.22

Measure the memory usage of the filtering operations in this section. You will benefit from the method given in Section A.6. Then, rearrange the code as explained in Section A.5, and measure the execution time of the convolution function. Comment on the results. Compare the results obtained here with the ones obtained in Exercises 2.16 and 2.19.

You should now understand how FIR filtering can be performed through direct implementation, CMSIS-DSP library convolution, and FIR filtering functions with the applications given in this section. Moreover, the execution time and memory usage values calculated in this section will aid in selecting the right implementation method.

IIR Filter Implementation

As mentioned in Section 2.5, the input–output relationship of an IIR filter can be represented by a constant coefficient difference equation. This equation can be coded directly to implement the IIR filter described by its parameters.

We demonstrate the direct implementation of IIR filtering in Listing 2.13. The complete project related to this code can be found in `Online_Student_Resources\Lab2\Direct_IIR_Filtering`. As in the FIR filtering case, we use the sum of the sinusoidal signals obtained in Task 2.6 as the input in this section. We provide the filter coefficients within the code. We will explain how these filter coefficients are obtained in Chapter 6.

```
#include "arm_math.h"

void IIR_filter (float b[], float a[], float x[], float y[],
int size []){
        int n,k,l;
        float sum;
        for(n=0;n<size [0];n++)
        {
```

```
            sum=0.0;
            for(k=0;k<size [1];k++)
            if(k<(n+1)) sum+=b[k]*x[n-k];
            for(l=1;l<size [2];l++)
            if(l<(n+1)) sum -=a[l]*y[n-l];
            y[n]=sum;
        }
}

#define N 512
#define K 3
#define L 3

int n;
int size[3];
float x[N],y[N];

float b[K] = {0.002044, 0.004088, 0.002044};
float a[L] = {1, -1.819168, 0.827343};

int main(void){
    for(n=0;n<N;n++)
    x[n] = arm_sin_f32(PI*n/128)+arm_sin_f32(PI*n/4);

    size [0]=N;
    size [1]=K;
    size [2]=L;

    IIR_filter (b,a,x,y,size);

    return(0);
}
```

Listing 2.13 IIR filtering with direct implementation.

Task 2.23

Run the code in Listing 2.13, and plot the output. What do you observe?

Task 2.24

Replace the filter coefficients in Listing 2.13 with the ones given in Listing 2.14. Run the code in Listing 2.13 again, and plot the output. What do you observe this time?

```
float b[K] = {0.705514, -1.414028, 0.705514};
float a[L] = {1, -1.359795, 0.462261};
```

Listing 2.14 New filter coefficients.

Task 2.25

Measure the memory usage of the filtering operations in this section. You will benefit from the method given in Section A.6. Then, rearrange the code as explained in Section A.5, and measure the execution time of the IIR filtering function. Comment on the results. Compare the results obtained from FIR and IIR filtering.

You should now understand the implementation details of IIR filtering, as well as the differences between FIR and IIR filtering operations.

3

The Z-Transform

Contents

3.1 Introduction

The Z-transform represents a discrete-time signal or system in the complex domain, which means it provides information that is not clearly evident in the time domain. This can aid in analyzing the signal or system from a different perspective. In this chapter, we will explore the Z-transform and its properties. First, we demonstrate the calculation of the Z-transform. Second, we introduce the Z-transform's properties. Third, we introduce methods for calculating the inverse Z-transform. Then, we apply the Z-transform to an

LTI system described by a constant coefficient difference equation. This will lead to a broader and stronger knowledge of the analysis of LTI systems. Related to this, we also introduce the block diagram representation of an LTI system.

3.2 Definition of the Z-Transform

The Z-transform of a discrete-time signal, $x[n]$, is defined as

$$X(z) = \sum_{n=-\infty}^{\infty} x[n]z^{-n} \tag{3.1}$$

where z is a complex variable.

Eqn. 3.1 indicates that $X(z)$ is the infinite sum of $x[n]$ multiplied by z^{-n}. This infinite sum is only valid if it converges. z values that satisfy this ***convergence*** constraint correspond to a part of the ***complex plane*** called the Region of Convergence (ROC). Thus, any Z-transform has an accompanying ROC.

> **convergence**
> *The property of approaching a limit value.*

> **complex plane**
> *A geometrical representation of complex numbers established using a real axis and a perpendicular imaginary axis.*

We can calculate the Z-transform of discrete-time signals using Eqn. 3.1. Let us start with the discrete unit impulse signal.

Example 3.1 The discrete unit impulse signal

The discrete unit impulse signal has the Z-transform

$$X(z) = \sum_{n=-\infty}^{\infty} \delta[n]z^{-n} = 1 \tag{3.2}$$

As can be seen in Eqn. 3.2, because the unit impulse is equal to zero for any non-zero value of n, the Z-transform of the discrete unit impulse signal is equal to one. This result will be useful in obtaining the impulse response of an LTI system. Here, the ROC is the whole complex plane because the Z-transform is independent of the value of z.

Example 3.2 The unit step signal

The Z-transform of the unit step signal is

$$X(z) = \sum_{n=-\infty}^{\infty} u[n]z^{-n} \tag{3.3}$$

$$= \sum_{n=0}^{\infty} z^{-n} \tag{3.4}$$

$$\tag{3.5}$$

Comparing this with the well-known closed form of a geometric series

$$\sum_{n=0}^{\infty} p^n = \frac{1}{1-p}, |p| < 1 \tag{3.6}$$

Eqn. 3.4 may be expressed as

$$\sum_{n=0}^{\infty} z^{-n} = \frac{1}{1-z^{-1}}, |z^{-1}| < 1 \tag{3.7}$$

Significantly, there is a constraint on the values of z for which this is valid, and this constraint defines the ROC for the Z-transform.

Example 3.3 The exponential signal

The Z-transform of the exponential signal is

$$X(z) = \sum_{n=-\infty}^{\infty} a^n u[n]z^{-n} \tag{3.8}$$

$$= \sum_{n=0}^{\infty} (a^{-1}z)^{-n} \tag{3.9}$$

$$= \frac{1}{1-az^{-1}}, |z| > |a| \tag{3.10}$$

3.3 Z-Transform Properties

The Z-transform has several properties. In this section, we will consider four important properties of the Z-transform. We selected these four properties because they will be used in derivations in further sections of this book.

3.3.1 Linearity

Assume that $X_1(z)$ and $X_2(z)$ are the Z-transforms of the discrete-time signals $x_1[n]$ and $x_2[n]$, respectively. Say we have a discrete-time signal $x_T[n] = \gamma_1 x_1[n] + \gamma_2 x_2[n]$, where γ_1 and γ_2 are constants.

Derivation of the linearity property is straightforward. Applying the Z-transform formula to $x_T[n]$, we obtain

$$X_T(z) = \sum_{n=-\infty}^{\infty} x_T[n]z^{-n} \tag{3.11}$$

$$= \gamma_1 \sum_{n=-\infty}^{\infty} x_1[n]z^{-n} + \gamma_2 \sum_{n=-\infty}^{\infty} x_2[n]z^{-n} \tag{3.12}$$

$$= \gamma_1 X_1(z) + \gamma_2 X_2(z) \tag{3.13}$$

We can thus clearly see that the Z-transform is linear.

3.3.2 Time Shifting

Derivation of this property is as follows. Assume that we shift $x[n]$ by n_0 to obtain $x_2[n] = x[n-n_0]$. The Z-transform of $x_2[n]$ is

$$X_2(z) = \sum_{n=-\infty}^{\infty} x_2[n]z^{-n} \tag{3.14}$$

$$= \sum_{n=-\infty}^{\infty} x[n-n_0]z^{-n} \tag{3.15}$$

Applying the change of variables $m = n - n_0$, Eqn. 3.15 becomes

$$X_2(z) = \sum_{m=-\infty}^{\infty} x[m]z^{-(m+n_0)} \tag{3.16}$$

$$= z^{-n_0}X(z) \tag{3.17}$$

This derivation states that if we shift a discrete-time signal in the time domain by n_0 samples, its Z-transform is multiplied by z^{-n_0} in the complex domain.

3.3.3 Multiplication by an Exponential Signal

Say that the Z-transform of a discrete-time signal, $x[n]$, is $X(z)$. Assume that we multiply $x[n]$ by an exponential signal, a^n, to obtain $x_2[n] = a^n x[n]$. The Z-transform of $x_2[n]$

will be

$$X_2(z) = \sum_{n=-\infty}^{\infty} x_2[n]z^{-n} \tag{3.18}$$

$$= \sum_{n=-\infty}^{\infty} x[n]a^n z^{-n} \tag{3.19}$$

$$= \sum_{n=-\infty}^{\infty} x[n](a^{-1}z)^{-n} \tag{3.20}$$

$$= X(a^{-1}z) \tag{3.21}$$

This property states that if we multiply a discrete-time signal by an exponential signal in the time domain, its Z-transform is scaled by the exponent term in the complex domain.

3.3.4 Convolution of Signals

The most important Z-transform property is the convolution property. Let us derive this property stepwise. The convolution sum of the signals $x[n]$ and $h[n]$ is

$$y[n] = \sum_{k=-\infty}^{\infty} x[k]h[n-k] \tag{3.22}$$

Taking the Z-transform of $y[n]$, we obtain

$$Y(z) = \sum_{n=-\infty}^{\infty} y[n]z^{-n} \tag{3.23}$$

$$= \sum_{n=-\infty}^{\infty} \sum_{k=-\infty}^{\infty} x[k]h[n-k]z^{-n} \tag{3.24}$$

Changing the order of summations and applying the time shifting property, we then obtain

$$Y(z) = \sum_{k=-\infty}^{\infty} x[k] \sum_{n=-\infty}^{\infty} h[n-k]z^{-n} \tag{3.25}$$

$$= \sum_{k=-\infty}^{\infty} x[k]H(z)z^{-k} \tag{3.26}$$

$H(z)$ is independent of the sum term, so it can be removed from the summation. This gives us

$$Y(z) = H(z)X(z) \tag{3.27}$$

We can summarize the result as follows. Convolving two discrete-time signals in the time domain results in multiplying their Z-transform in the complex domain. This result is extremely useful in obtaining the relationship between the difference equation representing a discrete-time system and its impulse response. We will also use this property in the filtering operation we introduce in Section 4.6.

3.4 Inverse Z-Transform

The formula for the inverse Z-transform is

$$x[n] = \frac{1}{2\pi j}\oint_C X(z)z^{n-1}dz \tag{3.28}$$

where C represents the closed contour within the ROC of the Z-transform. The integral in Eqn. 3.28 should be taken in the complex domain aided by contour integration methods. In other words, we cannot calculate it using Riemann integration methods, which are based on approximating the area under a given curve with infinitesimal rectangles. This is discussed in almost all engineering-based digital signal processing books [1, 2, 3, 4, 5, 6, 7].

In practical applications, one can find the inverse Z-transform without taking the complex integral in Eqn. 3.28. There are three methods to achieve this, which are summarized below.

3.4.1 Inspection

The idea behind this method is based on representing the Z-transform in terms of known Z-transform pairs. The first step is to partition the Z-transform, which can be performed using partial fraction expansion [1]. The inverse Z-transform of each extracted fractional part should be obtainable from Z-transform tables by inspection. The Z-transform properties can be used to obtain the known Z-transform pairs.

The inspection method is extremely powerful when the Z-transform can be expressed in terms of the ratio of two polynomials, which is always the case for LTI systems. This will be explained in Section 3.5. We present a simple example on the usage of the inspection method to calculate the inverse Z-transform.

Example 3.4 Inverse Z-transform by inspection

Consider a causal IIR filter with Z-transfer function

$$H(z) = \frac{1}{1 - 0.8z^{-1} + 0.15z^{-2}} \tag{3.29}$$

Eqn. 3.29 can be written as

$$H(z) = \frac{1}{(1 - 0.3z^{-1})(1 - 0.5z^{-1})} \tag{3.30}$$

Using partial fraction expansion, Eqn. 3.30 can be written as

$$H(z) = \frac{2.5}{1-0.5z^{-1}} - \frac{1.5}{1-0.3z^{-1}} \tag{3.31}$$

We know that

$$H_a(z) = \frac{1}{1-az^{-1}} \tag{3.32}$$

has the inverse Z-transform $h_a[n] = a^n u[n]$. Hence, the inverse Z-transform of Eqn. 3.31, and the impulse response of the filter, is

$$h[n] = (2.5(0.5)^n - 1.5(0.3)^n)u[n] \tag{3.33}$$

3.4.2 Long Division

The long division method is only applicable when the Z-transform is expressed as the ratio of two polynomials in z; the inverse Z-transform is calculated through polynomial division. The division will be in the form of Eqn. 3.1. Therefore, the inverse Z-transform of the signal can be obtained from the coefficients of the z^{-n} terms. The closed form representation of the signal cannot be obtained with the long division method. Below, we apply the long division method to Example 3.5.

Example 3.5 Inverse Z-transform by long division

The $H(z)$ from Example 3.5 can be written as

$$H(z) = \frac{z^2}{z^2 - 0.8z + 0.15} \tag{3.34}$$

As can be seen in Eqn. 3.34, $H(z)$ is the ratio of two polynomials, so we can apply division to obtain

$$H(z) = 1 + 0.8z^{-1} + 0.49z^{-2} + \cdots \tag{3.35}$$

$h[n]$ then becomes

$$h[n] = \delta[n] + 0.8\delta[n-1] + 0.49\delta[n-2] + \cdots \tag{3.36}$$

Although the closed form expression of $h[n]$ cannot be obtained through long division, the impulse response of the system can still be obtained from Eqn. 3.36.

3.4.3 Mathematical Program Usage

Another method for calculating the inverse Z-transform is using a mathematical program. MATLAB's symbolic math toolbox, for instance, can be used to derive forward and inverse Z-transforms.

3.5 Z-Transform and LTI Systems

The Z-transform can be used to represent a discrete-time LTI system in the complex domain. This will allow us to analyze it from a different perspective. To demonstrate this, we will make use of the constant coefficient difference equation of the LTI system discussed in Section 2.5.

3.5.1 From Difference Equation to Impulse Response

The Z-transform can be used to obtain the impulse response of an LTI system represented by its constant coefficient difference equation. Let us start by taking the Z-transform of Eqn. 2.27.

$$y[n] + \alpha_1 y[n-1] + \cdots + \alpha_{L-1} y[n-L+1] = \beta_0 x[n] + \cdots + \beta_{K-1} x[n-K+1] \quad (2.27)$$

We have

$$X(z) \sum_{k=0}^{K-1} \beta_k z^{-k} = Y(z) \sum_{l=0}^{L-1} \alpha_l z^{-l} \quad (3.37)$$

where $\alpha_0 = 1$. As can be seen here, the difference equation in the time domain became an algebraic equation in the complex domain.

Remember that we have $Y(z) = H(z)X(z)$ from Eqn. 3.27. We can obtain $H(z)$ from Eqn. 3.37 as

$$H(z) = \frac{Y(z)}{X(z)} = \frac{\sum_{k=0}^{K-1} \beta_k z^{-k}}{\sum_{l=0}^{L-1} \alpha_l z^{-l}} \quad (3.38)$$

Eqn. 3.38 clearly indicates that, for an LTI system, $H(z)$ will be the ratio of two polynomials. The numerator polynomial holds the coefficients of the $x[n]$ terms in the constant coefficient difference equation. The denominator polynomial holds the coefficients of the $y[n]$ terms.

An IIR filter has both $x[n]$ and $y[n]$ terms in its difference equation. Thus, it will have both numerator and denominator polynomials as in Eqn. 3.38. The degree of the denominator polynomial is called the order of the filter. Because an FIR filter has only $x[n]$ terms in its difference equation, its $H(z)$ will only have a numerator polynomial. The degree of the numerator polynomial is called the order of the filter.

One can obtain the time domain representation of an impulse response by taking the inverse Z-transform of Eqn. 3.38. The Z-transform allows us to derive the impulse response of an LTI system from its difference equation. Let us consider two examples on obtaining the impulse response of a system from its difference equation.

Example 3.6 Obtaining the impulse response of an IIR filter from its difference equation

Consider an IIR filter with the difference equation

$$\begin{aligned} y[n] = {} & 0.002044x[n] + 0.004088x[n-1] + 0.002044x[n-2] + 1.819168y[n-1] \\ & - 0.827343y[n-2] \end{aligned} \tag{3.39}$$

This is a *lowpass filter*.

> lowpass filter
> *A filter that attenuates the high frequency components of a signal.*

Representing the difference equation in Eqn. 3.39 in the z-domain, we obtain

$$\begin{aligned} Y(z) = {} & X(z)(0.002044 + 0.004088z^{-1} + 0.002044z^{-2}) \\ & - Y(z)(-1.819168z^{-1} + 0.827343z^{-2}) \end{aligned} \tag{3.40}$$

Rearranging the terms in Eqn. 3.40, we can obtain the corresponding transfer function

$$H(z) = \frac{Y(z)}{X(z)} = \frac{0.002044 + 0.004088z^{-1} + 0.002044z^{-2}}{1 - 1.819168z^{-1} + 0.827343z^{-2}} \tag{3.41}$$

Because $H(z)$ is available, its time domain representation can be obtained in the way described in Section 3.4. Applying partial fraction expansion to $H(z)$, we obtain

$$H(z) = \frac{Y(z)}{X(z)} = 0.00247 + \frac{4.1757}{1 - 0.9106z^{-1}} + \frac{-4.1761}{1 - 0.9086z^{-1}} \tag{3.42}$$

The inverse Z-transform of $H(z)$ can be obtained as

$$h[n] = 0.00247\delta[n] + (4.1757(0.9106)^n - 4.1761(0.9086)^n)u[n] \tag{3.43}$$

Example 3.7 Obtaining the impulse response of an IIR filter from its difference equation

Next, let us consider an IIR filter with the difference equation

$$\begin{aligned} y[n] = {} & 0.705514x[n] - 1.414028x[n-1] + 0.705514x[n-2] \\ & + 1.3598y[n-1] - 0.462261y[n-2] \end{aligned} \tag{3.44}$$

This is a *highpass filter*.

> highpass filter
> *A filter that attenuates the low frequency components of a signal.*

The transfer function of the IIR filter represented by its difference equation in Eqn. 3.44 is

$$H(z) = \frac{Y(z)}{X(z)} = 1.5262 + \frac{29.3691}{1 - 0.6816z^{-1}} - \frac{30.1898}{1 - 0.6782z^{-1}} \tag{3.45}$$

Applying the inverse Z-transform to $H(z)$, we obtain the impulse response of the filter, which is

$$h[n] = 1.5262\delta[n] + (29.3691(0.6816)^n - 30.1898(0.6782)^n)u[n] \tag{3.46}$$

3.5.2 From Impulse Response to Difference Equation

The Z-transform can also be used to construct the constant coefficient difference equation of an LTI system from its impulse response. To do so, the first step is to take the Z-transform of $h[n]$ to obtain $H(z)$. Then, $H(z)$ can be employed in Eqn. 3.38, so an algebraic equation can be formed between the $X(z)$ and $Y(z)$ terms as in Eqn. 3.37. Finally, applying the inverse Z-transform to this equation leads to the constant coefficient difference equation representation of the LTI system in the time domain.

As discussed in Section 2.5, an IIR filter will have both $x[n]$ and $y[n]$ terms in its difference equation, and it is not possible to use the convolution sum formula on an IIR filter. The Z-transform provides us with the constant coefficient difference equation of a system from its impulse response. Hence, we can implement it; i.e., we can use the Z-transform to implement an IIR filter from its impulse response. This is one of the most important usage areas of the Z-transform in LTI systems. Now we provide two examples of obtaining the difference equation of a system from its impulse response.

Example 3.8 Obtaining the difference equation of an IIR filter from its impulse response

Assume that we have an IIR filter with impulse response

$$h[n] = 0.00247\delta[n] + (4.1757(0.9106)^n - 4.1761(0.9086)^n)u[n] \tag{3.47}$$

The z-domain representation of $h[n]$ is

$$H(z) = \frac{Y(z)}{X(z)} = 0.00247 + \frac{4.1757}{1 - 0.9106z^{-1}} + \frac{-4.1761}{1 - 0.9086z^{-1}} \tag{3.48}$$

We can represent $H(z)$ as the ratio of two polynomials as

$$H(z) = \frac{Y(z)}{X(z)} = \frac{0.002044 + 0.004088z^{-1} + 0.002044z^{-2}}{1 - 1.819168z^{-1} + 0.827343z^{-2}} \tag{3.49}$$

Rearranging the terms in Eqn. 3.49, we obtain

$$\begin{aligned} Y(z) = {} & X(z)(0.002044 + 0.004088z^{-1} + 0.002044z^{-2}) \\ & - Y(z)(-1.819168z^{-1} + 0.827343z^{-2}) \end{aligned} \tag{3.50}$$

We can apply the inverse Z-transform to this result to obtain the corresponding difference equation

$$\begin{aligned} y[n] = {} & 0.002044x[n] + 0.004088x[n-1] + 0.002044x[n-2] \\ & + 1.819168y[n-1] - 0.827343y[n-2] \end{aligned} \tag{3.51}$$

As can be seen in Eqn. 3.51, the difference equation obtained in this example is the same as the one given in Example 3.6.

Example 3.9 Obtaining the difference equation of an IIR filter from its impulse response

Consider the LTI system with impulse response

$$h[n] = 1.5262\delta[n] + (29.3691(0.6816)^n - 30.1898(0.6782)^n)u[n] \tag{3.52}$$

The z-domain representation of $h[n]$ is

$$H(z) = \frac{Y(z)}{X(z)} = 1.5262 + \frac{29.3691}{1 - 0.6816z^{-1}} - \frac{30.1898}{1 - 0.6782z^{-1}} \tag{3.53}$$

We can rearrange the terms in Eqn. 3.53 to obtain

$$\begin{aligned} Y(z) = {} & X(z)(0.705514 - 1.414028z^{-1} + 0.705514z^{-2}) \\ & - Y(z)(-1.3598z^{-1} + 0.462261z^{-2}) \end{aligned} \tag{3.54}$$

Applying the inverse Z-transform, we can obtain the corresponding difference equation in the time domain:

$$\begin{aligned} y[n] = {} & 0.705514x[n] - 1.414028x[n-1] + 0.705514x[n-2] \\ & + 1.3598y[n-1] - 0.462261y[n-2] \end{aligned} \tag{3.55}$$

As can be seen in Eqn. 3.55, the difference equation obtained in this example is the same as the one given in Example 3.7.

3.6 Block Diagram Representation of LTI Systems

The derivations in the previous section allow us to represent LTI systems in modular form. More specifically, we can rewrite Eqn. 3.37 as

$$Y(z) = X(z)\sum_{k=0}^{K-1} \beta_k z^{-k} - Y(z)\sum_{l=1}^{L-1} \alpha_l z^{-l} \tag{3.56}$$

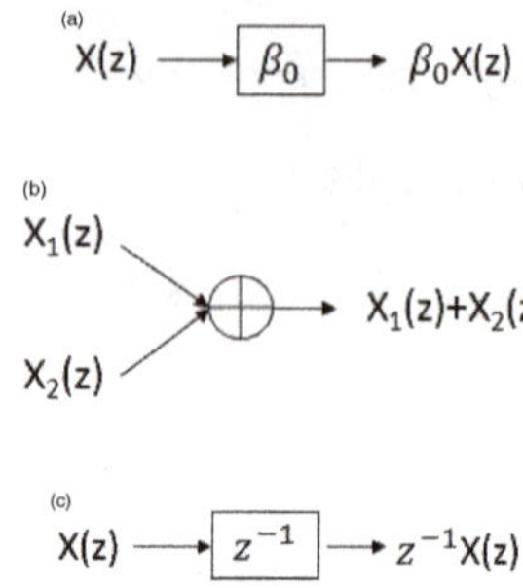

Figure 3.1 Block diagram representation of multiplication by a constant, addition, and a delay element. (a) Multiplication; (b) Addition; (c) Delay.

This leads us to the block diagram representation of an LTI system. Let us first focus on the basic building blocks used in Eqn. 3.56.

3.6.1 Building Blocks of a Block Diagram

There are three operations in Eqn. 3.56: multiplication by a constant, addition, and multiplication by z^{-1}. From the time shifting property of the Z-transform, we know that multiplying a signal by z^{-1} in the complex domain corresponds to shifting it by one sample in the time domain. This operation is called the delay operation. Any LTI system can be represented using only delay elements, multipliers, and adders. Each element can be represented schematically as in Figure 3.1.

3.6.2 From Difference Equation to Block Diagram

Eqn. 3.56 can be schematically described by the elements in Figure 3.1. This is called the block diagram representation of a discrete-time system. We provide the general LTI system in block diagram form in Figure 3.2.

This representation allows us to analyze discrete-time systems visually. Furthermore, it allows us to consider different structural forms of the same system, which we will discuss in Chapter 7. We now provide two examples on obtaining the block diagram representation of an IIR filter.

Example 3.10 Obtaining the block diagram representation of an IIR filter

Consider the IIR filter represented by the difference equation in Example 3.6. The block diagram representation of this IIR filter comprising both the feedforward and feedback terms can be seen in Figure 3.3.

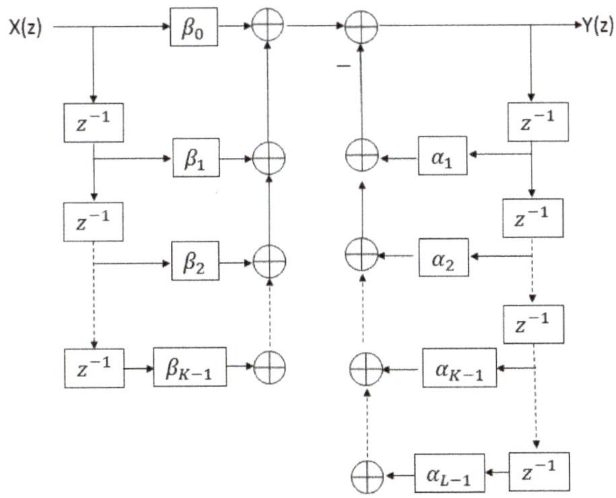

Figure 3.2 Block diagram representation of the general LTI system.

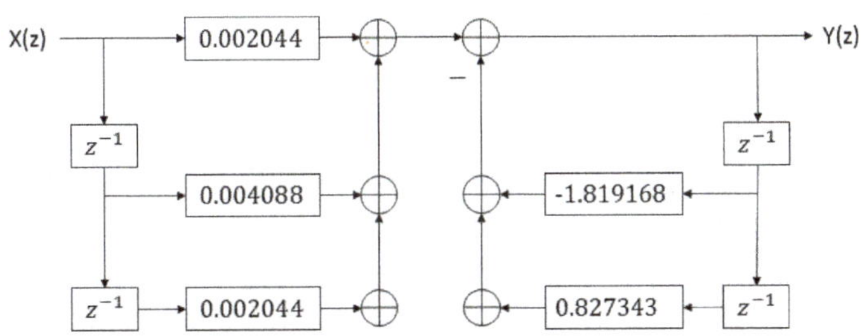

Figure 3.3 Block diagram representation of the IIR filter in Example 3.6.

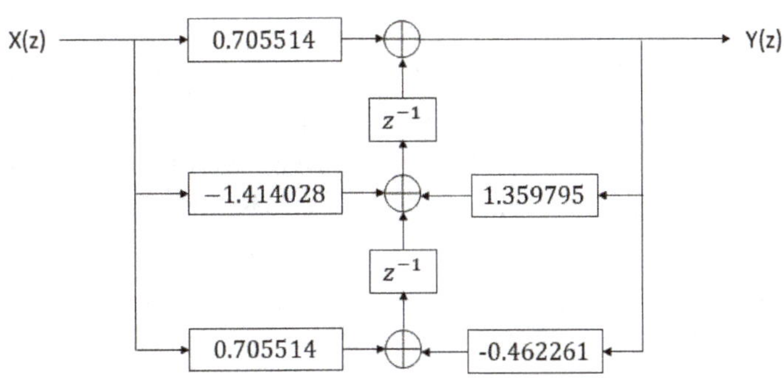

Figure 3.4 Block diagram representation of the IIR filter in Example 3.7.

Example 3.11 Obtaining the block diagram representation of an IIR filter

Consider the IIR filter represented by its difference equation in Example 3.7. The block diagram representation obtained from this IIR filter is shown in Figure 3.4.

3.6.3 From Block Diagram to Difference Equation

It is possible to obtain the difference equation of an LTI system (in the complex domain) from its block diagram representation. The main operation here is decomposing the block diagram in such a way that the output of each delay element is represented by a variable. These will be summed to obtain the final difference equation in the complex domain. This method will be of great use in obtaining the difference equation from a given block diagram representation. We now provide an example of this.

Example 3.12 Obtaining the difference equation from a block diagram

Consider the LTI system in Figure 3.3. We can obtain the transfer function (and the input–output relation) of the filter in the z-domain as

$$X(z)(0.002044+0.004088z^{-1}+0.002044z^{-2}) = Y(z)(1-1.819168z^{-1}+0.827343z^{-2}) \tag{3.57}$$

Rearranging the terms and taking the inverse Z-transform, we obtain the corresponding difference equation

$$\begin{aligned} y[n] &= 0.002044x[n]+0.004088x[n-1]+0.002044x[n-2] \\ &\quad +1.819168y[n-1]-0.827343y[n-2] \end{aligned} \tag{3.58}$$

3.7 Chapter Summary

The Z-transform represents a discrete-time signal or system in the complex domain. This opens up a new perspective for analyzing signals and systems. The Z-transform can be used to represent the convolution sum operation in the time domain as multiplication in the complex domain. Via this property, we can obtain the impulse response of an LTI system from its difference equation. Conversely, the difference equation can be obtained from the impulse response of the LTI system. The Z-transform also leads to the formation of the block diagram of an LTI system. This is a valuable tool that we will employ in Chapter 7. A close relationship exists between the Z-transform and discrete-time Fourier transform. We will benefit from this relationship in the next chapter.

3.8 Further Reading

The theoretical concepts considered in this chapter are explored further in a number of texts [1, 2, 3, 4, 5, 6, 7].

3.9 Exercises

3.9.1 Determine the Z-transform of the signal $x[n] = n(u[n] - u[n-20])$. What is the ROC for this transformation?

3.9.2 Determine the Z-transform of

(a) $x_1[n] = u[n] - u[n-10]$.

(b) $x_2[n] = x_1[n] * x_1[n]$.

3.9.3 Determine the signal $x[n]$ with Z-transform $X(z) = (1+z^{-1})(1-2z^{-2})(1+z^{-3})$.

3.9.4 A causal discrete-time system is represented as $H(z) = \frac{1-z^{-1}}{1-\alpha_0^2 z^{-2}}$.

(a) Comment on the stability of this system in terms of α_0.

(b) Find $h[n]$.

(c) Given the input $x[n] = \sin(\omega_0 n)u[n]$, find the corresponding output.

3.9.5 Find $y[n] = (-0.3)^n u[n] * (0.3)^n u[n]$ using the Z-transform.

3.9.6 Given the causal LTI system represented by $y[n] = 2y[n-1] - x[n-1]$, find $h[4]$ for this system using

(a) iterative calculation.

(b) the Z-transform.

3.9.7 Solve part (a) of Exercise 2.7.8 using the Z-transform.

3.9.8 Solve Exercise 2.7.9 using the Z-transform.

3.9.9 Given the input $x[n] = 0.5^n u[n]$, an LTI system has the output $Y(z) = \frac{2z^{-2}(1-0.5z^{-1})}{1+z^{-1}}$.

(a) What is $h[n]$ for this system?

(b) What is the difference equation representing this system?

3.9.10 Solve Exercise 2.7.11 using the Z-transform.

3.9.11 Given the input $x[n] = \frac{-1}{3}0.5^n u[n] - \frac{4}{3}2^n u[-n-1]$, a causal LTI system has the output $Y(z) = \frac{1+z^{-1}}{(1-z^{-1})(1+0.5z^{-1})(1-2z^{-1})}$.

(a) Find $X(z)$.

(b) What are $h[n]$ and $H(z)$ for this system?

(c) Is this system stable?

3.9.12 Given the input $x[n] = u[n]$, an LTI system has the output $y[n] = 0.5^{n-1}u[n+1]$.

(a) Find $H(z)$ and $h[n]$.

(b) Is this system stable?

(c) Is this system causal?

3.10 References

[1] Oppenheim, A.W. and Schafer, R.W. (2009) *Discrete-Time Signal Processing*, Prentice Hall, Third edn.
[2] Mitra, S.K. (2010) *Digital Signal Processing*, McGraw-Hill, Fourth edn.
[3] Proakis, J.G. and Manolakis, D.K. (1995) *Digital Signal Processing: Principles, Algorithms and Applications*, Prentice Hall, Third edn.
[4] Orfanidis, S. (1995) *Introduction to Signal Processing*, Prentice-Hall.
[5] McClellan, J.H., Schafer, R.W., and Yoder, M.A. (2003) *Signal Processing First*, Prentice-Hall.
[6] Manolakis, D.G. and Ingle, V.K. (2011) *Applied Digital Signal Processing*, Cambridge University Press.
[7] Oppenheim, A.W., Willsky, A.S., and Nawab, S.H. (1997) *Signals and Systems*, Prentice Hall, Second edn.

Note

This chapter does not include an accompanying lab.

4

Frequency Analysis of Discrete-Time Systems

Contents

4.1 Introduction

One of the most important tools in discrete-time signal processing is frequency analysis. As the name implies, the domain of interest is frequency. This domain provides valuable information that cannot be observed in the time domain. There are several methods for obtaining this information. This chapter explains these methods in detail. We will start with the discrete-time Fourier transform (DTFT), which can be used to calculate the Fourier transform of a discrete-time signal. The DTFT of a signal is a continuous function of frequency. Although this representation is mathematically tractable, it cannot be calculated or used in a digital system. This is because a continuous function cannot be represented fully in a digital system with limited resources. This limitation can be overcome by calculating samples of the DTFT in the frequency domain. This is called the discrete Fourier transform (DFT). As a result, the frequency domain characteristics of a discrete-time signal can be calculated for a digital system. However, the DFT requires the signal to be finite in time. We will explain why this is the case in this chapter. The DFT has a high computational load. Researchers have developed methods to decrease this computational load using an implementation technique called the Fast Fourier Transform (FFT). FFT is the main frequency analysis tool we will use in practical applications. The DFT and FFT are closely related to the discrete-time Fourier series expansion (DTFSE), which we will explore in this chapter. This will allow us to represent a discrete-time periodic signal as the sum of complex exponentials. The frequency analysis tools we have discussed before now do not allow us to observe the frequency content of a signal in terms of time. To do this, we can use the short-time Fourier transform (STFT), which we will introduce toward the end of this chapter. All these frequency domain representations may seem confusing, but they are all derived from the DTFT, which is analogous to the Fourier transform of a continuous-time signal. Based on this analogy, we expect this chapter will help you to grasp the main ideas behind discrete-time frequency analysis. To demonstrate how frequency analysis can be used in a practical application, we will explore signal filtering in the frequency domain. Finally, we will consider the linear phase concept, which is related to filtering.

4.2 Discrete-Time Fourier Transform

The discrete-time Fourier transform (DTFT) is the discrete version of the continuous-time Fourier transform. As a reminder, the Fourier transform of a continuous-time signal,

$x(t)$, is

$$X(j\Omega) = \int_{-\infty}^{\infty} x(t)e^{-j\Omega t}dt \tag{4.1}$$

The integral in Eqn. 4.1 must exist for the Fourier transform to be defined.

We begin this section with the definition of the DTFT. Then, we will explain the relationship between the DTFT and the Z-transform. Later, we will investigate the properties of the DTFT. We will also demonstrate the periodicity of the DTFT.

4.2.1 Definition of the DTFT

The DTFT can be used to represent a discrete-time signal in the frequency domain. The DTFT of a discrete-time signal, $x[n]$, is defined as

$$X(e^{j\omega}) = \sum_{n=-\infty}^{\infty} x[n]e^{-j\omega n} \tag{4.2}$$

$X(e^{j\omega})$ is a continuous and complex function of normalized frequency ω. Being a complex function, $X(e^{j\omega})$ can be represented as

$$X(e^{j\omega}) = |X(e^{j\omega})|e^{j\sphericalangle X(e^{j\omega})} \tag{4.3}$$

where $|X(e^{j\omega})|$ is the magnitude and $\sphericalangle X(e^{j\omega})$ is the phase term of $X(e^{j\omega})$. The magnitude term is closely related to the power of the signal at a given frequency. Therefore, it indicates the strength of the signal with respect to frequency. The phase term is related to the time delay for the given frequency.

We can calculate the DTFT of discrete-time signals using Eqn. 4.2. Let us start with the discrete unit impulse signal.

Example 4.1 The discrete unit impulse signal

The DTFT of the discrete unit impulse signal is

$$X(e^{j\omega}) = \sum_{n=-\infty}^{\infty} \delta[n]e^{-j\omega n} = 1 \tag{4.4}$$

Example 4.2 The exponential signal

The DTFT of the right-sided exponential signal (for $|a| < 1$) can be calculated as

$$X(e^{j\omega}) = \sum_{n=-\infty}^{\infty} a^n u[n]e^{-j\omega n} \tag{4.5}$$

$$= \sum_{n=0}^{\infty} (a^{-1}e^{j\omega})^{-n} \tag{4.6}$$

$$= \frac{1}{1-ae^{-j\omega}} \tag{4.7}$$

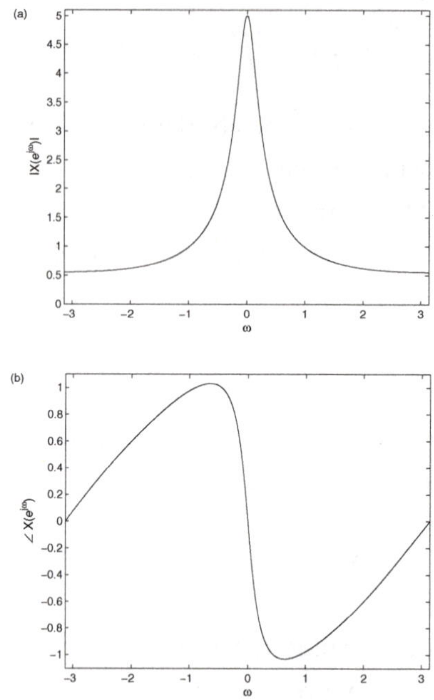

Figure 4.1 DTFT of the exponential signal. (a) Magnitude term; (b) Phase term.

Hence,

$$|X(e^{j\omega})| = \frac{1}{\sqrt{(1-a\cos(\omega))^2 + a^2\sin^2(\omega)}} \tag{4.8}$$

and

$$\sphericalangle X(e^{j\omega}) = -\arctan\left(\frac{a\sin(\omega)}{1-a\cos(\omega)}\right) \tag{4.9}$$

for the exponential signal.

We plot $|X(e^{j\omega})|$ and $\sphericalangle X(e^{j\omega})$ for $\omega \in [-\pi, \pi]$ and $a = 0.8$ in Figure 4.1. The magnitude term in this figure clearly indicates that the frequency content of the exponential signal is concentrated around low-frequency values. As the frequency value increases, the magnitude term decreases. The phase term shows almost piecewise linear characteristics with respect to frequency. We will explore such characteristics in Section 4.7.

Example 4.3 A finite length signal

The DTFT of a finite length signal $x[n] = \sum_{k=0}^{N-1} \beta_k \delta[n-k]$ is

$$X(e^{j\omega}) = \sum_{k=0}^{N-1} \beta_k e^{-j\omega k} \tag{4.10}$$

Let us take the finite signal $x[n] = \sin(\pi n/5)(u[n] - u[n-64])$. Eqn. 4.10 becomes

$$X(e^{j\omega}) = \sum_{k=0}^{63} \sin(\pi k/5) e^{-j\omega k} \tag{4.11}$$

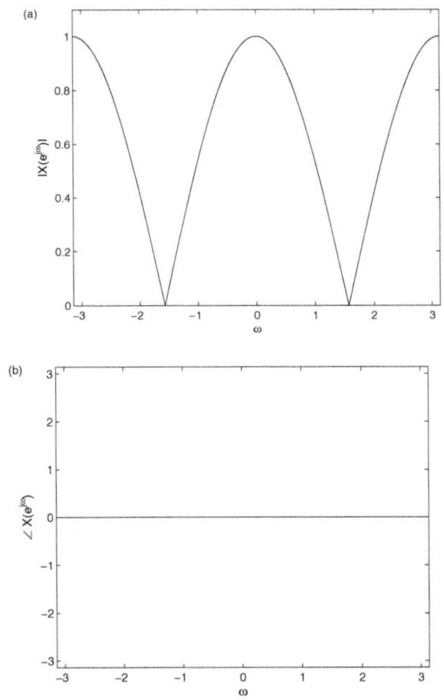

Figure 4.2 DTFT of the signal $x[n] = 0.5(\delta[n+1] + \delta[n-1])$. (a) Magnitude term; (b) Phase term.

The DTFT of a finite length signal is the sum of a finite number of complex exponentials. Now, we will explore two specific finite length signals.

Example 4.4 Two specific finite length signals

Let us consider two specific finite length signals. The first one is $x[n] = 0.5(\delta[n+1] + \delta[n-1])$. The DTFT of this signal is

$$X(e^{j\omega}) = \cos(\omega) \tag{4.12}$$

where $|X(e^{\omega})| = |\cos(\omega)|$ for $\omega \in [-\pi, \pi]$ and $\sphericalangle X(e^{j\omega}) = 0$ for $\omega \in [-\pi, \pi]$. We plot $|X(e^{j\omega})|$ and $\sphericalangle X(e^{j\omega})$ for $\omega \in [-\pi, \pi]$ in Figure 4.2.

The second signal is $x[n] = 0.5(\delta[n+1] - \delta[n-1])$. The DTFT of this signal is

$$X(e^{j\omega}) = j\sin(\omega) \tag{4.13}$$

where $|X(e^{j\omega})| = |\sin(\omega)|$ for $\omega \in [-\pi, \pi]$, $\sphericalangle X(e^{j\omega}) = -\pi/2$ for $\omega \in [-\pi, 0]$, and $\sphericalangle X(e^{j\omega}) = \pi/2$ for $\omega \in [0, \pi]$. We plot $|X(e^{j\omega})|$ and $\sphericalangle X(e^{j\omega})$ for $\omega \in [-\pi, \pi]$ in Figure 4.3.

As we will discuss further in Chapter 8, Figures 4.2 and 4.3 represent lowpass and highpass filters, respectively. As can be seen in Figure 4.2, the magnitude term has high values around 0 rad, which is a characteristic of a lowpass (or more specifically low-frequency pass) filter. Similarly, the magnitude term in Figure 4.3 has high values around 1.5 rad, which is a characteristic of a highpass (or more specifically high-frequency pass) filter.

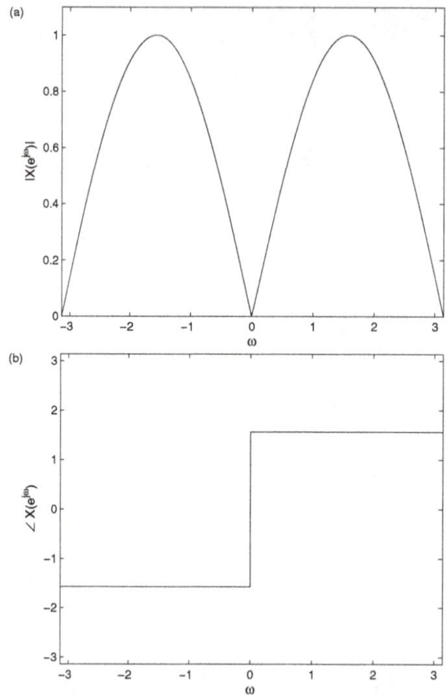

Figure 4.3 DTFT of the signal $x[n] = 0.5(\delta[n+1] - \delta[n-1])$. (a) Magnitude term; (b) Phase term.

4.2.2 Inverse DTFT

The inverse DTFT can be used to represent the frequency domain representation of a signal in the time domain. The inverse DTFT of $X(e^{j\omega})$ is

$$x[n] = \frac{1}{2\pi} \int_{-\pi}^{\pi} X(e^{j\omega}) e^{j\omega n} d\omega \tag{4.14}$$

4.2.3 The Relationship between DTFT and Z-transform

The complex entity z in the Z-transform can be represented as $z = re^{j\omega}$. If we take $r = 1$, we obtain

$$X(e^{j\omega}) = X(z)|_{z=e^{j\omega}} \tag{4.15}$$

Eqn. 4.15 shows that the DTFT is a special case of the Z-transform. That is, the DTFT is equal to the Z-transform evaluated at $|z| = 1$.

4.2.4 Frequency Response of an LTI System

The DTFT of the impulse response of an LTI system is called the frequency response. It explains how the system responds to a sinusoidal input signal (composed of complex exponentials) for a given frequency. Consider the frequency response of an LTI system with impulse response $h[n]$. Assume that the complex signal $x[n] = Ae^{\omega_0 n}$ is fed as an input to this system. Using the convolution sum formula, the output $y[n]$ can be calculated as

$$y[n] = h[n] * x[n] \tag{4.16}$$

$$= \sum_{k=-\infty}^{\infty} h[k]Ae^{j\omega_0(n-k)} \tag{4.17}$$

$$= Ae^{j\omega_0 n} \sum_{k=-\infty}^{\infty} h[k]e^{-\omega_0 k} \tag{4.18}$$

$$= H(e^{j\omega_0})x[n] \tag{4.19}$$

where $H(e^{j\omega_0})$ is the DTFT of $h[n]$ evaluated at frequency ω_0. This shows that the system responds to a complex exponential based on its frequency, which is the basis of filtering. We will explore this in more detail in Section 4.6.

4.2.5 Properties of the DTFT

All the Z-transform properties introduced in Section 3.3 also apply to the DTFT. As a reminder, they are listed below.

$$\alpha_1 x_1[n] + \alpha_2 x_2[n] \Longleftrightarrow \alpha_1 X_1(e^{j\omega}) + \alpha_2 X_2(e^{j\omega}) \tag{4.20}$$

$$x[n - n_0] \Longleftrightarrow e^{-j\omega n_0} X(e^{j\omega}) \tag{4.21}$$

$$a^n x[n] \Longleftrightarrow X(a^{-1} e^{j\omega}) \tag{4.22}$$

$$y[n] = x[n] * h[n] \Longleftrightarrow Y(e^{j\omega}) = X(e^{j\omega})H(e^{j\omega}) \tag{4.23}$$

There is one DTFT property that does not apply to the Z-transform. This is the periodicity of the DTFT. We now give a step by step explanation of this.

Take $\omega = \omega + 2\pi k$ from Eqn. 4.2 with $k \in \mathbb{I}$. We have

$$X\left(e^{j(\omega+2\pi k)}\right) = \sum_{n=-\infty}^{\infty} x[n]e^{-j(\omega+2\pi k)n} \tag{4.24}$$

$$= \sum_{n=-\infty}^{\infty} x[n]e^{-j\omega n} e^{-j2\pi kn} \tag{4.25}$$

$$= X\left(e^{j\omega}\right) \tag{4.26}$$

This derivation clearly indicates that the DTFT of a discrete-time signal is periodic with fundamental period 2π.

4.3 Discrete Fourier Transform

The DTFT of a signal $x[n]$, $X(e^{j\omega})$, is a continuous function of frequency ω as mentioned in Section 4.2. Although we cannot handle continuous representations in a digital system, one possible solution is to sample $X(e^{j\omega})$ in frequency and calculate these sampled values. The only constraint here is that $x[n]$ should be limited in time [1]. In other words, $x[n]$ should have at most N samples. This is always satisfied in real-life applications because the digital signal processing hardware can only handle a limited number of samples at once.

Let us take samples of $X(e^{j\omega})$ by setting $\omega = 2\pi k/N$ for $k = 0, 1, \cdots, N-1$. Substituting these values into Eqn. 4.2, we obtain

$$X[k] = \sum_{n=0}^{N-1} x[n] e^{-j\frac{2\pi k}{N}n} \tag{4.27}$$

$X[k]$ with $k = 0, 1, \cdots, N-1$ is called the discrete Fourier transform (DFT) of $x[n]$. We now present an example by calculating the DFT of the first signal.

Example 4.5 DFT of the finite length signal

The DTFT of the signal $x[n] = 0.5(\delta[n+1] + \delta[n-1])$ is $X(e^{j\omega}) = \cos(\omega)$. We can calculate the DFT of this signal with $N = 4$ as

$$X[k] = \cos(\pi k/2) \tag{4.28}$$

where $k = 0, 1, 2, 3$. We will use this result for calculating the inverse DFT as well.

The inverse DFT is defined as

$$x[n] = \frac{1}{N} \sum_{k=0}^{N-1} X[k] e^{j\frac{2\pi k}{N}n} \tag{4.29}$$

where $n = 0, 1, \cdots, N-1$. As can be seen in Eqn. 4.29, calculating the inverse DFT does not require any complex integration. Instead, a finite sum suffices. This will be useful in Section 4.3.2. For a detailed derivation of the inverse DFT, please see [1].

Eqn. 4.29 demonstrates another property of the DFT. As we have mentioned previously, the DFT requires the signal to be finite in time. In fact, the DFT assumes that the signal $x[n]$ is periodic with period N. We can prove the property as follows. Let us calculate $x[n+N]$ using Eqn. 4.29. We have

$$x[n+N] = \frac{1}{N}\sum_{k=0}^{N-1} X[k]e^{j\frac{2\pi k}{N}(n+N)} \tag{4.30}$$

$$= \frac{1}{N}\sum_{k=0}^{N-1} X[k]e^{j\frac{2\pi k}{N}n}e^{j2\pi k} \tag{4.31}$$

$$= \frac{1}{N}\sum_{k=0}^{N-1} X[k]e^{j\frac{2\pi k}{N}n} \tag{4.32}$$

$$= x[n] \tag{4.33}$$

In Eqn. 4.32, we used the property that $e^{j2\pi k} = 1$ for $k \in \mathbb{I}$. This derivation indicates that the DFT assumes the time limited signal $x[n]$ to be periodic with period N. Let us demonstrate this property through an example.

Example 4.6 Inverse DFT of a finite length signal

In Example 4.5, we obtained the DFT of the sample signal $x[n] = 0.5(\delta[n+1] + \delta[n-1])$ for $N = 4$ to obtain

$$X[k] = \cos(\pi k/2) \tag{4.34}$$

where $k = 0, 1, 2, 3$. We can calculate the inverse DFT of this signal using Eqn. 4.29 to obtain

$$x[n] = \frac{1}{4}(1 - e^{j\pi n}) \tag{4.35}$$

As can be seen in Eqn. 4.35, $x[1] = 0.5$ and $x[-1] = 0.5$ as expected. Moreover, the obtained $x[n]$ is clearly periodic with period $N = 4$.

4.3.1 Calculating the DFT

The DFT of a discrete-time signal can be calculated based on Eqn. 4.27. As $X[k]$ is complex-valued, it can be written as $X[k] = X_R[k] + jX_I[k]$. Using Euler's expansion formula, we can obtain the real and imaginary parts of $X[k]$ for a real-valued $x[n]$ as

$$X_R[k] = \sum_{n=0}^{N-1} x[n]\cos\left(\frac{2\pi kn}{N}\right) \tag{4.36}$$

$$X_I[k] = -\sum_{n=0}^{N-1} x[n]\sin\left(\frac{2\pi kn}{N}\right) \tag{4.37}$$

where $k = 0, 1, \cdots, N-1$.

This way, we can calculate the real and imaginary parts of $X[k]$ separately. It is important to note that the DFT defined in Eqn. 4.27 applies to complex-valued $x[n]$s, but Eqns. 4.36 and 4.37 are only applicable to real-valued $x[n]$s. We will explore this in detail in Lab 4.12.

4.3.2 Fast Fourier Transform

The DFT is extensively used to calculate the frequency content of discrete-time signals. However, it is computationally expensive. As can be seen in Eqn. 4.27, calculating each $X[k]$ requires N complex multiplications and $N-1$ complex additions. If $X[k]$ is to be calculated for N values, the total number of complex multiplications and additions will be N^2 and $N(N-1)$, respectively. A complex multiplication consists of four real multiplications and two real additions. Thus, there are $4N^2$ multiplication operations and $2N(N-1) = 2N^2 - 2N$ addition operations, which means that a total of $6N^2 - 2N$ operations are required. This value will become very large as N increases.

To decrease the computational cost of the DFT, we introduce the fast Fourier transform (FFT). FFT computation algorithms are based on the effective calculation of Fourier transform coefficients in an intelligent way. The widely used versions of the FFT include radix-2, radix-4, and radix-8, which are applicable when N is equal to integer powers of 2, 4, and 8, respectively. If N is not equal to such a value, the sequence of sample values to be transformed may be zero-padded to a suitable length.

FFT algorithms can reduce the total number of arithmetic operations required to calculate an N point DFT to $(N/2) \times \log_2(N)$ when N is a power of two [1, 2]. Let us analyze the computational gain when the FFT is used instead of the DFT. Assume that we want to calculate the DFT of a signal with $N = 256 = 2^8$. The DFT will require 2^{18} arithmetic operations. This value decreases to 2^{10} when FFT is used, which means the computational gain will be of order 2^8. We introduce FFT calculation methods in detail in Lab 4.12, where we will also compare the actual computational gain obtained when the FFT is used instead of the DFT.

4.4 Discrete-Time Fourier Series Expansion

A periodic discrete-time signal can be represented as the sum of complex exponential signals [1] called discrete-time Fourier series expansion (DTFSE). The formula for DTFSE of a periodic discrete-time signal with period N is

$$x[n] = \frac{1}{N} \sum_{k=-N+1}^{N-1} \tilde{X}[k] e^{j\frac{2\pi k}{N}n} \tag{4.38}$$

where the coefficients $\tilde{X}[k]$ for $k = 0, \cdots, N-1$ are defined as

$$\tilde{X}[k] = \sum_{n=0}^{N-1} x[n] e^{-j\frac{2\pi k}{N}n} \tag{4.39}$$

Comparing Eqns 4.27 and 4.39, we can see that $X[k]$ and $\tilde{X}[k]$ are the same. In other words, the coefficients of the DTFSE of a discrete-time signal are equal to its DFT. Moreover, if the period of a discrete-time signal is suitable for FFT calculations, then

DTFSE coefficients can be calculated in a fast manner. The sum in Eqn. 4.38 can be from $k=0$ to K, where $K \leq N-1$. In this case, we will have an approximation of the discrete-time signal via DTFSE.

The relationship between $X[k]$ and $\tilde{X}[k]$ also provides insight into the requirement for the periodicity of the discrete-time signal in the DFT calculation. We will explore how to calculate the DTFSE coefficients in Lab 4.12.

We can further simplify Eqn. 4.38 to obtain

$$x[n] = \frac{1}{N} \sum_{k=-N+1}^{N-1} |\tilde{X}[k]| \cos\left(\frac{2\pi k}{N} n + \sphericalangle \tilde{X}[k]\right) \tag{4.40}$$

Here, we benefit from $\tilde{X}[-k] = \tilde{X}^{*}[k]$ for real-valued periodic signals, where * corresponds to the complex conjugate operation [3]. As a result, the periodic signal is approximated by cosine signals with the phase term. Let us provide an example of approximating a periodic signal with DTFSE.

Example 4.7 Approximating a periodic signal with DTFSE

Let a periodic square signal $x[n]$ be defined within one period ($N=16$) as $x[n]=1$ for $n=0,\cdots,7$ and $x[n]=0$ for $n=8,\cdots,15$. We take the upper bound in Eqn. 4.38 as $K=1,5,15$ and provide the approximated signals in Figure 4.4. As can be seen in this figure, the approximation becomes better as K increases. We can apply DTFSE to other periodic discrete-time signals as well. We will provide such examples in Lab 4.12.

4.5 Short-Time Fourier Transform

DFT calculations take the whole time range into account. Thus, it is not possible to observe the frequency content of a signal around a certain time interval using the DFT. One solution to this problem is dividing the signal into windows in time. Then, the DFT can be taken for each window separately, and the frequency content of a signal around a certain time interval can be observed. This method is called the short-time Fourier transform (STFT).

STFT calculations can be performed in three steps. The first step is multiplying the signal by an appropriate window function, $v[n]$. This will partition the signal. Let us use the Hamming window for this operation.

Example 4.8 The Hamming window

The Hamming window with length M is defined as [1]

$$v[n] = (0.54 - 0.46\cos(2\pi n/M))\,(u[n] - u[n-M]) \tag{4.41}$$

The array $v[n]$ for $M=64$ is plotted in Figure 4.5.

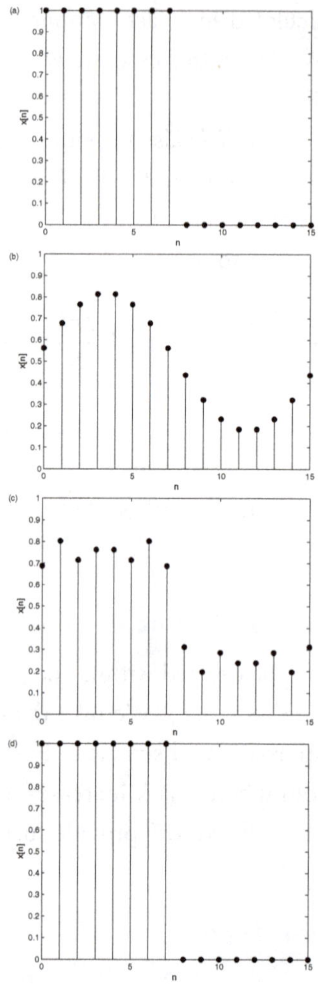

Figure 4.4 Approximating the discrete-time periodic signal in Exercise 4.7 using DTFSE. (a) Original signal; (b) Approximation for K=1; (c) Approximation for K=5; (d) Approximation for K=15.

As the Hamming window is defined, we can multiply the signal with it to obtain $s[m] = v[m] \times x[n-m]$. Here, the signal is shifted by n instead of the window. Hence, $s[m]$ will always be defined for $m = 0, 1, \cdots, M-1$.

The second step is applying the DFT to the windowed portion of the signal. The result should be stored in a matrix. The third step is shifting the window and calculating the DFT again. The overall STFT operation can be defined in mathematical terms as

$$X_{ST}[k,n] = \sum_{m=0}^{M-1} x[n-m]v[m]e^{-j\frac{2\pi k}{M}m} \tag{4.42}$$

where $k = 0, 1, \cdots, M-1$.

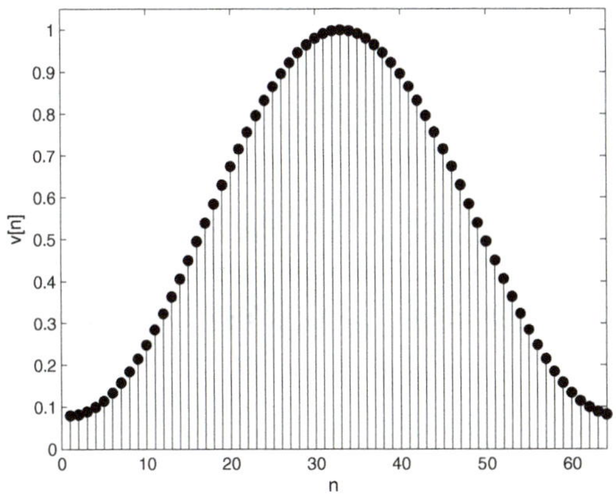

Figure 4.5 The Hamming window.

If $X_{ST}[k,n]$ is calculated for more than one n value, then it will be in the form of a two-dimensional matrix. Rows will represent frequency samples and columns will represent time values. Entries of this matrix will be complex-valued. Taking the magnitude of the entries will lead to the ***spectrogram*** of the signal. In general, the spectrogram is represented as a color (or grayscale) coded image.

> **spectrogram**
> *The visual representation of the spectrum of frequencies in a signal.*

There are several constraints in implementing the STFT. First, the window size M should be selected such that the local frequency content of the signal can be observed. Second, M should be selected such that the FFT can be used for each window to decrease the computation cost. Third, sampling in time should be such that the overlap in windows should be reasonable. Taking $n = 1$ in Eqn. 4.42 will lead to the best resolution in time. However, the computation cost and the memory limitations of the hardware for storing the $X_{ST}[k,n]$ values should be taken into account in implementation.

The STFT can be explained best when used on a ***chirp signal***, which we will now define.

> **chirp signal**
> *A chirp is a signal in which the frequency increases (up-chirp) or decreases (down-chirp) over time.*

Example 4.9 The chirp signal

The chirp is a signal defined as

$$x[n] = \cos\left(\omega_1 n + (\omega_2 - \omega_1)\frac{n^2}{2(N-1)}\right) \tag{4.43}$$

where ω_1 and ω_2 are the start and end frequency values of the signal, respectively, and N is the length of the signal.

We can rewrite Eqn. 4.43 as

$$x[n] = \cos(\omega_0[n]n) \tag{4.44}$$

where $\omega_0[n] = (\omega_1 + (\omega_2 - \omega_1)\frac{n}{2(N-1)})$. This representation clearly indicates that the frequency of the chirp signal changes linearly with time.

Let us generate a discrete-time chirp signal with $\omega_1 = 0$ and $\omega_2 = \pi$ rad. We plot the obtained chirp signal for 200 samples in Figure 4.6.

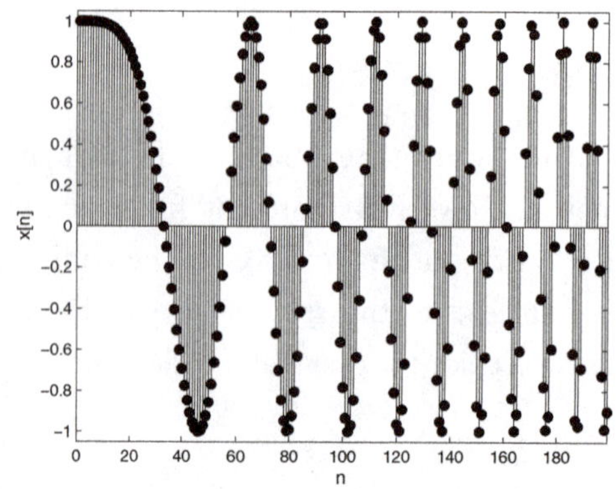

Figure 4.6 Discrete-time chirp signal.

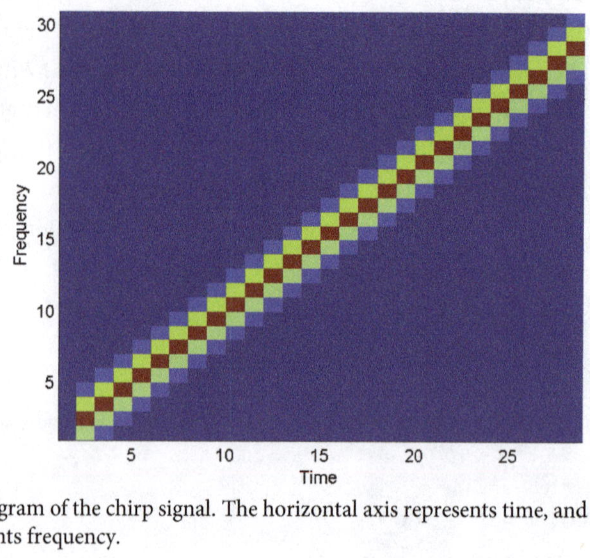

Figure 4.7 Spectrogram of the chirp signal. The horizontal axis represents time, and the vertical axis represents frequency.

We now calculate the STFT of the chirp signal with window size 64. We will provide the complete C code to calculate the STFT of a given signal in Lab 4.12. We provide the obtained spectrogram for the chirp signal in Figure 4.7. As can be seen in this figure, there is an increase in frequency in the spectrogram.

We also provide the application of the STFT in real-life signals in Lab 4.12. There, we emphasize the usefulness of the STFT in extracting information from a given signal.

4.6 Filtering in the Frequency Domain

Convolving two discrete-time signals in the time domain corresponds to multiplying their DTFT in the frequency domain as in Eqn. 4.23. One can use the polar form of complex functions $X(e^{\omega})$, $H(e^{j\omega})$, and $Y(e^{j\omega})$ to rewrite this equation as

$$|Y(e^{j\omega})|e^{j\sphericalangle Y(e^{j\omega})} = |X(e^{j\omega})|e^{j\sphericalangle X(e^{j\omega})}|H(e^{j\omega})|e^{j\sphericalangle H(e^{j\omega})} \tag{4.45}$$

where

$$|Y(e^{j\omega})| = |X(e^{j\omega})||H(e^{j\omega})| \tag{4.46}$$

$$\sphericalangle Y(e^{j\omega}) = \sphericalangle X(e^{j\omega}) + \sphericalangle H(e^{j\omega}) \tag{4.47}$$

As can be seen in Eqn. 4.46, the magnitude terms $|X(e^{j\omega})|$ and $|H(e^{j\omega})|$ are multiplied in the frequency domain to form $|Y(e^{j\omega})|$, and phase terms are summed as in Eqn. 4.47. Hence, we can adjust the frequency content of the input signal by the impulse response of the system, which is the basis of the filtering operation. Subsequently, $y[n]$ can be obtained from $Y(e^{j\omega})$ through the inverse DTFT.

4.6.1 Circular Convolution

Although Eqn. 4.45 allows us to apply the convolution operation in the frequency domain, it is valid in theory as $X(e^{j\omega})$ and $H(e^{j\omega})$ are continuous functions of frequency ω. Therefore, they cannot be processed by a digital system.

We can construct Eqn. 4.23 using the DFT as long as a certain condition is met. Take $h[n]$ as an FIR filter with length L. The input signal should be finite in time with length K. Let us take $N = \max(L,K)$ and append the shorter signal with zeros to obtain $x[n]$ and $h[n]$ of equal length N. Using the convolution sum formula, we can calculate the output of the system from input $x[n]$ as $y[n] = x[n] * h[n]$. Here, we can use the inverse

DFT representation of $x[n]$ given in Eqn. 4.29 to obtain

$$y[n] = \sum_{l=0}^{N-1} h[l] \frac{1}{N} \sum_{k=0}^{N-1} X[k] e^{j\frac{2\pi k}{N}(n-l)} \tag{4.48}$$

$$= \frac{1}{N} \sum_{k=0}^{N-1} X[k] e^{j\frac{2\pi n}{N}k} \sum_{l=0}^{N-1} h[n] e^{-j\frac{2\pi k}{N}l} \tag{4.49}$$

$$= \frac{1}{N} \sum_{k=0}^{N-1} X[k] H[k] e^{j\frac{2\pi n}{N}k} \tag{4.50}$$

From Eqn. 4.29, we know that

$$y[n] = \frac{1}{N} \sum_{k=0}^{N-1} Y[k] e^{j\frac{2\pi n}{N}k} \tag{4.51}$$

Using Eqns. 4.50 and 4.51 together, we can deduce that

$$Y[k] = X[k]H[k] \tag{4.52}$$

Therefore, using the DFT, we can replace $X(e^{j\omega})$ and $H(e^{j\omega})$ by $X[k]$ and $H[k]$, respectively, and Eqn. 4.45 still holds for these. Importantly, signals corresponding to $X[k]$ and $H[k]$ are periodic in the time domain as discussed in Section 4.3. Hence, convolution of these signals is defined as

$$y[n] = x[n] \circledast h[n] \tag{4.53}$$

where $\circledast$ represents circular convolution. This operation will provide a different result compared to the previously defined convolution operation. The reason for this is the periodicity of the signals $x[n]$ and $h[n]$.

Based on the circular convolution operation, we can rewrite Eqn. 4.23 as

$$y[n] = x[n] \circledast h[n] \Longleftrightarrow Y[k] = X[k]H[k] \tag{4.54}$$

Hence, Eqn. 4.45 becomes

$$|Y[k]|e^{j\sphericalangle Y[k]} = |X[k]|e^{j\sphericalangle X[k]}|H[k]|e^{j\sphericalangle H[k]} \tag{4.55}$$

where

$$|Y[k]| = |X[k]||H[k]| \tag{4.56}$$

$$\sphericalangle Y[k] = \sphericalangle X[k] + \sphericalangle H[k] \tag{4.57}$$

4.6.2 Linear Convolution

As there are two convolution operations, we should discriminate them. We will call the convolution operation that we have been using up to now linear convolution. Now, the problem we have is obtaining $y[n] = x[n] * h[n]$ from $y[n] = x[n] \circledast h[n]$. To note here, we will not calculate the circular convolution in the time domain. Instead, we will use the relationship given in Eqn. 4.55 and apply the inverse DFT to $Y[k]$ to obtain $y[n]$.

In order to calculate the linear convolution from the inverse DFT of $Y[k]$, we must do the following. Assume that the length of $x[n]$ is N and the length of $h[n]$ is M. We should first append zeros to $x[n]$ and $h[n]$ such that each has length $N+M-1$. Then, we should calculate $X[k]$ and $H[k]$. Furthermore, remember we must satisfy the FFT length constraints for fast calculation. Then, we can use Eqn. 4.55 to apply the inverse DFT to $Y[k]$ to obtain $y[n]$.

In some practical applications, filtering can be applied in the frequency domain instead of applying the convolution operation. We will see the advantages of this approach in Lab 4.12.

4.7 Linear Phase

The final topic to be considered in frequency domain analysis is linear phase. We know that phase components are added in the DFT representation of the convolution operation (as given in Eqn. 4.47). Thus, they have an effect on the output signal. To reflect that this effect is undesired, it is sometimes called phase distortion.

Minimum distortion is achieved when the phase term of $h[n]$ is a linear function of frequency. We call such filters linear phase filters. FIR filter coefficients can be adjusted to obtain a linear phase response. This can be done by selecting the filter coefficients to be symmetrical around a constant index. Then, phase distortion will become a time delay at the output signal. We will further consider the linear phase concept in Lab 4.12. Now, we explain the effect of linear phase through a simple example.

Example 4.10 Linear phase example

Assume that an FIR filter is defined as $h[n] = a_0\delta[n] + a_1\delta[n-1] + a_0\delta[n-2]$. The frequency response of this filter is

$$H(e^{j\omega}) = (2a_0\cos(\omega) + a_1)\,e^{-j\omega} \tag{4.58}$$

Assume that the input to this filter is $x[n] = \cos(\omega_0 n)$. Using Euler's formula, we can rewrite this input as

$$x[n] = \frac{1}{2}\left(e^{j\omega_0 n} + e^{-j\omega_0 n}\right) \tag{4.59}$$

Using Eqn. 4.19, we can write the output of the filter as

$$y[n] = \frac{1}{2}\left(H(e^{j\omega_0})e^{j\omega_0 n} + H(e^{-j\omega_0})e^{-j\omega_0 n}\right) \tag{4.60}$$

If $(2a_0\cos(\omega_0) + a_1) > 0$, the phase term becomes $\sphericalangle H(e^{j\omega}) = -\omega$. We can rewrite Eqn. 4.60 as

$$y[n] = |2a_0\cos(\omega_0) + a_1|\cos(\omega_0(n-1)) \tag{4.61}$$

Hence, the phase term only adds a one sample delay to the output.

4.8 Chapter Summary

Analyzing the characteristics of discrete-time signals in the frequency domain provides valuable information. In this chapter, we focused on Fourier transform techniques. We started with the DTFT, which can be linked to the Z-transform introduced in the previous chapter. Although the DTFT is a powerful tool for obtaining the frequency content of a discrete-time signal, it cannot be implemented on a digital system. This leads to the DFT, which is in fact the sampled version of the DTFT in the frequency domain. The DFT has a heavy computational load, so we introduced the FFT to overcome this. The DFT and FFT are also closely related to DTFSE. Therefore, we introduced DTFSE to represent periodic discrete-time signals using complex exponential functions. We also introduced the STFT to analyze the frequency content of a discrete-time signal in terms of time. Throughout this chapter, we emphasized that all these Fourier transforms originate from the DTFT, which is very similar to the Fourier transform for continuous-time signals. We also explored the filtering operation in the frequency domain using the DFT. Finally, we evaluated the linear phase concept in the filtering operation. All these frequency domain analysis tools will be extremely useful in understanding and designing digital filters.

4.9 Further Reading

Frequency analysis in discrete time is a vast topic. We suggest that you refer to the wider literature on this topic if you want to delve further into the details [1, 3, 4, 5, 6, 7, 8]. If you do not feel comfortable with heavy mathematical derivations, we suggest referring to Smith's book [9].

4.10 Exercises

4.10.1 Determine the DTFT of the following signals.

(a) $x_1[n] = \cos(\pi n/7)$.

(b) $x_2[n] = \delta[n+14] + \delta[n-14]$.

(c) $x_3[n] = e^{-5n}u[n]$.

(d) $x_4[n] = x_2[n] * x_3[n]$.

(e) $x_5[n] = x_2[n]x_3[n]$.

4.10.2 Determine the DTFT of the following discrete-time signals.

(a) $x[n] = \alpha^{|n|}$, where $|\alpha| < 1$.

(b) $x[n] = A\alpha^n \cos(\omega_0 n + \phi)u[n]$, where $|\alpha| < 1$ and A, α, ω_0, and ϕ are real-valued.

4.10.3 Determine the DTFT of the discrete-time signals given in Exercise 2.7.1.

4.10.4 Determine the DTFT of the discrete-time signal given in Exercise 3.9.3.

4.10.5 Determine the inverse DTFT of the following.

(a) $X(e^{j\omega}) = \sin(4\omega + \pi/8)$.

(b) $X(e^{j\omega}) = \cos(-2\omega + 1/2)$.

4.10.6 Given the causal LTI system represented by $y[n] = 2y[n-1] - x[n-1]$, find

(a) the frequency response of this system.

(b) the output obtained from this system when the input is $x[n] = \cos(n/8)$.

4.10.7 Let $H(e^{j\omega}) = \frac{1-e^{j2\omega}}{1-e^{-j2\omega}}$ represent an LTI system in the frequency domain.

(a) What is $h[n]$?

(b) What will be the output of this system given input signals $x[n] = \cos(\pi n/2)$ and $x[n] = \sin(\pi n/2)$?

4.10.8 For the signals in Exercise 2.7.2,

(a) obtain their DFT by writing the C code.

(b) plot the magnitude of the obtained DFTs.

4.10.9 An FIR filter is represented as $h[n] = \sum_{k=1}^{5} \frac{1}{1+|3-k|}\delta[n-k]$.

(a) Find $H(e^{j\omega})$.

(b) What is the eight-point DFT of $h[n]$?

(c) If we apply the inverse DFT to the result in part (b), what signal will be the obtained?

4.10.10 A periodic discrete-time signal is represented as $x[n] = \cos(\pi n/7) - \sin(2\pi n/8)$. Find the DTFSE of this signal.

4.10.11 A periodic discrete-time signal is represented within one period as $x[n] = \sin(\pi n/8)/(\pi n)$ for $n = 0,\cdots,31$.

(a) Write a C code to calculate the DTFSE of this periodic signal.

(b) Use the generated C code to approximate $x[n]$ with 6, 16, and 30 discrete-time Fourier series elements.

4.10.12 Comment on the linear phase property of the following systems.

(a) $y[n] = 2y[n-1] - x[n-1]$.

(b) $y[n] = -2x[n] + x[n-1] + x[n+1]$.

4.10.13 Two impulse responses of FIR filters are $h_1[n] = \cos(\pi n/6)$ and $h_2[n] = \sin(\pi n/6)$ for $n = -6,\cdots,6$.

(a) Determine the frequency response of both filters.

(b) Plot the phase component of the obtained frequency responses.

(c) Check whether these filters satisfy the linear phase condition.

4.10.14 For the given FIR filters in Exercise 4.10.13,

(a) write a C code to calculate the output of these filters given the input $x[n] = \cos(\pi n/6)$ for $n = 0,\cdots,60$.

(b) comment on the phase shift after the filtering operation.

4.11 References

[1] Oppenheim, A.W. and Schafer, R.W. (2009) *Discrete-Time Signal Processing*, Prentice Hall, Third edn.
[2] Kuo, S.M., Lee, B.H., and Tian, W. (2013) *Real-Time Digital Signal Processing: Fundamentals, Implementations and Applications*, Wiley, Third edn.
[3] Proakis, J.G. and Manolakis, D.K. (1995) *Digital Signal Processing: Principles, Algorithms and Applications*, Prentice Hall, Third edn.
[4] Mitra, S.K. (2010) *Digital Signal Processing*, McGraw-Hill, Fourth edn.
[5] Orfanidis, S. (1995) *Introduction to Signal Processing*, Prentice-Hall.
[6] McClellan, J.H., Schafer, R.W., and Yoder, M.A. (2003) *Signal Processing First*, Prentice-Hall.
[7] Manolakis, D.G. and Ingle, V.K. (2011) *Applied Digital Signal Processing*, Cambridge University Press.
[8] Oppenheim, A.W., Willsky, A.S., and Nawab, S.H. (1997) *Signals and Systems*, Prentice Hall, Second edn.
[9] Smith, S.W. (1997) *The Scientist and Engineer's Guide to Digital Signal Processing*, California Technical Pub.

4.12 Lab 4

4.12.1 Introduction

This lab is about the frequency domain analysis of discrete-time signals and systems using Fourier transforms. We will begin with discrete Fourier transform (DFT) calculations. Then, we will consider the fast Fourier transform (FFT) using CMSIS-DSP library functions. Next, we will provide examples on the usage of discrete-time Fourier series expansion. Furthermore, we will explore filtering in the frequency domain. Finally, we will consider the linear phase property of LTI filters.

4.12.2 Calculating the DFT

In this section, we will explore DFT implementation. The aim here is to demonstrate how to calculate the DFT in C. Moreover, we will discuss how to speed up calculations using CMSIS-DSP library functions.

Using C Functions

We can implement the DFT using C functions in the `math.h` header file. In Listing 4.1, we provide a C function for calculating the DFT of a given signal. Here, the code block is not optimized. It is provided to show the basic setup of the DFT operation.

```
void DFT(float x[], float XR[], float XI[], int size)
{
int n,k;
float sumR,sumI;

for(k=0;k<size;k++)
{
sumR=0.0;
sumI=0.0;
for(n=0;n<size;n++)
{
sumR+=x[n]*cos(2*PI*k*n/(float)size);
sumI+=x[n]*sin(2*PI*k*n/(float)size);
}
XR[k]=sumR;
XI[k]=-sumI;
}
}
```

Listing 4.1 C function for calculating the DFT.

The signal used in this section is the sum of two sinusoids given in Section 2.9.2. The normalized angular frequencies of these sinusoidal signals are $w_1 = \pi/128$ rad/sample and $w_2 = \pi/4$ rad/sample, respectively.

The C code for DFT calculations is given in Listing 4.2. Here, the sample size N is set to 256. The complete project related to this code can be found in `Online_Student_Resources\Lab4\DFT_calculations`.

```
#include "math.h"
#include "time.h"

#define PI 3.14159265358979
#define N 256

```

```
int n;
float s1[N],s2[N],s[N];
float s_complex[2*N];
float w1=PI/128;
float w2=PI/4;
float XR[N],XI[N],Mag[N];

unsigned int CoreClock = 16000000;
unsigned int CycleCount;
float ExecutionTime ;

void DFT(float x[], float XR[], float XI[], int size){
	int n,k;
	float sumR ,sumI;
	for(k=0;k<size;k++)
	{
		sumR=0.0;
		sumI=0.0;
		for(n=0;n<size;n++)
		{
		sumR+=x[2*n+0]*cos(2*PI*k*n/(float)size)
		+x[2*n+1]*sin(2*PI*k*n/(float)size);
		sumI+=-x[2*n+1]*cos(2*PI*k*n/(float)size)
		+x[2*n+0]*sin(2*PI*k*n/(float)size);
		}
		XR[k]=sumR;
		XI[k]=-sumI;
	}
}

int main(void)
{

	//Sinusoidal signals
	for(n=0;n<N;n++){
		s1[n]=sin(w1*(float)n);
		s2[n]=sin(w2*(float)n);
		s[n]=s1[n]+s2[n];
	}

	//Complex sum of sinusoidal signals
	for(n=0;n<N;n++){
		s_complex[2*n]=s[n];
		s_complex[2*n+1]=0;
	}

	StartTiming ();

	DFT(s_complex,XR,XI,N);

	//Magnitude calculation
	for(n=0;n<N;n++)
	Mag[n]=sqrt(pow(XR[n],2)+pow(XI[n],2));

	CycleCount = StopTiming();
	ExecutionTime = CalculateTime (CycleCount,CoreClock);

	return(0);
}
```

Listing 4.2 DFT calculations.

Task 4.1

Run the code in Listing 4.2. Plot the real and imaginary parts of the obtained DFT using the arrays XR and XI. Plot the magnitude of the DFT using the Mag array. Comment on the results. Calculate the total time required to calculate the DFT without setting the core clock of the CPU.

Task 4.2

Now, change the frequency value of the second signal in Listing 4.2 to $w_2 = \pi/5$ rad/sample. Again, set the sample size N to 256. Run the modified code. Plot the `XR`, `XI`, and `Mag` arrays again. What do you observe this time? Write the observed frequency components of the signal. Are these values correct? If they are not, why?

Using CMSIS-DSP Library Functions

There is no dedicated DFT function in the CMSIS-DSP library. However, we can calculate DFT values using the available CMSIS-DSP library functions. To do so, we can use the functions `arm_sin_f32` and `arm_cos_f32` instead of `sin` and `cos`, respectively. We can also compute the magnitude of the DFT using the `arm_cmplx_mag_f32` function. Do not forget to make the necessary adjustments given in Section A.7.3 before using these functions.

Task 4.3

Use the mentioned CMSIS-DSP library functions to calculate the DFT of the signal given in Section 4.12.2. Compare the execution time of the direct and CMSIS-DSP library-based implementations. Which one is faster?

Task 4.4

Change the sample size to 512. Run the modified code. Compare the execution time for sample sizes 256 and 512. How does sample size affect the execution time?

You should now know how to find the frequency components of a given signal using the DFT. Moreover, you should know how to implement the DFT in C. Observe that time consuming operations can be reduced by using optimized CMSIS-DSP library functions. Through experiments, you should have observed that increasing the sample size increases the calculation time.

4.12.3 Fast Fourier Transform

In this book, we use the CMSIS-DSP library for FFT calculations. This library contains two different function sets for calculating the FFT. The first set is for complex FFT calculations, and the second set is for real FFT calculations.

Complex FFT

Complex FFT functions can be used for both complex and real valued signals. However, real valued signals must be represented as if they are complex valued. There are five predefined complex FFT functions in the CMSIS-DSP library. These are as follows.

```
arm_cfft_f32
arm_cfft_radix2_init_f32
arm_cfft_radix2_f32
arm_cfft_radix4_init_f32
arm_cfft_radix4_f32
```

All these functions can be used for both FFT and inverse FFT calculations. However, the last four of these are slower and less general than the `arm_cfft_f32` function. Therefore, we do not consider them here. The `arm_cfft_f32` function uses a mixed-radix algorithm. In other words, it performs the FFT using multiple radix-8 stages and one radix-2 or radix-4 stage. We demonstrate all the parameters used in `arm_cfft_f32` in Listing 4.3.

```
void arm_cfft_f32(const arm_cfft_instance_f32 *S, float32_t *p1, uint8_t ifftFlag,
uint8_t bitReverseFlag);
/*
S: the instance of FFT/IFFT
p1: Complex input/output buffer with a size of twice FFT/IFFT length
ifftFlag: the flag to indicate the type of FFT computed
bitReverseFlag: the flag to indicate the order of the output
*/
```

Listing 4.3 Parameters used in the `arm_cfft_f32` function.

In Listing 4.3, the instance of `S` stores information about the FFT/IFFT structure, such as the length of the FFT/IFFT; ***pointers*** for ***twiddle factor*** and ***bit reversal*** tables; and bit reversal table length. There are nine predefined instances for nine supported FFT lengths. These are 16, 32, 64, 128, 256, 512, 1024, 2048, and 4096. The instance structure is `arm_cfft_sR_f32_len`*`FFTLength`*. The `ifftFlag` parameter indicates whether the FFT or inverse FFT is computed. The inverse FFT is calculated when this flag is set, and the FFT is calculated when it is reset. The `bitReverseFlag` parameter indicates whether the output is in normal or bit reversed order. The output is in normal order when this flag is set and in reverse order when it is reset.

> **pointer**
> *A data type in a programming language that contains the address information of another value.*

> **twiddle factor**
> *Trigonometric constant coefficients multiplied by data in the course of FFT algorithm.*

> **bit reversal**
> *Reverse the order of the bits in a digital word.*

The CMSIS-DSP library can also be used to calculate the magnitude of the complex FFT. The function for this operation is `arm_cmplx_mag_f32`. We define it in Listing 4.4.

```
arm_cmplx_mag_f32(float32_t *pSrc, float32_t *pDst, uint32_t numSamples);
/*
pSrc: the pointer for the complex input
pDst: the pointer for the real output
numSamples: the number of complex samples in the input array
*/
```

Listing 4.4 The `arm_cmplx_mag_f32` function.

The FFT of the signal in Section 4.12.2 can be obtained using CMSIS-DSP library complex FFT functions as in Listing 4.5. The complete project related to this code can be found in `Online_Student_Resources\Lab4\FFT_calculations`.

```
#include "arm_math.h"
#include "arm_const_structs.h"
#include "time.h"

#define N 256

int n;
float s1[N],s2[N],s[N];
float s_complex[2*N];
float w1=PI/128;
float w2=PI/4;
float Mag[N];

unsigned int CoreClock = 16000000;
unsigned int CycleCount;
float ExecutionTime ;

int main(void)
{

	//Sinusoidal signals
	for(n=0;n<N;n++){
		s1[n]=arm_sin_f32(w1*(float)n);
		s2[n]=arm_sin_f32(w2*(float)n);
		s[n]=s1[n]+s2[n];
	}

```

```
        //Complex sum of sinusoidal signals
        for(n=0;n<N;n++){
            s_complex[2*n]=s[n];
            s_complex[2*n+1]=0;
        }

        StartTiming ();

        //CFFT calculation
        arm_cfft_f32(&arm_cfft_sR_f32_len256, s_complex, 0, 1);

        //Magnitude calculation
        arm_cmplx_mag_f32(s_complex, Mag, N);

        CycleCount = StopTiming();
        ExecutionTime = CalculateTime (CycleCount,CoreClock);

        return(0);
}
```

Listing 4.5 FFT calculations using CMSIS-DSP library functions.

Task 4.5

Run the code in Listing 4.5. Plot the `Mag` array. Is the obtained result the same as the one in Section 4.12.2? Compute the execution time of the FFT process and compare it with the DFT.

Now, we will find the FFT of real-world signals. First, we will use the accelerometer signal. The C code for acquiring the accelerometer data as a signal is given in Listing 4.6. The complete project related to this code can be found in `Online_Student_Resources\Lab4\FFT_accelerometer`. Here, the sample size is set as 512. The sampling period for the accelerometer signal is set to 10 ms. The acquired signal is stored in memory, and then its FFT is calculated.

```
#include "arm_math.h"
#include "arm_const_structs.h"

#include "stm32f4xx_hal.h"
#include "STM32F4-Discovery\stm32f4_discovery_accelerometer.h"
#include "hal_config.h"

#define N 512

int16_t n;
int16_t acc_data[3] = {0}; // x-axis, y-axis, z-axis
float32_t X[2*N] = {0};
float32_t Mag[N];

int main(void)
{
        HAL_Init();
        SystemClock_Config();
        BSP_ACCELERO_Init();

        for(n=0;n<N;n++){
            BSP_ACCELERO_GetXYZ(acc_data);
```

```
            X[2*n]=acc_data[0];
            X[2*n+1]=0;
            HAL_Delay(10);
        }

        //CFFT calculation
        arm_cfft_f32(&arm_cfft_sR_f32_len512, X, 0, 1);

        //Magnitude calculation
        arm_cmplx_mag_f32(X, Mag, N);

        return(0);
}
```

Listing 4.6 Calculating the FFT of the accelerometer signal using CMSIS-DSP library functions.

Task 4.6

Make the necessary adjustments, as given in Section 1.4.6, to execute the project. In order to acquire the accelerometer data in action, move the ST Discovery kit back and forth five times with a 1-s delay each time. To note here, the direction of the movement should be along the widest side of the ST Discovery kit with the mini USB connector facing forward. Plot the acquired accelerometer data. Calculate the FFT of the signal and plot its magnitude. Compare the results with the ones obtained in the previous section.

Task 4.7

Now, we will find the FFT of the audio signal captured by the AUP audio card when the letter "E" is spelled. Make the necessary adjustments to capture the signal from the AUP audio card as explained in Section 1.4.7. Set the sampling rate to 8 kHz and the buffer size to 1024. Plot the acquired audio data. Calculate the FFT of the signal, and plot its magnitude. What do you observe?

To note here, the maximum FFT calculation size is 4096 in the CMSIS-DSP library. If the signal at hand has more than 4096 samples, then the only way to find the frequency spectrum is to use the DFT or write your own FFT code. Moreover, HAL libraries and CMSIS-DSP library FFT functions cannot be compiled owing to its code limitation.

Here, you should observe that the FFT has a clear advantage in terms of computation time. Further, the usage of the CMSIS-DSP library simplifies the code. If the sample size of the signal at hand exceeds 1024, then the DFT must be used owing to Keil's code size limitation.

4.12.4 Discrete-Time Fourier Series Expansion

The aim in this section is to demonstrate that a signal can be approximated using Discrete-Time Fourier Series Expansion (DTFSE). Detailed information on DTFSE has been provided in Section 4.4. We provide the C function to calculate the DTFSE approximation in Listing 4.7.

```
void DTFSE(float X[], float xc[], int size, int Kx){
	int n,k;
	float sumR, P, A;
	for(n=0;n<size;n++){
		sumR=0.0;
		for(k=0;k<=Kx;k++){
			A = sqrt(xc[2*k]*xc[2*k]+xc[2*k+1]*xc[2*k+1]);
			P = atan2(xc[2*k+1],xc[2*k]);
			sumR += A*arm_cos_f32((2*PI*k*n/(float)size)+P)/N;
		}
		X[n]=sumR;
	}
}
```

Listing 4.7 C function to calculate the DTFSE.

Let a periodic square signal be defined within one period ($N = 16$) as $x[n] = 1$ for $n = 0, \cdots, 7$ and $x[n] = 0$ for $n = 8, \cdots, 15$. The C code for the DTFSE calculations on this square signal is given in Listing 4.8. The complete project related to this code can be found in `Online_Student_Resources\Lab4\DTFSE_calculations`.

```
#include "arm_math.h"
#include "arm_const_structs.h"
#include "time.h"

#define N 16
#define K 1

unsigned int CoreClock = 16000000;
unsigned int CycleCount;
float ExecutionTime;

int n;
float s_complex[2*N],DTFSEcoef[2*N];
float s_real[N];
float s_imag[N];
float y_real[N];

void DTFSE(float X[], float xc[], int size, int Kx){
	int n,k;
	float sumR, P, A;
	for(n=0;n<size;n++){
		sumR=0.0;
		for(k=0;k<=Kx;k++){
			A = sqrt(xc[2*k]*xc[2*k]+xc[2*k+1]*xc[2*k+1]);
			P = atan2(xc[2*k+1],xc[2*k]);
			sumR += A*arm_cos_f32((2*PI*k*n/(float)size)+P)/N;
		}
		X[n]=sumR;
	}
}

int main(){
	//square signal
	for(n=0;n<N;n++){
		if(n<N/2) s_real[n]=1;
		else s_real[n]=0;
```

```
            s_imag[n]=0.0;
            s_complex[2*n+0]=s_real[n];
            s_complex[2*n+1]=s_imag[n];
        }

    arm_copy_f32(s_complex,DTFSEcoef,2*N);

    //Coefficient calculation with CFFT function
    arm_cfft_f32(&arm_cfft_sR_f32_len16, DTFSEcoef, 0, 1);

    StartTiming ();

    //DTFSE
    DTFSE(y_real, DTFSEcoef, N, K);

    CycleCount = StopTiming();
    ExecutionTime = CalculateTime (CycleCount,CoreClock);

    return(0);
}

```

Listing 4.8 DTFSE calculations.

Task 4.8

Run the code in Listing 4.8 by setting K as 1, 5, and 15. Plot the approximated signals. Calculate the execution time for each approximation. What do you observe?

Task 4.9

We will now use the periodic triangle signal generated in Section 2.9.4. Modify the code in Listing 4.8 by replacing the square signal with this triangle signal (having a period of 16 samples), and rerun the code. First, plot the obtained triangle signal. Then, plot the approximated signals by setting K as 2, 5, 8, and 15. Comment on the results.

You should observe that if the DFT coefficients of a periodic signal are at hand, the signal can be approximately or fully generated by using them in the DTFSE framework. The degree of approximation directly affects the computation time. This may affect the system performance when the periodic signal is long, so the choice of approximation degree is a trade-off between quality and speed.

4.12.5 Short-Time Fourier Transform

In this section, we focus on the short-time Fourier transform (STFT). We provide the STFT calculation and spectrogram formation of the chirp signal in Listing 4.9. The complete project related to this code can be found in `Online_Student_Resources\Lab4\STFT_calculations`.

```
#include "arm_math.h"
#include "arm_const_structs.h"

#define N 1024 // Sample number
#define M 64 // window size

float32_t x[N]; // chirp signal
float32_t w1=0; // Starting frequency of the chirp signal
float32_t w2=PI; // Ending frequency of the chirp signal

float32_t XST[4*(N/M-1)*M/2]; // STFT saved in an array
float32_t s[2*M]; // subwindow
float32_t v[M]; // Hamming window
float32_t FFTOutput[M]; // FFT Output buffer

int32_t n,m; // Counters
int32_t nind=0;

int main(void)
{
        //chirp signal
        for(n=0;n<N;n++)
        x[n]=arm_cos_f32(w1*n+(w2-w1)*n*n/(2*(N-1)));

        //Hamming window
        for (m=0;m<M;m++)
        v[m]=0.54-0.46*cos(2*PI*m/M);

        for (n=M;n<N;n=n+M/2){

                //Select subwindow
                for (m=0;m<M;m++){
                        s[2*m]=v[m]*x[n-m];
                        s[2*m+1]=0;
                }

                //Finding the FFT of window
                arm_cfft_f32(&arm_cfft_sR_f32_len64, s, 0, 1);
                arm_cmplx_mag_f32 (s, FFTOutput, M);

                arm_copy_f32(FFTOutput, &XST[nind*M], M);

                nind++;
        }

        return(0);
}
```

Listing 4.9 STFT calculation of the chirp signal.

The window size for STFT calculation is set as $M = 64$ in Listing 4.9. Spectrogram values are stored in the array XST. This array should be converted to matrix form when displayed. It is kept as an array so the Keil interface can be used to save it.

Task 4.10

Run the code in Listing 4.9. Plot the generated spectrogram. Then, change the STFT window size to $M = 16$. Rerun the code. Compare the two generated spectrograms. What do you observe?

We will now apply the STFT to real-world signals. We will first use the accelerometer data as a signal for this purpose. The C code for acquiring the accelerometer signal is given in Listing 4.10. The complete project related to this code can be found in `Online_Student_Resources\Lab4\STFT_accelerometer`.

```
#include "arm_math.h"
#include "arm_const_structs.h"

#include "stm32f4xx_hal.h"
#include "STM32F4-Discovery\stm32f4_discovery_accelerometer.h"
#include "hal_config.h"

#define N 1024
#define M 64 // window size

int16_t acc_data[3]; // x-axis, y-axis, z-axis
float32_t x[N];

float32_t XST[4*(N/M-1)*M/2]; // STFT saved in an array
float32_t s[2*M]; // subwindow
float32_t v[M]; // Hamming window
float32_t FFTOutput[M]; // FFT Output buffer
int32_t n,m; // Counters
int32_t nind=0;

void STFT(){
        //Hamming window
        for (m=0;m<M;m++)
        v[m]=0.54-0.46*cos(2*PI*m/M);

        for (n=M;n<N;n=n+M/2){

                //Select subwindow
                for (m=0;m<M;m++){
                        s[2*m]=v[m]*x[n-m];
                        s[2*m+1]=0;
                }

                //Finding the FFT of window
                arm_cfft_f32(&arm_cfft_sR_f32_len128, s, 0, 1);
                arm_cmplx_mag_f32 (s, &XST[nind*M], M);
                nind++;
        }
}

int main(void)
{
        HAL_Init();
        SystemClock_Config();
        BSP_ACCELERO_Init();

        for(n=0;n<N;n++){
                BSP_ACCELERO_GetXYZ(acc_data);
                x[n]=acc_data[0];
                HAL_Delay(10);
        }

        STFT();

        return(0);
}
```

Listing 4.10 Calculating the STFT of the accelerometer signal.

Task 4.11

Run the code in Listing 4.10. Set the sample size to 1024, and set the sampling period for the accelerometer signal to 10 ms. In order to see the accelerometer in action, move the ST Discovery kit back and forth five times with a two second delay each time. To note here, the direction of the movement should be along the widest side of the ST Discovery kit with the mini USB connector facing forward. Plot the obtained signal and its spectrogram. Comment on the results.

Task 4.12

Finally, we use the STFT to analyze an audio signal. To do so, make the necessary adjustments as discussed in Section 1.4.7 to capture the audio signal from the AUP audio card when the word "HELLO" is spelled. Set the sampling rate to 8 kHz and buffer size to 8000. Use an STFT window size of $M = 128$. Plot the obtained audio signal and its spectrogram. Comment on the results.

4.12.6 Filtering in the Frequency Domain

We have already implemented the filtering operation in the time domain in Chapter 2.9. Filtering can also be done in the frequency domain. We have provided the theoretical basis for this operation in Section 4.6. We will now provide practical examples of this.

Refer to the sinusoidal signal given in Section 4.12.2. Let us filter out the sinusoidal signal component with frequency $\pi/4$ rad/sample. The C code for this application is given in Listing 4.11. The complete project related to this code can be found in `Online_Student_Resources\Lab4\FiF_calculations`.

```
#include "arm_math.h"
#include "arm_const_structs.h"
#include "FIR_lpf_coefficients.h"
#include "time.h"

#define N 512

unsigned int CoreClock = 16000000;
unsigned int CycleCount;
float ExecutionTime ;

int n;
float s1[N],s2[N],s[N];
float s_complex[2*N];
float w1=PI/128;
float w2=PI/4;

float df_complex[2*N];
float result_complex[2*N];
float result[N];

int main(void){

	//Sinusoidal signals
	for(n=0;n<N;n++){
```

```
            s1[n]=arm_sin_f32(w1*(float)n);
            s2[n]=arm_sin_f32(w2*(float)n);
            s[n] = s1[n] + s2[n];
      }

      //Complex sum of sinusoidal signals
      for(n=0;n<N;n++){
            s_complex[2*n]=s[n];
            s_complex[2*n+1]=0;
      }

      // Complex impulse response of filter
      for(n=0;n<K;n++){
            df_complex[2*n+0]=h[n];
            df_complex[2*n+1]=0;}

      // Finding the FFT of the filter
      arm_cfft_f32(&arm_cfft_sR_f32_len512, df_complex, 0, 1);

      StartTiming ();

      // Finding the FFT of the input signal
      arm_cfft_f32(&arm_cfft_sR_f32_len512, s_complex, 0, 1);

      // Filtering in the frequency domain
      arm_cmplx_mult_cmplx_f32(s_complex, df_complex, result_complex, 2*N);

      // Finding the complex result in time domain
      arm_cfft_f32(&arm_cfft_sR_f32_len512, result_complex, 1, 1);

      // Constructing the result from the real part of the complex result
      for(n=0;n<N;n++)
      result[n]=result_complex[2*n];

      CycleCount = StopTiming();
      ExecutionTime = CalculateTime (CycleCount,CoreClock);

      return(0);
}
```

Listing 4.11 Filtering in the frequency domain.

Task 4.13

Run the code in Listing 4.11. Plot the obtained output. Calculate the total time required without setting the core clock. Compare this result with the one obtained in Task 2.17. Are the results the same? Which filtering operation is faster?

Here, you should have learned how to apply filtering in the frequency domain. You should also be aware that filtering in the frequency domain gives the same results as in the time domain, except the transient response of the filter is different. Filtering in the frequency domain can be faster if the filter length is larger than 64.

4.12.7 Linear Phase

Linear phase is an important filter property to be explored. We have covered the theoretical aspects of linear phase in Section 4.7. We now provide examples on filters with and without linear phase.

We provide the C code to plot the phase component of a generic FIR filter (represented by its coefficients) in Listing 4.12. The complete project related to this code can be found in `Online_Student_Resources\Lab4\Linear_phase_calculations`. We provide the filter coefficients to test this code in the same folder in the file `linear_phase_FIR_coefficients.h`.

```
#include "math.h"
#include "arm_math.h"
#include "arm_const_structs.h"
#include "linear_phase_FIR_coefficients.h"

int n;

float h_complex[2*K];
float h_r[K], h_i[K], Phase[K], Mag[K];

int main(void){

        // Complex impulse response of filter
        for(n=0;n<K;n++){
                h_complex[2*n+0]=h[n];
                h_complex[2*n+1]=0;
        }

        // Finding the FFT of the filter
        arm_cfft_f32(&arm_cfft_sR_f32_len64, h_complex, 0, 1);

        // Finding the magnitude of the filter
        arm_cmplx_mag_f32 (h_complex, Mag , K);

        // Finding the phase of the filter
        for(n=0;n<K;n++){
                h_r[n]=h_complex[2*n+0];
                h_i[n]=h_complex[2*n+1];
                Phase[n]=atan2(h_r[n],h_i[n]);
        }

        return(0);
}
```

Listing 4.12 Phase response of a generic FIR filter.

Task 4.14

Run the code in Listing 4.12. Plot the `Mag` and `Phase` arrays. What do you observe?

Task 4.15

Replace the header file in Listing 4.12 with the file `FIR_lpf_coefficients.h` given in `Online_Student_Resources\Lab4\Linear_phase_calculations`. Rerun the code. Plot the `Mag` and `Phase` arrays. What do you observe this time?

Here, you should observe the difference between filters with and without the linear phase property. To note here, linear phase FIR filters can be obtained when the filter coefficients have a certain structure as mentioned in Section 4.7. It is not easy to obtain an IIR filter with the linear phase property.

5

Conversion between Continuous-Time and Discrete-Time Signals

Contents

5.1 Introduction

Most signals we encounter in real-life applications are in continuous-time form. In order to process these using a digital system, we must convert them to discrete time first. After processing a discrete-time signal, it should be converted back to continuous-time. To understand how these operations can be performed in an effective manner, the theory behind conversion between continuous-time and discrete-time signals must be learned, which is explained in this chapter. Explaining conversion operations requires mathematical derivations. Although we aim to keep mathematical usage at a minimal level, it is not possible to eliminate it completely. Thus, we ask the reader to persevere with this chapter to appreciate the importance of these derivations. To explain conversion concepts, we will first explore continuous-time to discrete-time conversion from a theoretical perspective. In practice, we will consider the analog to digital conversion (ADC) corresponding to this operation. We will next deal with discrete-time to continuous-time conversion from a theoretical perspective. We will explore digital to analog conversion (DAC) as the practical counterpart of this operation. Finally, we will consider downsampling and interpolation operations.

5.2 Continuous-Time to Discrete-Time Conversion

The first step in processing a continuous-time signal in a digital system is converting it to discrete-time. The sampling theorem provides the theoretical background for this operation. Below, we explore it in both the time and frequency domains.

5.2.1 Sampling Theorem in the Time Domain

Assume that there is a continuous-time signal $x(t)$ and we want to obtain its samples in time. In mathematical terms, we can perform this operation by multiplying $x(t)$ with an ***impulse train*** as

$$x(nT_s) = x(t) \sum_{n=-\infty}^{\infty} \delta(t - nT_s) \tag{5.1}$$

where T_s is the period of the impulse train. Using the sifting theorem

$$x(t)\delta(t - t_0) = x(t_0)\delta(t - t_0) \tag{5.2}$$

we can rewrite Eqn. 5.1 as

$$x(nT_s) = \sum_{n=-\infty}^{\infty} x(nT_s)\delta(t - nT_s) \tag{5.3}$$

impulse train
An input signal that consists of an infinite series of impulse units equally separated in time.

Eqn. 5.3 indicates that samples of the continuous-time signal $x(t)$ will appear as the coefficient of impulse signals periodically located in time by T_s. The signal $x(nT_s)$ in Eqn. 5.3 is still in continuous-time. If we focus solely on the coefficient of the impulse signals with their location as an index, we obtain

$$x[n] = x(nT_s) \tag{5.4}$$

In Eqn. 5.4, $x[n]$ represents the discrete-time signal obtained from $x(t)$. $x[n]$ does not hold any information on the sampling period. Therefore, we should keep T_s or the sampling frequency, $f_s = 1/T_s$, to associate $x[n]$ with $x(nT_s)$.

Example 5.1 Sampling a tone signal in the time domain

Consider a continuous-time tone signal $x(t) = \cos(\Omega_M t)$ with angular frequency Ω_M rad/s. We can sample it by T_s to obtain the corresponding discrete-time signal $x[n] = \cos(\Omega_M T_s n)$. Assume that $\Omega_M = 1000\pi$ rad/s. We can take the sampling period as $T_s = 10^{-4}$ s. After sampling, the discrete-time signal will be $x[n] = \cos(\pi n/10)$. We provide part of the continuous-time signal and its samples in Figure 5.1. The samples contain information about the original continuous-time signal.

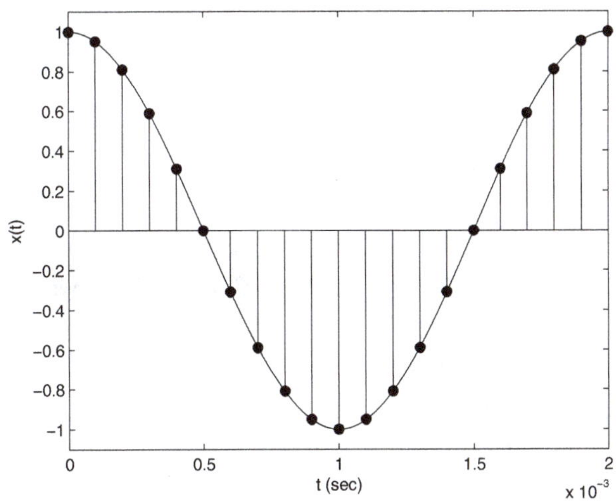

Figure 5.1 Part of the tone signal and its samples.

We must select an appropriate sampling frequency for the signal. The sampling theorem states what this frequency should be so no information is lost in the sampling process. We can summarize the sampling theorem without proof as follows. Assume that the continuous-time signal to be sampled is bandlimited with maximum frequency Ω_M rad/s. The sampling frequency for this signal should be greater than twice this value (i.e., $2\pi f_s > 2\Omega_M$) so that we do not lose any information in sampling.

Based on the sampling theorem, f_s should be larger than 1000 Hz in Example 5.1. In other words, T_s should be smaller than 0.001 s. The sampling theorem can be proven in the frequency domain, which we will now demonstrate.

5.2.2 Sampling Theorem in the Frequency Domain

We can use continuous-time signal properties to prove the sampling theorem in the frequency domain. The Fourier transform of the impulse train is

$$\sum_{k=-\infty}^{\infty} \delta(t - kT_s) \Longleftrightarrow 2\pi f_s \sum_{k=-\infty}^{\infty} \delta(j(\Omega - k2\pi f_s)) \tag{5.5}$$

where $f_s = 1/T_s$ and Ω represent frequency values of continuous-time signals. Multiplying two continuous-time signals in the time domain corresponds to their convolution in the frequency domain [1]. This property allows us to represent the signal $x(nT_s)$ in the frequency domain as

$$x(t) \sum_{k=-\infty}^{\infty} \delta(t - kT_s) \Longleftrightarrow X(j\Omega) * 2\pi f_s \sum_{k=-\infty}^{\infty} \delta(j(\Omega - k2\pi f_s)) \tag{5.6}$$

The convolution operation in Eqn. 5.6 is a special type because $X(j\Omega)$ is convolved with impulses. Convolving a signal with an impulse shifts the signal to the location of the impulse signal [1]. Using this property, we can rewrite Eqn. 5.6 as

$$x(nT_s) \Longleftrightarrow 2\pi f_s \sum_{k=-\infty}^{\infty} X(j(\Omega - k\Omega_s)) \tag{5.7}$$

where $\Omega_s = 2\pi f_s$.

Eqn. 5.7 states that the Fourier transform of $x(nT_s)$ is periodic with period $\Omega_s = 2\pi f_s$ rad/s. In Section 5.2.1, we assumed that $x(t)$ was a ***bandlimited signal***. In other words, $X(j\Omega) = 0$ when $|\Omega| > \Omega_M$ rad/s. In order to obtain $x(t)$ from $x(nT_s)$ (to be explained in more detail in Section 5.4), frequency components of $x(t)$ must not be distorted during the sampling process. In simple terms, replicas of $X(j\Omega)$ should not overlap in Eqn. 5.7. Hence, we should have at least $\Omega_s - \Omega_M > \Omega_M$ or $\Omega_s > 2\Omega_M$. This proves the sampling theorem in the frequency domain. Now, let's reconsider Example 5.1 in the frequency domain.

> **bandlimited signal**
> *A signal with power spectral density that is bound within a limited range of frequencies.*

Example 5.2 Sampling the tone signal in the frequency domain

The Fourier transform of the continuous-time tone signal $x(t) = \cos(\Omega_M t)$ with angular frequency Ω_M rad/s is

$$X(j\Omega) = \pi(\delta(\Omega + \Omega_M) + \delta(\Omega - \Omega_M)) \tag{5.8}$$

This transform can be represented in the frequency domain as in Figure 5.2.

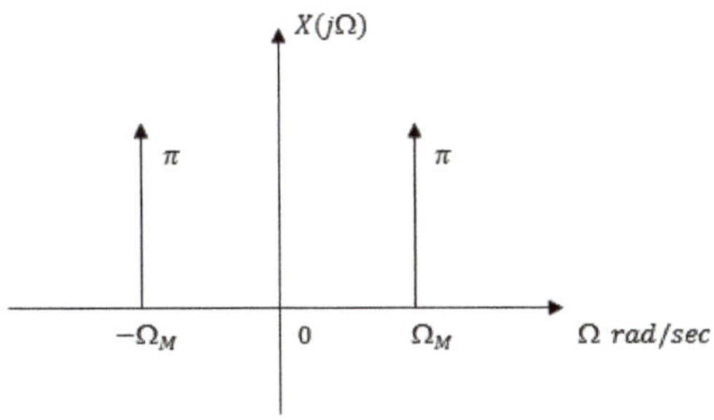

Figure 5.2 Fourier transform of the tone signal from Exercise 5.2 in the frequency domain.

After sampling, the Fourier transform of the sampled signal $x(nT_s)$ will be as in Figure 5.3. As can be seen in this figure, replicas of $X(j\Omega)$ do not overlap because the sampling theorem is satisfied.

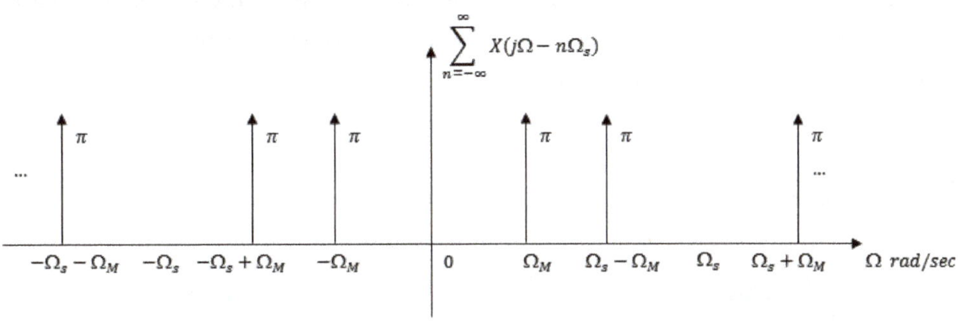

Figure 5.3 Fourier transform of the sampled tone signal from Exercise 5.2 in the frequency domain.

Eqn. 5.7 also provides insight on the periodicity of the frequency components of discrete-time signals. We will apply normalization to Eqns. 5.4 and 5.7 by setting $T_s = 1$ to show this property. After normalization, we will have $x[n] = x(n)$. Moreover, frequency components of the sampled signal will be located periodically every 2π rad in the frequency domain. This is the same as for the DTFT in Eqn. 4.24. Therefore, we can demonstrate the periodicity of frequency components of the discrete-time signal $x[n]$ using normalization.

5.2.3 Aliasing

What happens when the sampling theorem is not satisfied in the sampling process? Or, what happens when $\Omega_s < 2\Omega_M$? The answer is that replicas of $X(j\Omega)$ will overlap. We can observe this effect on the Fourier transform of the sampled tone signal in Figure 5.4.

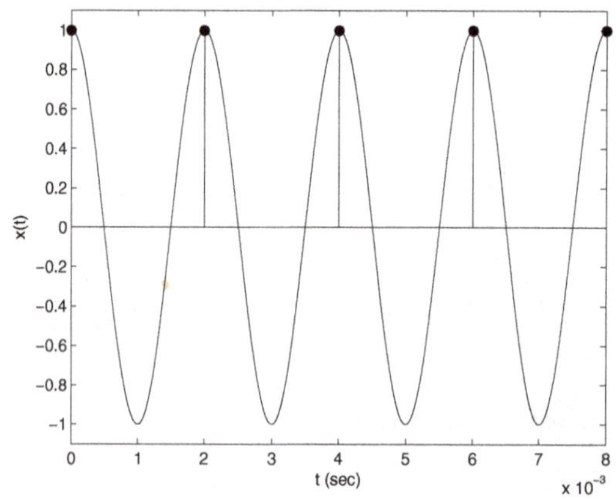

Figure 5.4 Fourier transform of the sampled tone signal in the frequency domain when there is aliasing.

Comparing Figures 5.3 and 5.4, it can be seen that the neighboring frequency content overlaps in Figure 5.4, while the frequency content of the original signal is unchanged. This is called ***aliasing***. Now, we focus on the effect of aliasing in the time domain on the tone signal by reconsidering Example 5.1.

> **aliasing**
> *The distortion of a sampled signal resulting from sampling a signal at a frequency that is too low.*

Example 5.3 The effect of aliasing while sampling the tone signal

As explained in Section 5.2.1, the tone signal should be sampled by a sampling frequency greater than twice its maximum frequency. Let us reconsider Example 5.1 by keeping $\Omega_M = 1000\pi$ rad/s. We provide the continuous time signal in Figure 5.5 (a). In order to observe the effect of aliasing, let us take the sampling period as $T_s = 0.002$ s. Therefore, the corresponding sampling frequency will be $\Omega_s = 1000\pi$ rad/s, which is half of the required sampling frequency. After sampling, the discrete-time signal will be $x[n] = \cos(2\pi n)$ or $x[n] = 1$ for all n. We plot these samples in Figure 5.5 (b).

As can be seen in Figure 5.5 (b), the sampled signal does not provide any information on the corresponding continuous-time signal it originated from because of aliasing. Referring to Figure 5.1, in which there was no aliasing, it can be seen that it was possible to obtain the original continuous-time signal from those samples.

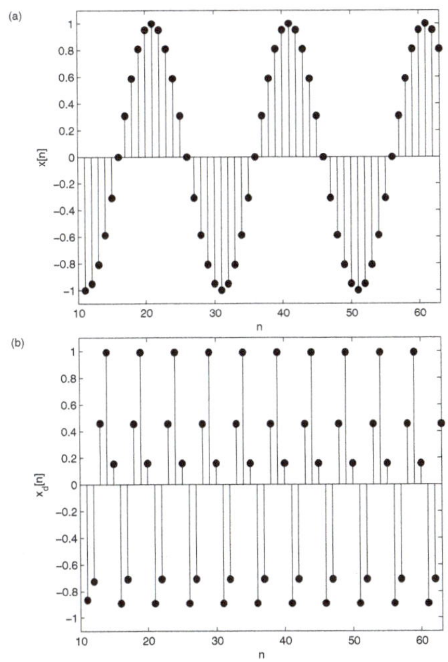

Figure 5.5 Part of the tone signal and its aliased samples. (a) Continuous-time tone signal; (b) Sampled tone signal with aliasing

The effect of aliasing can be minimized by filtering the continuous-time signal before sampling. The lowpass filter used for this operation is called an anti-aliasing filter. The bandwidth of the anti-aliasing filter should be at most half of the sampling frequency. This filtering operation discards some frequency components of the original signal, which leads to information loss. However, this operation prevents contamination of frequency components via aliasing. In practical applications, there are no bandlimited signals [2], so the anti-aliasing filter is a necessary step before the sampling operation. Now, we will focus on practical issues related to sampling.

5.3 Analog to Digital Conversion

In practice, the sampling operation is performed by an analog to digital converter (ADC). Please see [3] for the hardware setup of an actual ADC module. ADC not only samples the signal in time but also quantizes the analog value. Hence, the result can be represented as a proper value in the digital system.

The quantization operation is performed as follows. First, the minimum and maximum values of the input signal should be set, which we will call V_{min} and V_{max}, respectively. Next, we set the quantization level, which represents how many bits will be assigned per sample. Let us call this value N. Finally, the resulting digital value for the analog input V_{in} is calculated as

$$Q = \left\lfloor \frac{2^N (V_{in} - V_{min})}{(V_{max} - V_{min})} \right\rfloor \tag{5.9}$$

where $\lfloor \cdot \rfloor$ is the *floor function*. Please note that quantization yields an integer value between 0 and $2^N - 1$. We can use the value in two different ways. First, we can convert it to a float and perform floating-point operations as we have done previously. Second, we can convert the integer value to a fixed-point value and use it accordingly. We will see how to do this in Chapter 10.

> **floor function**
> *A function that maps a real number to the largest previous or the smallest following integer.*

The difference (in terms of the analog value) between two successive quantization levels is called the resolution. A higher quantization level corresponds to a better resolution. Before going further, we will introduce the successive approximation register (SAR) architecture as a sample ADC hardware. The SAR ADC includes an SAR logic circuitry, an N-bit DAC, a comparator, and a sample and hold circuitry as shown in Figure 5.6. In simple terms, it performs a binary search. The analog voltage is sampled and held for one sampling period. The DAC output is first set to $V_{ref}/2$. Then, the DAC output voltage and sampled input voltage are compared. If the input voltage (V_{in}) is greater than the DAC output, the comparator output is set to logic level 1, and the MSB bit of the N-bit SAR remains at logic level 1. Otherwise, the comparator output is set to logic level 0, and the MSB bit of the N-bit SAR is cleared to logic level 0. Then, SAR logic moves to the next bit. This procedure continues until all N-bits are calculated.

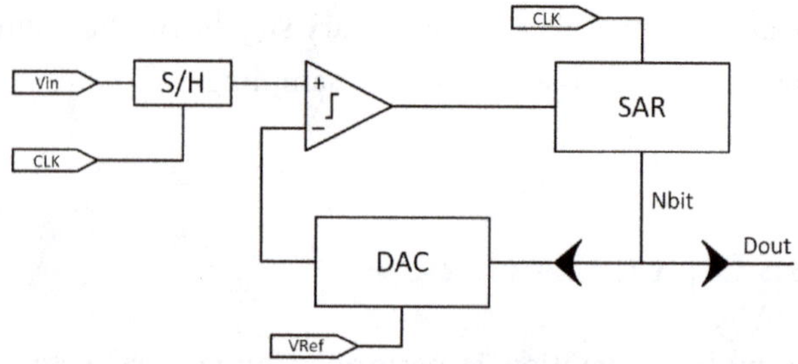

Figure 5.6 Composition of the SAR ADC.

5.4 Discrete-Time to Continuous-Time Conversion

The processed digital signal must be converted to continuous-time. This operation is called reconstruction. In this section, we analyze it from both frequency and time domain perspectives. Here, we assume that the original bandlimited continuous-time signal $x(t)$ is sampled with sampling frequency $f_s = 1/T_s$ Hz and the bandwidth of $x(t)$ is Ω_M rad/s.

5.4.1 Reconstruction in the Frequency Domain

We now explain the reconstruction operation in the frequency domain. To do so, we start with the discrete-time signal $x[n]$. Considering the sampling period, we know that $x[n] = x(nT_s)$. Moreover, the Fourier transform of $x(nT_s)$ is periodic with period $\Omega_s = 2\pi f_s$ rad/s as in Eqn. 5.7. If the sampling operation is performed properly (by taking $\Omega_s > 2\Omega_M$), there will be no aliasing. As a result, the frequency content of the original signal $x(t)$ is in keeping with the frequency content of the sampled signal $x(nT_s)$, which can be seen by focusing on the copy of the frequency content by setting $k = 0$ in Eqn. 5.7. This part corresponds to the original signal. If this part can be filtered, then we can reconstruct the original signal $x(t)$. In theory, this can be performed by selecting an ideal lowpass filter with the pass-band slightly higher than Ω_M rad/s. Hence, we can reconstruct the original continuous-time signal from its samples. Let us explain this operation on the sampled tone signal in Example 5.1.

Example 5.4 Reconstructing the tone signal in the frequency domain

Let us first focus on the frequency domain representation of the sampled tone signal in Figure 5.3. As can be seen in this figure, the frequency content of the original signal is available around the ***center frequency***. This part can be extracted using an ideal lowpass filter with frequency response $H_{ILPF}(j\Omega)$ and pass-band $\Omega_M + \epsilon$ rad/s. Here, ϵ represents a value slightly larger than zero. We provide the reconstructed signal in the frequency domain in Figure 5.7. As can be seen in this figure, the reconstruction operation keeps only one copy of the periodic frequency content of the discrete-time signal. This copy corresponds to the frequency content of the original continuous-time signal.

> **center frequency**
> *The measure of a central frequency, between the upper and lower cut-off frequencies.*

5.4.2 Zero-Order Hold Circuit for Reconstruction

The impulse response of an ideal lowpass filter with ***cut-off frequency*** $\omega_c = \Omega_M$ rad/s is

$$h_{ILPF}(t) = \frac{\omega_c T_s}{\pi} sinc(\omega_c t) \tag{5.10}$$

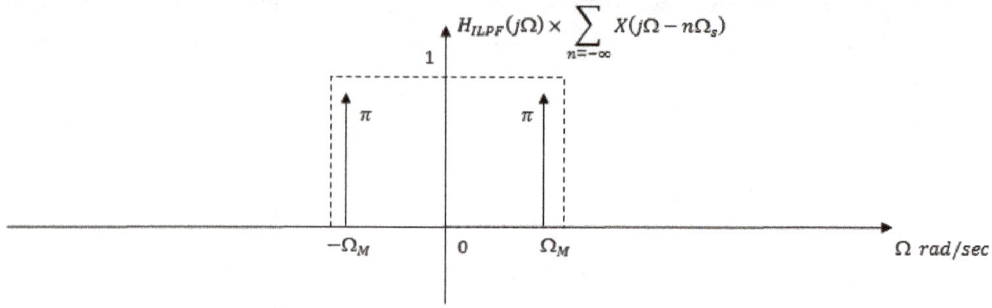

Figure 5.7 Reconstructing the tone signal from its samples in the frequency domain.

where $sinc(t) = \sin(t)/t$ [1]. In Eqn. 5.10, $h_{ILPF}(t)$ is defined for $t \in (-\infty, \infty)$. Hence, it is noncausal and not realizable. A noncausal system may process future input values to generate a current output value. On the contrary, a causal system does not anticipate future input values for operation.

> **cut-off frequency**
> *The frequency in which the output power of the system decreases to a given proportion of the power in the passband.*

Causal and realizable hold circuits can be used instead of the ideal lowpass filter. In this book, we will only consider the zero-order hold (ZOH) circuit for this purpose. As its name implies, the circuit keeps its input value at its output for T_d s. Therefore, the impulse response of the ZOH circuit can be written as $h_{ZOH}(t) = u(t) - u(t - T_d)$. The Fourier transform of $h_{ZOH}(t)$ is

$$H_{ZOH}(j\Omega) = T_d sinc\left(\frac{\Omega T_d}{2}\right) e^{-j\frac{\Omega T_d}{2}} \tag{5.11}$$

The magnitude of $H_{ZOH}(j\Omega)$ becomes

$$|H_{ZOH}(j\Omega)| = T_d \left| sinc\left(\frac{\Omega T_d}{2}\right) \right| \tag{5.12}$$

Example 5.5 Magnitude of the frequency response of the ZOH circuit when $T_d = 1$

We plot $|H_{ZOH}(j\Omega)|$ with $T_d = 1$ in Figure 5.8. As can be seen in the figure, the ZOH circuit can be taken as a lowpass filter. The first frequency value for which the magnitude of this filter drops down to zero is an important value. In this case, the important frequency value is $2\pi T_d$ rad/s. We can take $T_d = T_s$, so the ZOH circuit can be used as a practical lowpass filter in reconstruction. This filter does not provide perfect reconstruction due to its *sidelobes* beyond the cut-off frequency.

sidelobes
In a frequency graph, the lobes that represent secondary frequencies.

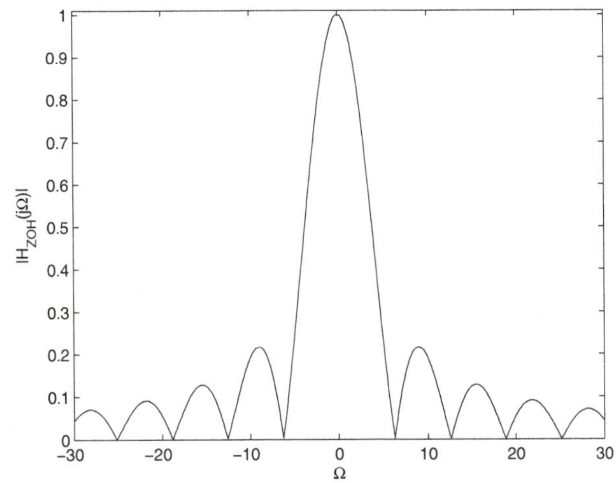

Figure 5.8 Magnitude of the frequency response of the ZOH circuit.

5.4.3 Reconstruction in the Time Domain

The reconstruction operation can also be explained in the time domain. To do so, we start with reconstruction using the ideal lowpass filter. As mentioned in the previous section, reconstruction in the frequency domain corresponds to multiplication of the frequency response of the sampled signal and the ideal lowpass filter. We know that multiplication in the frequency domain corresponds to convolution in the time domain. Hence, we can write the reconstructed signal $x_r(t)$ in the time domain as

$$x_r(t) = x(nT_s) * h_{ILPF}(t) \tag{5.13}$$

$$= \sum_{n=-\infty}^{\infty} x(nT_s)\delta(t - nT_s) * \frac{\omega_c T_s}{\pi} sinc(\omega_c t) \tag{5.14}$$

$$= \frac{\omega_c T_s}{\pi} \sum_{n=-\infty}^{\infty} x(nT_s) sinc\left(\omega_c(t - nT_s)\right) \tag{5.15}$$

Eqn. 5.15 indicates that every sample is represented by a weighted sinc function in reconstruction. If the sampling period satisfies the sampling theorem, we will have perfect reconstruction (i.e., $x_r(t) = x(t)$) using the sum of sinc functions.

Reconstruction using the ZOH circuit is similar. Now, we will convolve $x(nT_s)$ with h_{ZOH}. We set $T_d = T_s$ to obtain

$$x_r(t) = x(nT_s) * h_{ZOH}(t) \tag{5.16}$$

$$= \sum_{n=-\infty}^{\infty} x(nT_s)\delta(t-nT_s) * (u(t) - u(t-T_s)) \tag{5.17}$$

$$= \sum_{n=-\infty}^{\infty} x(nT_s)\,(u(t-nT_s) - u(t-(n+1)T_s)) \tag{5.18}$$

The reconstruction operation using the ZOH circuit will yield a continuous-time signal with staircase form. We now provide an example of this operation.

Example 5.6 Reconstructing the sampled tone signal using the ZOH circuit

We can reconstruct the sampled tone signal from Example 5.1. Let us take $T_d = T_s$ in the ZOH circuit. Part of the reconstructed signal is shown in Figure 5.9. As can be seen in this figure, the reconstructed signal is in staircase form. However, it still has an approximately sinusoidal shape.

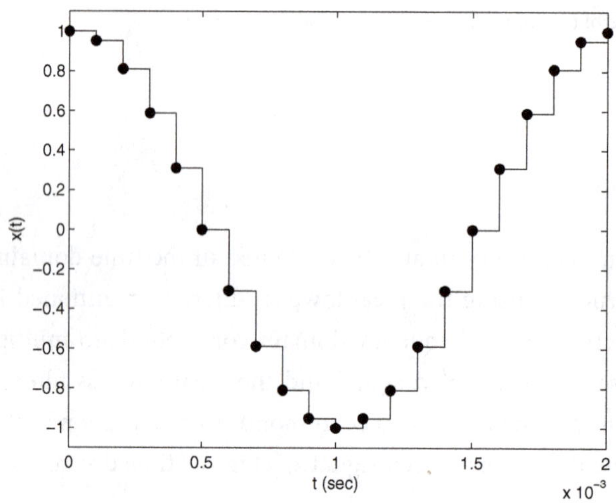

Figure 5.9 Reconstructing the sampled tone signal using the ZOH circuit.

5.5 Digital to Analog Conversion

A digital to analog converter (DAC) module is used to convert a digital value back to analog form. There are several DAC architectures. These can be categorized as Kelvin divider, segmented, binary weighted, R-2R ladder, multiplying, sigma-delta, or PWM [4]. Each DAC architecture has its advantages and disadvantages. We will not go into the details of the DAC module here. Instead, we ask that you consult the accompanying lab for the details of the DAC module used in practice.

The theoretical derivation in the previous section states that perfect reconstruction is possible after sampling a signal, but this is not the case in practice. First, we have the ADC operation to perform the sampling operation, but the ADC module quantizes the amplitude of each sample besides sampling in the time domain, which is an irreversible process. Second, a continuous-time signal cannot be bandlimited in real life, which means it must be filtered by an anti-aliasing filter before the sampling operation. This filter discards some high-frequency components of the original signal. These distortions occur before the reconstruction operation. Third, we know that the ideal lowpass filter used to reconstruct the sampled signal cannot be realized. Instead, a non-ideal lowpass filter must be used (such as a ZOH circuit). This also distorts the frequency content of the signal, so there is no such thing as a perfect reconstruction in practice. We handle practical digital to analog conversion concepts in Lab 5.11.

As a sample DAC architecture, we use the R/2R ladder. As shown in Figure 5.10, the N-bit R/2R ladder DAC includes a termination resistor, N 2R valued resistors, $N - 1$ R valued resistor, and an amplifier. The termination resistor is connected to ground and ensures the Thevenin resistance of the resistor network is R regardless of the number of bits in the network. Digital to analog conversation is performed by switching the 2R resistors to the reference voltage (V_{Ref}) or ground according to digital word information. If all inputs are connected to ground, the output is 0 volts. If all inputs are connected to V_{Ref}, the output voltage approaches V_{Ref}. If some inputs are connected to ground and others connected to V_{Ref}, then the output will be between 0 volts and V_{Ref}. The output range is independent of the number of bits. It is only limited by V_{Ref}. The number of bits only increases the resolution of the DAC. The operational amplifier at the output acts as a buffer and ensures there are output loads without affecting the ladder network.

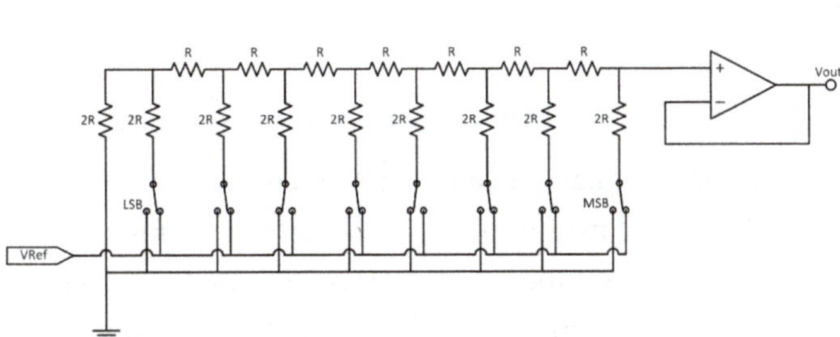

Figure 5.10 Composition of the N-bit R/2R ladder DAC.

5.6 Changing the Sampling Frequency

The sampling frequency of a discrete-time signal can be altered in two ways. If the original continuous-time signal is available, we can apply the sampling operation with the desired sampling frequency. However, if the original continuous-time signal is not available, this operation is not possible. The only option left is to alter the digital signal characteristics using mathematical calculations. In this section, we will explain how this is done.

5.6.1 Downsampling

The number of samples to be processed in the discrete-time signal may be more than necessary, or the original sampling frequency may be higher than needed. Therefore, we may want to decrease the number of samples in the discrete-time signal. This is called downsampling. The downsampled signal can be represented in the time domain as

$$x_d[n] = x[nM] \tag{5.19}$$

where $M \in \mathbb{I}$ is the downsampling factor.

The downsampling operation affects the frequency content of a discrete-time signal. We can see this in the frequency domain as

$$X_d(e^{j\omega}) = \frac{1}{M}\sum_{i=0}^{M-1} X(e^{j(\omega/M - 2\pi i/M)}) \tag{5.20}$$

Eqn. 5.20 indicates that the frequency content of the original signal is expanded by a factor of M after downsampling [2]. Therefore, the original signal components with frequency values lower than π/M rad will expand to π rad and further, which may cause aliasing. To avoid this, we must apply prefiltering before the downsampling operation, where the aim is to discard the original signal components with frequency values higher than π/M rad.

Example 5.7 Downsampling the sampled tone signal

Let us use the sampled tone signal in Example 5.1 and downsample it by a factor of four. We provide the original and downsampled signals in the time domain in Figure 5.11. As can be seen in this figure, the number of samples is decreased by a factor of four.

We also provide the FFT magnitude of the original and downsampled tone signals in Figure 5.12. The FFT magnitude of the original tone signal extends from sample 0 to 255. Because the downsampling operation is performed by a factor of four in this example, the FFT magnitude of the downsampled tone signal shrinks down to the range 0 to 63 in part (b) of the figure. Here, you should keep in mind that the FFT is calculated on the downsampled signal. This is why the number of FFT samples is also decreased. However, it is still possible to see the effect of downsampling in the expanded frequency range in this figure.

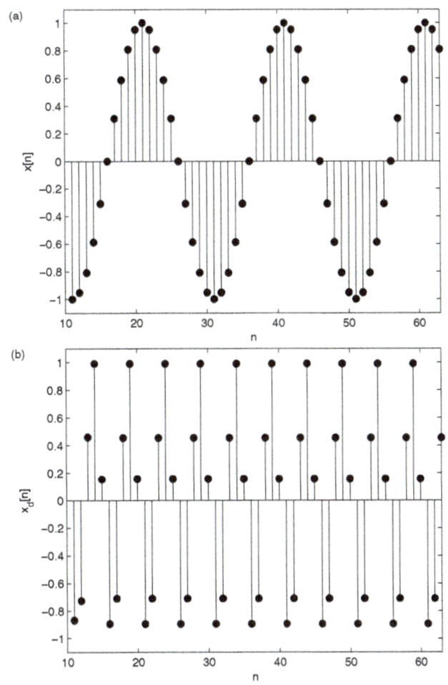

Figure 5.11 Downsampling the sampled tone signal. (a) Original signal; (b) Downsampled signal.

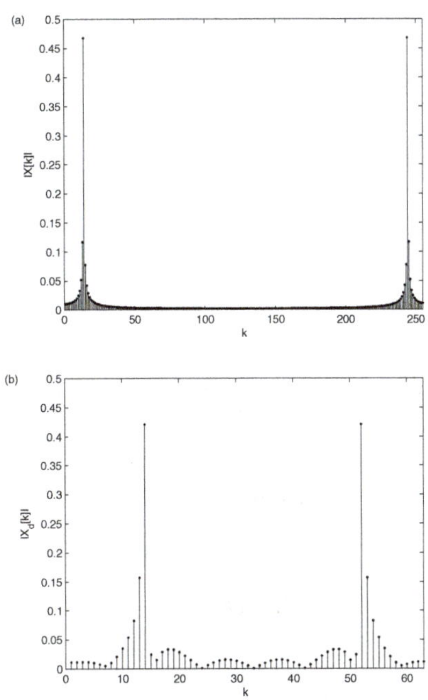

Figure 5.12 The effect of downsampling in the frequency domain. Magnitude of the FFT of the original and downsampled tone signals. (a) Original signal; (b) Downsampled signal.

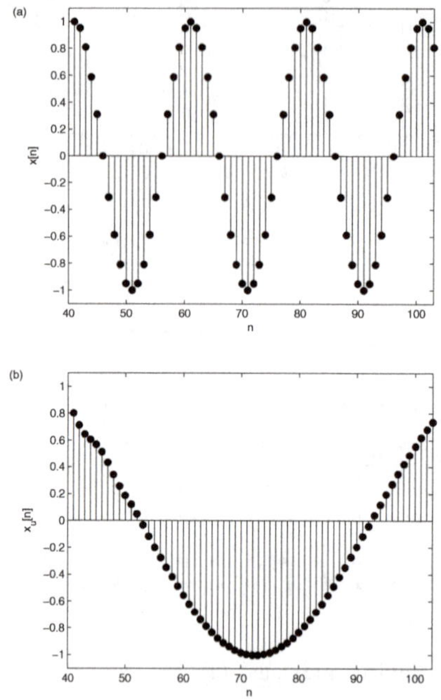

Figure 5.13 Interpolating the sampled tone signal. (a) Original signal; (b) Interpolated signal.

5.6.2 Interpolation

Similar to downsampling, the number of samples to be processed in the discrete-time signal may be fewer than necessary, or the original sampling frequency may be lower than needed. Therefore, we may want to increase the number of samples in the discrete-time signal. This is called interpolation, and it is performed in two steps. First, zeros are inserted between discrete-time signal samples. Second, interpolation is used to set the values of the intermediate samples.

Inserting zeros between discrete-time signal samples can be represented mathematically as

$$x_u[n] = \begin{cases} x[n/L] & n/L \in \mathbb{I} \\ 0 & \text{otherwise} \end{cases} \tag{5.21}$$

where L is the interpolation factor. We can see the effect of this operation in the frequency domain as

$$X_u(e^{j\omega}) = X(e^{j\omega L}) \tag{5.22}$$

Eqn. 5.22 indicates that the frequency content of the original signal is compressed by a factor of L after adding zeros [2]. Therefore, the original signal components with frequency values in the range π to $L\pi$ rad will be located between π/L and π rad in the interpolated signal. As a result, multiple copies of the original signal will be inserted in the original frequency range in the interpolated signal. In order to eliminate these undesired copies, we should apply a lowpass filter after interpolation. This filter should discard the frequency content of the interpolated signal between π/L and π rad. This operation corresponds to smoothing in the time domain.

Example 5.8 Interpolating the sampled tone signal

Let us take the sampled tone signal in Example 5.1 and interpolate it by a factor of four. We provide the original and interpolated signals in the time domain in Figure 5.13. As can be seen in this figure, the interpolated signal keeps the original signal shape, but the number of samples is increased by a factor of four.

We also provide the FFT magnitude of the original and interpolated tone signals in Figure 5.14. The FFT magnitude of the original tone signal extends from sample 0 to 255. Because the interpolation operation is performed by a factor of four in this example, the FFT magnitude of the interpolated tone signal expands up to the range 0 to 1023 in part (b) of the figure. You should keep in mind that the FFT is calculated on the interpolated signal, which is why the number of FFT samples also increased. However,

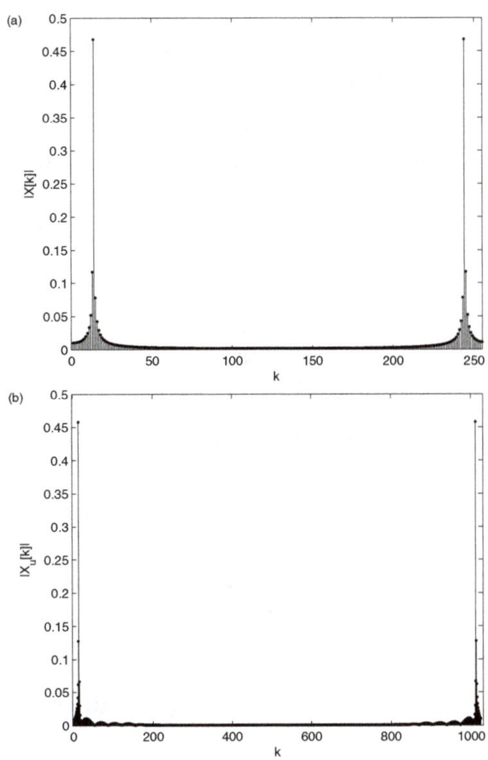

Figure 5.14 The effect of interpolation in the frequency domain. Magnitude of the FFT of the original and interpolated signals. (a) Original signal; (b) Interpolated signal.

it is still possible to see the effect of interpolation in the compressed frequency range in this figure.

5.7 Chapter Summary

In order to process a continuous-time signal in a digital system, the first step is converting it to digital form. The basis of this operation is the sampling theorem. In this chapter, we explored the theorem in both time and frequency domains. The practical module performing the operation in a microcontroller is the ADC. We introduced the general properties of this module to emphasize its practical constraints and limitations. As the sampled signal is processed in a digital system, it must be converted back to analog form. This operation is called reconstruction, which we explored from both theoretical and practical perspectives. Lastly, we introduced mathematical methods to change the sampling frequency of a discrete-time signal. These topics will be extensively used in practical applications. Therefore, we strongly suggest that you master them.

5.8 Further Reading

As mentioned in the introduction, using mathematics was unavoidable in this chapter. You might want to refer to these books for more background [1, 2, 5, 6, 7, 8, 9]. As in previous chapters, if you do not feel comfortable with the heavy mathematical derivations in these books, we suggest referring to Smith's book [10].

5.9 Exercises

5.9.1 An unknown continuous-time signal is sampled without aliasing. As a result, four seconds of the signal is represented by 85000 samples. What is the highest frequency component that could be available in the continuous-time signal?

5.9.2 A continuous-time sinusoidal signal $x(t) = \sin(\Omega_0 t)$ is sampled with sampling period T_s. The obtained discrete-time signal is represented as $x[n]$.

- **(a)** What is the fundamental period of $x[n]$ if $\Omega_0 = 20$ rad/s and $T_s = \pi/8$ s?
- **(b)** Is there a general constraint on T_s such that there is no aliasing after sampling; $x[n]$ being a periodic signal?

5.9.3 The continuous-time signal $x(t) = \cos(\pi t/10)$ is sampled with sampling frequency f_s.

- **(a)** What should the minimum value of f_s be to avoid aliasing?
- **(b)** Plot 32 samples of the obtained discrete-time signal with $f_s = 10$ Hz. Will there be aliasing for this sampling frequency?

(c) Plot 32 samples of the obtained discrete-time signal with $f_s = 0.1$ Hz. Will there be aliasing for this sampling frequency?

(d) Find another continuous-time signal (with a specified sampling frequency) such that the discrete-time signal in part (c) is obtained.

5.9.4 A continuous-time signal is represented in the time domain as $x(t) = sinc(\pi t)$. Find the frequency domain representation of the sampled signal when

(a) the sampling frequency is set to the lowest possible value obtained from the sampling theorem.

(b) the sampling frequency is set to twice the value obtained from the sampling theorem.

5.9.5 A continuous-time signal is represented in the time domain as $x(t) = u(t+10) - u(t-10)$. This signal is sampled with sampling frequency $f_s = 20$ Hz. Plot the magnitude of the frequency domain representation of the sampled signal.

5.9.6 A continuous-time signal is represented in the frequency domain as $X(j\Omega) = (u(\Omega+10) - u(\Omega - 10))e^{-j\Omega}$. This signal is sampled with sampling frequency $\Omega_s = 20$ rad/s. Plot the magnitude of the frequency domain representation of the sampled signal.

5.9.7 A continuous-time signal is sampled with an appropriate sampling frequency. The ADC module used in operation provides a 12-bit output. This module has an input voltage range in the range 0–3 V. Sampled signal values are stored in an integer array. Based on these values, what will the resolution of the sampled signal be?

5.9.8 We want to reconstruct the signal $x(t) = \cos(\pi t/10)$ with sampling frequency $f_s = 10$ Hz. Plot the reconstructed signal when

(a) an ideal LPF with an appropriate cut-off frequency is used.

(b) a ZOH circuit with an appropriate T_d value is used.

5.9.9 The ZOH circuit is not the only option for interpolation. Find the time and frequency domain characteristics of the

(a) first-order hold circuit.

(b) second-order hold circuit.

5.9.10 The continuous-time signal $x(t) = \cos(\pi t/10) + \sin(\pi t)$ is sampled with sampling frequency $f_s =$ 10 Hz.

(a) Plot 128 samples of the discrete-time signal obtained by the sampling operation.

(b) Apply downsampling to the discrete-time signal by a factor of three. Plot 128 samples of the downsampled signal.

(c) Apply interpolation to the discrete-time signal by a factor of three. Plot 128 samples of the interpolated signal.

(d) Using downsampling and interpolation operations, how can we change the sampling frequency of the original signal to $f_s = 35$ Hz?

5.10 References

[1] Oppenheim, A.W., Willsky, A.S., and Nawab, S.H. (1997) *Signals and Systems*, Prentice Hall, Second edn.
[2] Oppenheim, A.W. and Schafer, R.W. (2009) *Discrete-Time Signal Processing*, Prentice Hall, Third edn.
[3] Ünsalan, C. and Gürhan, H.D. (2013) *Programmable Microcontrollers with Applications: MSP430 LaunchPad with CCS and Grace*, McGraw-Hill.
[4] Kester, W. (2004) *The Data Conversion Handbook*, Elsevier.
[5] Mitra, S.K. (2010) *Digital Signal Processing*, McGraw-Hill, Fourth edn.
[6] Proakis, J.G. and Manolakis, D.K. (1995) *Digital Signal Processing: Principles, Algorithms and Applications*, Prentice Hall, Third edn.
[7] Orfanidis, S. (1995) *Introduction to Signal Processing*, Prentice-Hall.
[8] McClellan, J.H., Schafer, R.W., and Yoder, M.A. (2003) *Signal Processing First*, Prentice-Hall.
[9] Manolakis, D.G. and Ingle, V.K. (2011) *Applied Digital Signal Processing*, Cambridge University Press.
[10] Smith, S.W. (1997) *The Scientist and Engineer's Guide to Digital Signal Processing*, California Technical Pub.

5.11 Lab 5

5.11.1 Introduction

This lab introduces the basic concepts of conversion between continuous-time and discrete-time signals from a practical perspective. We will begin with analog to digital conversion. Then, we will focus on digital to analog conversion. Finally, we will explain the interpolation and decimation operations in a practical example. As these operations depend on the ADC and DAC modules on the STM32F4 Discovery kit, you should review the relevant sections of Chapter 1. Moreover, you will need an oscilloscope, a ***signal generator***, and suitable connection cables to carry out the applications introduced in this chapter. Operations to be performed in this chapter are explained in detail in previous sections.

> **signal generator**
> *A laboratory instrument commonly used to generate repeating or non-repeating electronic signals in either the analog or the digital domain.*

5.11.2 Analog to Digital Conversion

Analog to digital conversion is the first step in processing a continuous-time signal in a digital system. The theoretical background of analog to digital conversion is given in Section 5.3. Here, we will start with the basic setup to adjust the ADC module of the STM32F407VGT6 microcontroller. Then, we will analyze the effect of aliasing on a sampled signal.

Basic Setup

We will sample an analog sinusoidal signal generated from a signal generator in order to show how the ADC module of the STM32F407VGT6 microcontroller works. Adjust the signal generator to obtain an analog sinusoidal signal with a 2-kHz frequency, 3-V peak to peak amplitude, and 1.5-V *offset* value. Connect the output of the signal generator to the oscilloscope and observe the generated *analog signal*. You should see a similar signal to that shown in Figure 5.15.

> **voltage offset**
> *A DC voltage value added to an AC waveform.*

> **analog signal**
> *A continuous-time signal.*

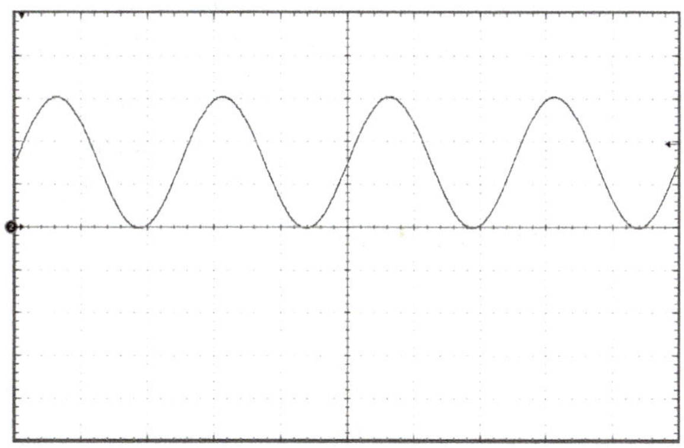

Figure 5.15 An analog sinusoidal signal generated from a signal generator.

Now, we will feed this analog sinusoidal signal to the ADC module of the STM32F407VGT6 microcontroller. An input voltage value higher than 3 V may damage the STM32F407VGT6 microcontroller. Therefore, make sure that the generated sinusoidal signal does not exceed this value. The sampling frequency will be set to 16 kHz in analog to digital conversion. The timer of the microcontroller should trigger its internal ADC module every 62.5 μs. You should review the working principles of the ADC module in Section 1.4.4. We provide a sample project for the initial setup of the sampling operation in `Online_Student_Resources\Lab5\sampling_initial_setup`.

Task 5.1

Connect the signal generator output to the PA1 pin of the STM32F407VGT6 microcontroller. Execute the project given in `Online_Student_Resources\Lab5\sampling_initial_setup`. Plot the sampled signal. What do you observe?

Aliasing

The sampling frequency of the previous example is specifically selected based on the sampling theorem. As a reminder, this theorem tells us that the minimum sampling frequency for an analog signal should be selected to be more than twice its maximum frequency. If the sampling frequency does not satisfy this constraint, aliasing will occur.

Task 5.2

In order to observe aliasing, change the sampling frequency to 8, 4, and 3.2 kHz in the project given in `Online_Student_Resources\Lab5\sampling_initial_setup`. Rerun the code for each sampling frequency and plot the sampled signals. When do you expect to observe aliasing? Comment on the sampled signals.

This section has covered analog to digital conversion with a particular focus on the working principles of the ADC module of the STM32F407VGT6 microcontroller. We expect you to master this concept by converting an analog signal to digital form via practical applications. While doing this, the effect of aliasing can also be observed. Related to this, we expect you to now understand how the sampling frequency of the ADC module should be selected to avoid aliasing.

5.11.3 Digital to Analog Conversion

A digital signal can be represented in analog form using a digital to analog converter. We discussed digital to analog conversion in Section 5.5. Here, we will provide examples of analog signal generation from a digital representation using a microcontroller. Specifically, we will generate analog sinusoidal, square, triangle, and sawtooth signals. To do so, we will use the periodic signal generation method given in Section 2.9.4 to construct one period of the desired digital signal. We will use the DAC module of the STM32F407VGT6 microcontroller to generate analog signals. Details on this DAC module can be found in Section 1.4.4.

We will first generate periodic sinusoidal and square signals with a period of 160 samples. If the sampling frequency of the DAC is set to 16 kHz, the generated analog signals will have a 16000/160 = 100 Hz frequency. The generated analog signals will have a 3-V peak to peak amplitude and a 1.5-V offset value.

We will use the ST Discovery kit's onboard push-button to switch between digital signals fed to the DAC module. Initially, the sinusoidal signal is fed to the DAC. We provide a sample project for this application in `Online_Student_Resources\Lab5\analog_signal_generation`.

Task 5.3

Connect the PA4 pin of the STM32F407VGT6 microcontroller to the oscilloscope. Execute the project given in `Online_Student_Resources\Lab5\analog_signal_generation`. Observe the periodic sinusoidal and square analog signals using the push-button. Comment on the generated analog signals.

Task 5.4

Change the code given in `Online_Student_Resources\Lab5\analog_signal_generation` to generate triangle and sawtooth signals with the same specifications. Rerun the code and observe the analog signals on the oscilloscope. Comment on the generated analog signals.

Task 5.5

Now, repeat Task 5.4, but change the sampling frequency to 1.6 kHz and change the period of digital signals to 16 samples. Run the code and observe each analog signal on the oscilloscope. What has changed in the generated analog signals?

As a result of this section, you should understand the usage of the DAC module in the STM32F407VGT6 microcontroller. Moreover, you should observe how the sampling frequency of the DAC module affects the generated analog signal.

5.11.4 Changing the Sampling Frequency

In Section 5.6, we explained how the sampling frequency of a discrete-time signal can be changed using decimation or interpolation operations. Here, we will apply these operations to a tone signal generated in the microcontroller. Then, we will feed the decimated and interpolated signals to the AUP audio card. We will be able to hear the changes in the decimated and interpolated tone signal.

Generating the Original Tone Signal

We will first generate a 500-Hz tone signal with 8000 samples. This signal will be fed to the AUP audio card. The sampling frequency for this operation is set to 8 kHz. Hence, the duration of the tone signal will be 1 s. In order to hear this signal continuously, we will send the samples of the tone signal to the AUP audio card repeatedly. The complete project for this application is given in `Online_Student_Resources\Lab5\tone_generation`.

Task 5.6

Run the code in `Online_Student_Resources\Lab5\tone_generation` and listen the tone signal. Plot the original tone signal.

Decimation

Below, we provide the C function for performing the downsampling operation on a given digital signal. You should be able to grasp the basic mechanism behind downsampling by inspecting it.

```
//Variables: Input array, output array, dummy array, prefilter array,
//state array for filter, sample size, filter size, donwsampling coefficient
void decimate(float32_t *i, float32_t *o, float32_t *d, float32_t *f,
float32_t *fState, uint32_t s_size, uint32_t t_size, uint32_t d_coef){
arm_fir_init_f32(&S, t_size, f, fState, s_size);
arm_fir_f32(&S, i, d, s_size);

for(n=0;n<(s_size/d_coef);n++)
*(o+n) = *(d+(d_coef*n));
}
```

The downsampling operation is extensively used in digital signal processing. For this reason, the CMSIS-DSP library has predefined functions, which are given below. Note that the downsampling operation is called decimation in this library. The filter to be used in the downsampling operation should be provided by the user for the CMSIS-DSP library function.

```
arm_status arm_fir_decimate_init_f32(arm_fir_decimate_instance_f32 *S,
uint16_t numTaps,uint8_t M,float32_t *pCoeffs,float32_t *pState, uint32_t blockSize)
/*
S: points to an instance of the floating-point FIR decimator structure.
numTaps: number of coefficients in the filter.
M: decimation factor.
pCoeffs: points to the filter coefficients.
pState: points to the state buffer.
blockSize: number of input samples to process per call.
*/

void arm_fir_decimate_f32(const arm_fir_decimate_instance_f32 *S,
float32_t *pSrc, float32_t *pDst, uint32_t blockSize)
/*
S: points to an instance of the floating-point FIR decimator structure.
pSrc: points to the block of input data.
pDst: points to the block of output data.
blockSize: number of input samples to process per call.
*/
```

Let us apply decimation to a signal using the CMSIS-DSP library downsampling functions. The decimation factor is set to 6 for the following exercise. The complete

Keil project for this exercise is given in `Online_Student_Resources\Lab5\decimation_example`.

Task 5.7

This example concerns downsampling a signal by a factor of 6. In other words, we change its sampling rate from $fs = 48$ kHz to $fs' = fs/6 = 8$ kHz.

Consider the continuous-time signal

$$x(t) = 8000\sin(1000\pi t) + 2400\cos(12000\pi t + 0.3) \tag{5.23}$$

This is a signal comprised of two sinusoidal components with frequencies 500 Hz and 6000 Hz, respectively. Sampling at a frequency of $fs = 48$ kHz yields the discrete-time signal represented by the sequence

$$x(n) = 8000\sin(1000\pi n/48000) + 2400\cos(12000\pi n/48000 + 0.3) \tag{5.24}$$

192 samples of this signal are shown in Figure 5.16.

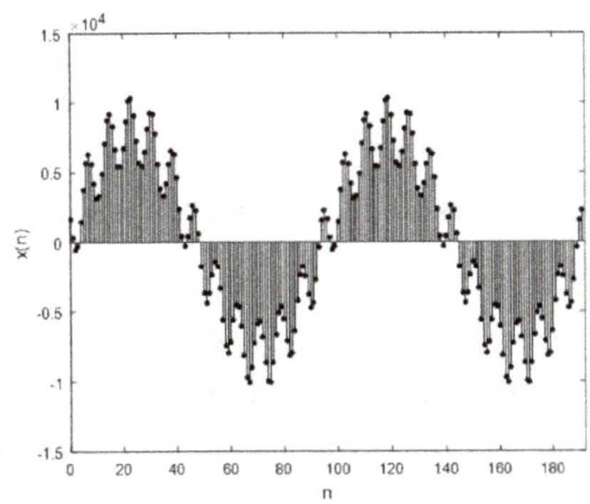

Figure 5.16 192 samples of the discrete-time signal from Task 5.7.

Because the sampling frequency is an integer multiple of both 500 Hz and of 6000 Hz, $x[n]$ is periodic with a period of 96 samples. Furthermore, as the frequencies 500 Hz (fs/96) and 6000 Hz (fs/8) are both less than the Nyquist frequency (fs/2), aliasing is not an issue. We would expect to be able to reconstruct the signal $x(t)$ correctly from its samples, $x[n]$. We will consider the higher frequency component of the signal to be an important high-frequency detail.

Figure 5.17 shows the continuous-time analog output signal produced by writing the sample sequence $x[n]$ to the DAC in the TLV320AIC23B codec. This result was achieved using the program `stm32f4_downsample_x6_A_intr.c` and may be considered as confirmation that the signal $x(t)$ can be reconstructed correctly from its samples. The 500 Hz sinusoidal component and the smaller amplitude 6000 Hz sinusoidal component are clearly visible in the oscilloscope trace.

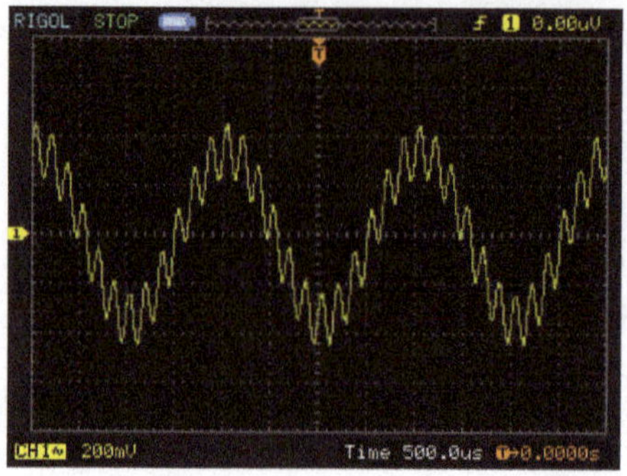

Figure 5.17 Continuous-time signal reconstructed from discrete-time signal $x[n]$ using the TLV320AIC23B DAC with sampling frequency $fs = 48$ kHz.

The digital to analogue reconstruction process is represented in block diagram form in Figure 5.18.

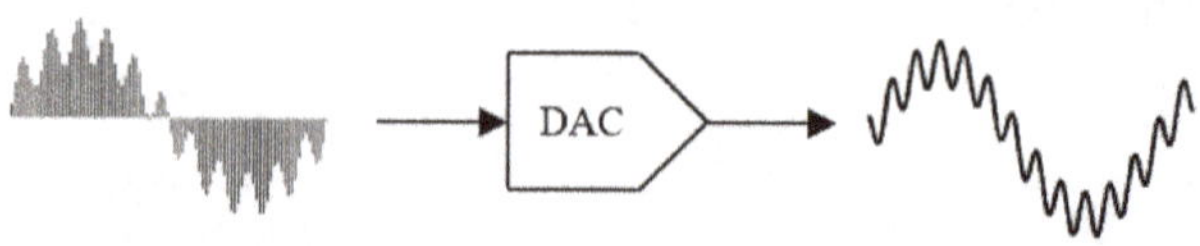

Figure 5.18 Block diagram representation of the reconstruction of a continuous-time signal from the discrete-time signal $x[n]$ using the TLV320AIC23B DAC with sampling frequency $fs = 48$ kHz.

A discrete-time signal sampled at $fs' = 8$ kHz can only represent sinusoidal components correctly at frequencies less than $fs'/2 = 4$ kHz. For this reason, before sampling a signal, it should be passed through a lowpass antialiasing filter with a cut-off frequency of less than 4 kHz. This requirement holds whether the signal to be sampled is a continuous-time signal or, as in this example, a discrete-time signal.

The program `stm32f4_downsample_x6_B_intr.c` applies an antialiasing filter to the signal considered in the program `stm32f4_downsample_x6_A_intr.c`. The antialiasing filter is implemented in MATLAB's `FDATool` as a 64-coefficient lowpass FIR filter with a cut-off frequency of $fs/12 = 4$ kHz ($fs = 48$ kHz). Figure 5.19 shows the magnitude frequency response of this filter as predicted by `FDATool`.

Figure 5.20 shows the result of passing the first 192 samples of signal $x[n]$ through the filter. Starting with zero initial conditions, the filter output exhibits an initial transient response (affecting the first 64 output samples) before settling to give a periodic output waveform in response to its periodic input signal. As expected, the $fs/8 = 6$ kHz sinusoidal component of the input signal is not present in the filtered signal. Further, as expected, the $fs/96 = 500$ Hz component of the input signal passes through the antialiasing filter.

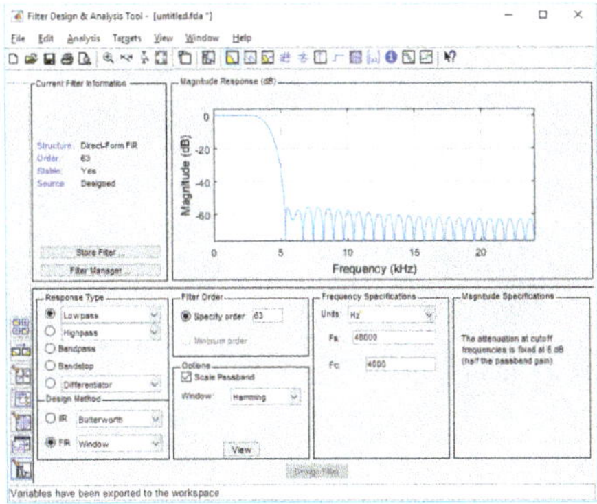

Figure 5.19 64-coefficient FIR antialiasing filter designed using `FDATool`.

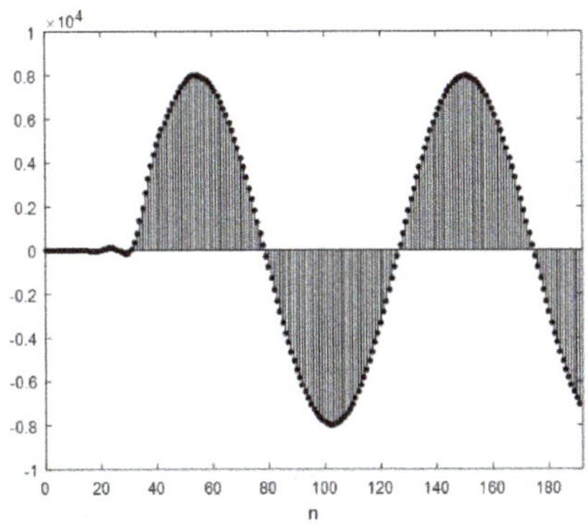

Figure 5.20 Result of passing the discrete-time signal $x[n]$ through an antialiasing filter.

The program `stm32f4_downsample_x6_B_intr.c` then samples the filtered signal at $fs' = 8$ kHz. In other words, it writes every sixth sample value to the TLV320AIC23B DAC and discards the rest at a sampling rate of $fs/6 = fs' = 8$ kHz. Figures 5.21 and 5.22 show the downsampled signal written to the DAC and the continuous-time analogue signal reconstructed by the DAC, respectively.

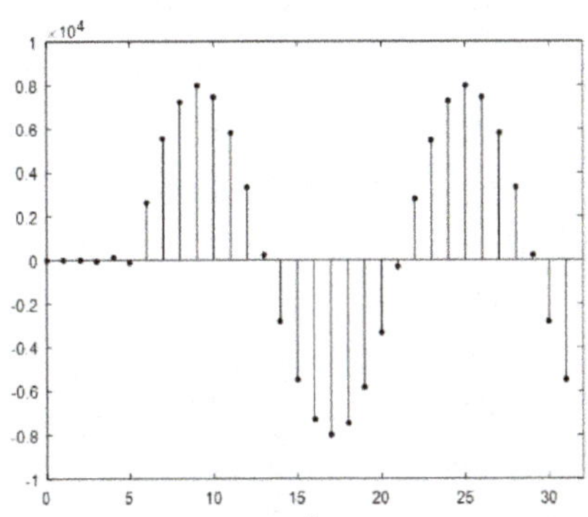

Figure 5.21 Filtered signal downsampled by a factor of 6.

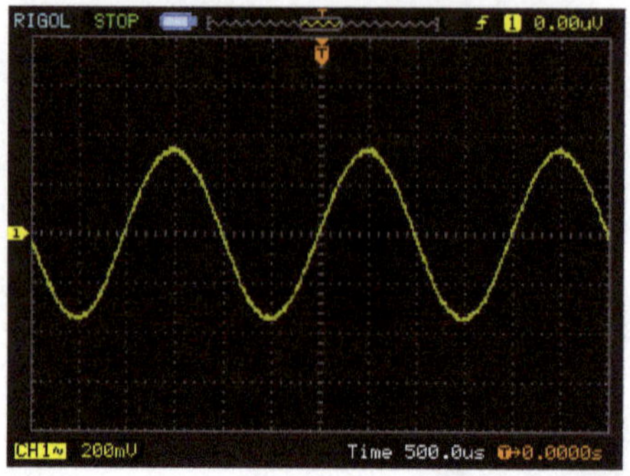

Figure 5.22 Continuous-time signal reconstructed from a downsampled signal using the TLV320AIC23B DAC with sampling frequency $fs = 8$ kHz.

The filtering, downsampling, and digital to analog reconstruction processes are represented in block diagram form in Figure 5.23.

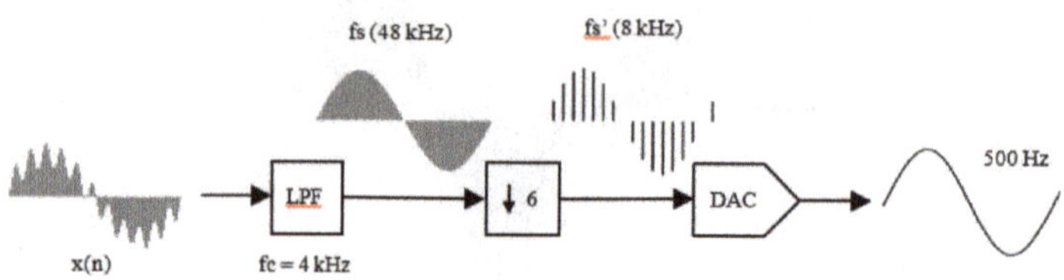

Figure 5.23 Block diagram representation of reconstruction of a continuous-time signal from the discrete-time signal $x[n]$, following filtering and downsampling, using the TLV320AIC23B DAC with sampling frequency $fs = 8$ kHz.

To summarize, to correctly downsample or decimate a discrete-time signal, a lowpass filter must be applied before sampling at a lower rate. Consequently, high-frequency details that are present in the original signal are necessarily lost in the downsampling or decimation process.

To emphasize the necessity of the antialiasing filter implemented by the program `stm32f4_downsample_x6_B_intr.c`, the user may toggle between enabling and disabling the filter by pressing the blue user push button on the Discovery board.

With the filter disabled, every sixth sample of $x[n]$ is written to the DAC. This downsampled signal and the corresponding DAC output are shown in Figures 5.24 and 5.25, respectively. Significantly, in these signals, the $fs/8 = 6$ kHz component of the discrete-time signal $x[n]$ appears as an aliased signal component at frequency $((fs/6) - (fs/8)) = fs/24 = 2$ kHz.

In the program `stm32f4_downsample_x6_B_intr.c`, the FIR antialiasing filter is specified by its coefficients, which are read from the header file `downsample_x6.h`. As shown in Figure 5.19, these coefficients yield a lowpass magnitude frequency response that is 6 ***decibel*** (dB) down at $fs'/2 = 4$ kHz and rolls off steeply at higher frequencies to more than 60 dB down at frequencies greater than 5.5 kHz. Consider

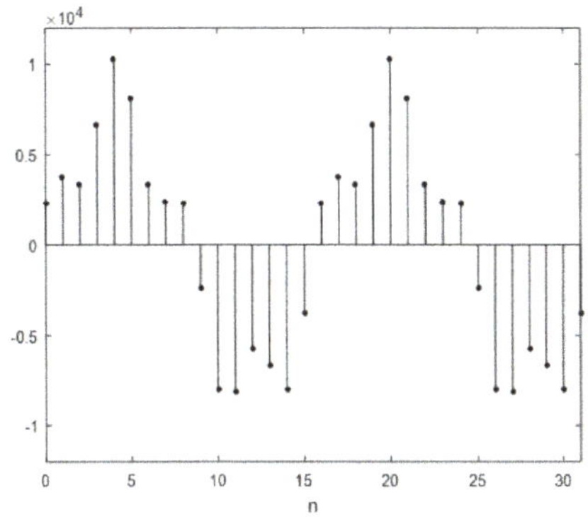

Figure 5.24 Signal $x[n]$ downsampled by a factor of 6 without having first been passed through an antialiasing filter.

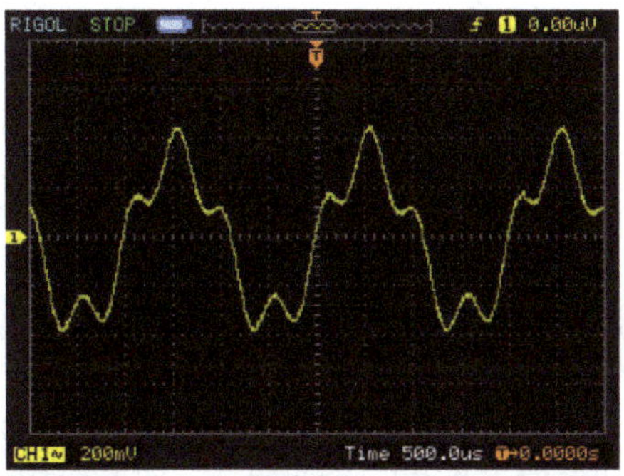

Figure 5.25 Continuous-time signal reconstructed from downsampled signal using the TLV320AIC23B DAC with sampling frequency $fs = 8$ kHz.

the lower order antialiasing filter shown in Figure 5.26. The gain of this 9-coefficient FIR filter is 6 dB down at $fs'/2 = 4$ kHz but rolls off far less steeply than that of the 64-coefficient filter considered earlier.

> **decibel**
> *A logarithmic unit of the ratio between the power of two signals.*

This filter can be tested in the program `stm32f4_downsample_x6_B_intr.c` by changing the pre-processor command

```
#include "downsample_x6_order_63.h"
```

to

```
#include "downsample_x6_order_8.h"
```

Listen to the downsampled signal using headphones, or observe it using an oscilloscope, and you should be able to detect a low-amplitude unwanted aliased 2 kHz component.

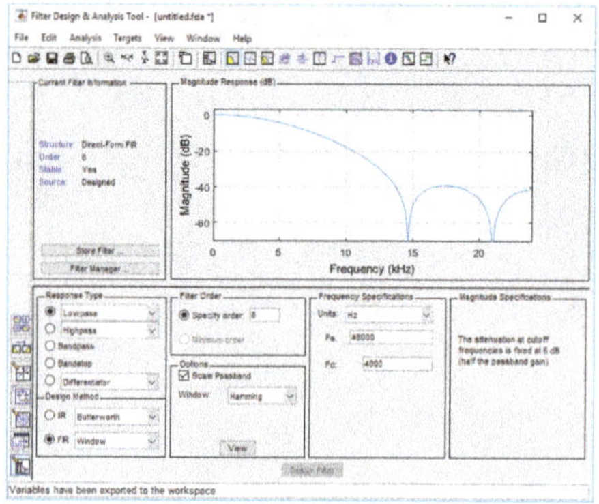

Figure 5.26 9-coefficient FIR antialiasing filter designed using `FDATool`.

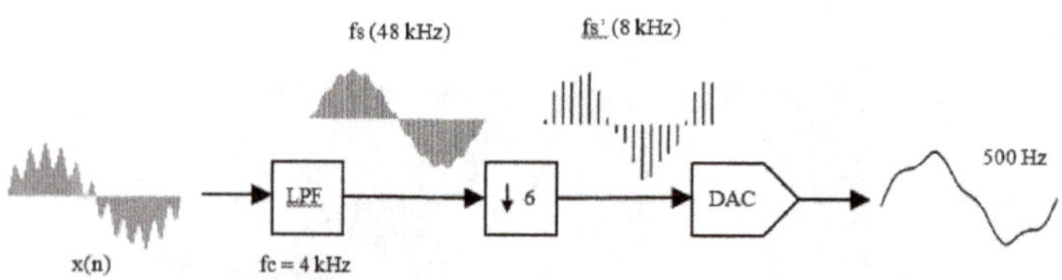

Figure 5.27 Block diagram representation of the reconstruction of a continuous-time signal from the discrete-time signal $x[n]$, following filtering and downsampling, using the TLV320AIC23B DAC with sampling frequency $fs = 8$ kHz, where a lower order and less effective antialiasing filter has been used.

The CMSIS DSP library function `arm_fir_decimate_f32()` implements downsampling and antialiasing as described above. The coefficients of the FIR antialiasing filter used by this function are specified by the user.

The program `stm32f4_downsample_x6_CMSIS_intr.c` is functionally equivalent to the program `stm32f4_downsample_x6_B_intr.c` (with antialiasing filter enabled) but uses the CMSIS function to implement the downsampling operation. It also stores the first 32 sample values that are output by the function, which has zero initial conditions, in the array `buffer[]`. The contents of this array can be saved to a file and plotted using a MATLAB function. These contents also illustrate the startup transient response shown earlier in Figure 5.21.

Interpolation

As with decimation, we provide the C function to perform the interpolation operation on a given digital signal below. We expect you to grasp the basic mechanism behind interpolation by inspecting it.

```
//Variables : Input array, output array, dummy array, filter array, state array
//for filter, sample size, filter size, interpolator coefficient
void interpolate(float32_t *i, float32_t *o, float32_t *d, float32_t *f,
float32_t *fState, uint32_t s_size, uint32_t t_size, uint32_t i_coef){
for(n=0;n<s_size;n++)
```

```
*(d+(i_coef*n)) = *(i+n);

arm_fir_init_f32(&S, t_size, f, fState, s_size*i_coef);
arm_fir_f32(&S, d, o, s_size*i_coef);
}
```

The CMSIS-DSP library has predefined functions for the interpolation operation as well. We provide them below. The filter to be used in the interpolation operation should be provided by the user.

```
arm_status arm_fir_interpolate_init_f32(arm_fir_interpolate_instance_f32 *S,
uint8_t L, uint16_t numTaps, float32_t *pCoeffs, float32_t *pState,
uint32_t blockSize)
/*
S: points to an instance of the floating-point FIR interpolator structure.
L: upsample factor.
numTaps: number of filter coefficients in the filter.
pCoeffs: points to the filter coefficient buffer.
pState: points to the state buffer.
blockSize: number of input samples to process per call.
*/

void arm_fir_interpolate_f32(const arm_fir_interpolate_instance_f32 *S,
float32_t *pSrc,float32_t *pDst,uint32_t blockSize)
/*
S: points to an instance of the floating-point FIR interpolator structure.
pSrc: points to the block of input data.
pDst: points to the block of output data.
blockSize: number of input samples to process per call.
*/
```

Let us apply interpolation to the generated tone signal using CMSIS-DSP library upsampling functions. The interpolation factor for this example is 4. The complete Keil project for this application is given in `Online_Student_Resources\Lab5\interpolation_example`.

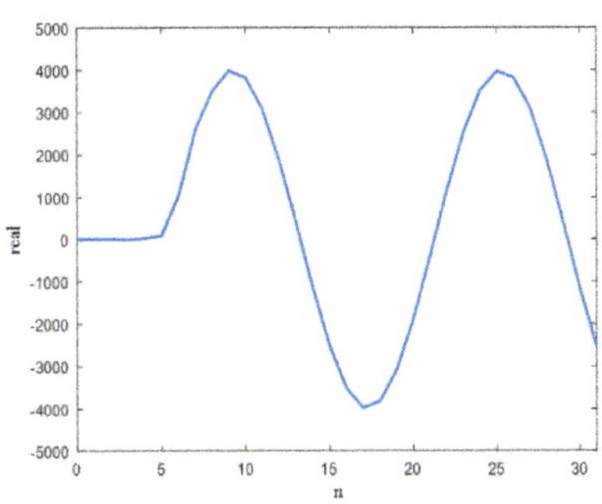

Figure 5.28 Initial 32 sample values obtained using the program `stm32f4_downsample_x6_CMSIS_intr.c`, which uses the CMSIS function `arm_fir_decimate_f32()` to filter and downsample the signal $x[n]$, showing the transient response of the 64-coefficient FIR filter with zero initial conditions.

Task 5.8

Run the code in `Online_Student_Resources\Lab5\interpolation_example`, and listen to the tone signal. What do you hear? In addition, plot the interpolated tone signal.

Operations explained in this section are also described as sample rate conversions. They are specifically used when the sampling frequency of the digital signal at hand must be altered using only software tools so that the user does not need to revisit the original analog signal.

6

Digital Processing of Continuous-Time Signals

Contents

6.1 Introduction

Consider a continuous-time LTI system with input $x(t)$ and output $y(t)$. The aim of this chapter is to replace this continuous-time system with its digital equivalent. The first step is to use an ADC module to extract the digital form of $x(t)$ as $x[n]$. Then, $x[n]$ will be processed by the discrete-time system to obtain the output $y[n]$. This signal will be written to a DAC to obtain the corresponding continuous-time signal $y(t)$. We introduced ADC and DAC operations in Chapter 5. Here, we will focus on the discrete-time system.

We will start by considering frequency mapping between continuous and discrete-time representations. Then, we will evaluate different methods of converting a continuous-time system to discrete-time system. We use the same simple continuous-time system throughout this chapter to compare different methods.

6.2 Frequency Mapping

The result of applying the DTFT to a sequence $x[n]$ (i.e., transforming a discrete-time signal into the frequency domain) is

$$X(e^{-j\omega}) = \sum_{n=-\infty}^{\infty} x[n]e^{-j\omega n} \tag{6.1}$$

$X(e^{-j\omega})$ is a periodic and continuous function of normalized frequency ω with period 2π rad, which means only one cycle of $X(e^{-j\omega})$ must be considered in order to fully represent the signal in the frequency domain.

It is conventional to use the range $[-\pi, \pi]$ for ω, and this corresponds to a range of actual frequencies $[-\Omega_s/2, \Omega_s/2]$ rad/s, where $\Omega_s = 2\pi/T_s$ is the sampling frequency of the system. In other words, an actual frequency of Ω rad/s will map to a normalized frequency of $\omega = \Omega T_s$ rad.

Example 6.1 Frequency mapping

Consider the continuous-time signal comprised of three different frequency components,

$$x(t) = A_1\cos(\Omega_1 t) + A_2\cos(2\Omega_1 t) + A_3\cos(3\Omega_1 t) \tag{6.2}$$

If $x(t)$ is sampled at frequency $\Omega_s = 9\Omega_1$, then $T_s = 2\pi/9\Omega_1$, and the discrete-time signal $x[n]$ will have frequency components at normalized frequencies $\Omega_1 T_s = 2\pi/9$ rad, $2\Omega_1 T_s = 4\pi/9$ rad, and $3\Omega_1 T_s = 6\pi/9$ rad.

$$x[n] = x(nT_s) = A_1\cos\left(\Omega_1 n\frac{2\pi}{9\Omega_1}\right) + A_2\cos\left(2\Omega_1 n\frac{2\pi}{9\Omega_1}\right) + A_3\cos\left(3\Omega_1 n\frac{2\pi}{9\Omega_1}\right) \tag{6.3}$$

The discrete-time signal $x[n]$ is comprises frequency components at normalized frequencies ω_1, $2\omega_1$, and $3\omega_1$ rad, where $\omega_1 = \Omega_1 T_s$.

$$x[n] = A_1 \cos(n\omega_1) + A_2 \cos(2n\omega_1) + A_3 \cos(3n\omega_1) \tag{6.4}$$

We can apply the same reasoning in reconstructing a continuous-time signal from its discrete-time representation. Then,

$$\Omega = \frac{\omega}{2\pi}\Omega_s \tag{6.5}$$

where $\Omega \in [-\Omega_s/2, \Omega_s/2)$ rad/s and $\omega \in [-\pi, \pi)$ rad.

6.3 A Simple Continuous-Time System

Consider a lowpass filter consisting of a resistor and a *capacitor* as shown in Figure 6.1.

> **capacitor**
> *A passive electrical component that is designed to accumulate a charge and store electrical energy.*

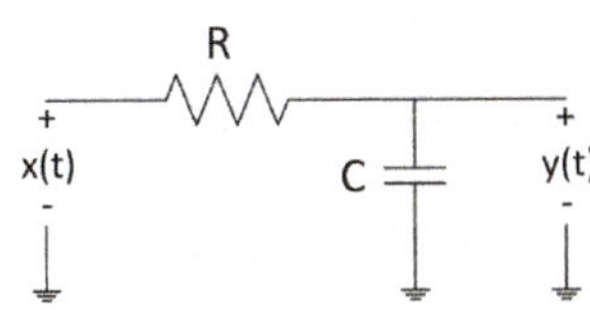

Figure 6.1 A first-order continuous-time RC filter.

The input signal is voltage $x(t)$ and the output signal is voltage $y(t)$. Using Kirchhoff's current and voltage laws [1], the differential equation representing the RC filter can be shown to be

$$\frac{dy(t)}{dt} + \alpha y(t) = \alpha x(t) \tag{6.6}$$

where $\alpha = 1/RC$.

Assuming zero initial conditions, the *Laplace transform* [2] of Eqn. 6.6 is

$$sY(s) + \alpha Y(s) = \alpha X(s) \tag{6.7}$$

corresponding to the transfer function

$$H(s) = \frac{Y(s)}{X(s)} = \frac{\alpha}{s+\alpha} \tag{6.8}$$

> **Laplace transform**
> *A transformation method used to map a continuous function of time to a continuous function of complex frequency.*

6.4 Representing Continuous-Time Systems in Discrete Time

Several methods of representing a continuous-time system in discrete time exist. In this section, we will consider the backward difference, bilinear transformation, bilinear transformation with prewarping, and impulse invariance methods.

6.4.1 Backward Difference

As the name implies, this approach approximates the derivative terms in the differential equation with backward difference at time $t = nT_s$ as

$$\frac{dy(t)}{dt} \approx \frac{y(nT_s) - y((n-1)T_s)}{T_s} \tag{6.9}$$

where n is an integer representing the sample number, and T_s is the sampling period. Because we approximate the differential equation at time $t = nT_s$, $x(t) = x(nT_s)$ and $y(t) = y(nT_s)$. The differential equation in Eqn. 6.6 then becomes

$$y(nT_s) = y((n-1)T_s) - \alpha T_s(y(nT_s) - x(nT_s)) \tag{6.10}$$

Normalizing the time index (by setting $T_s = 1$) in Eqn. 6.10, we obtain

$$y[n] = y[n-1] - \alpha T_s(y[n] - x[n]) \tag{6.11}$$

Rearranging Eqn. 6.11, we obtain

$$y[n] = \frac{1}{1+\alpha T_s} y[n-1] + \frac{\alpha T_s}{1+\alpha T_s} x[n] \tag{6.12}$$

Assuming zero initial conditions and taking the Z-transform of Eqn. 6.11, we obtain Z-transfer function

$$\frac{Y(z)}{X(z)} = H(z) = \frac{\alpha}{\frac{1-z^{-1}}{T_s} + \alpha} \tag{6.13}$$

This is equivalent to substituting

$$s = \frac{1-z^{-1}}{T_s} \tag{6.14}$$

into Eqn. 6.8. This conversion works for all LTI continuous-time systems [3].

To summarize, the backward difference method can be described as

$$H(z) = H(s)\big|_{s=\frac{1-z^{-1}}{T_s}} \tag{6.15}$$

The backward difference method distorts the frequency content during transformation. To show how this distortion occurs, let us start with the transformation formula in Eqn. 6.14. Because we focus on the frequency content, we will have $s = j\Omega$ and $z^{-1} = e^{-j\omega}$. In other words, we analyze Eqn. 6.14 around the imaginary axis in continuous-time. Then, we obtain

$$j\Omega = \frac{1 - e^{-j\omega}}{T_s} \tag{6.16}$$

If we apply Euler's formula to $e^{-j\omega}$, we obtain

$$j\Omega T_s = 1 - \cos(\omega) + j\sin(\omega) \tag{6.17}$$

From this equation, we can obtain $\cos(\omega) = 1$ and $\sin(\omega) = \Omega T_s$. This leads to $\tan(\omega) = \Omega T_s$. Hence,

$$\omega = \arctan(\Omega T_s) \tag{6.18}$$

Eqn. 6.18 clearly indicates that the continuous-time frequency content is transformed to discrete-time frequency by a nonlinear arctangent function, which will have an effect on the representation of a continuous-time filter in discrete time. We will observe this effect in Lab 6.10.

6.4.2 Bilinear Transformation

The bilinear transformation method is based on approximating the integral terms in a differential equation by trapezoids. First, we must first convert the differential equation representing a continuous-time system to the corresponding integral equation. Then, integral terms are approximated. Referring back to the differential equation in Eqn. 6.6, we can take the integral of both sides in a certain time interval to obtain

$$\int_{(n-1)T_s}^{nT_s} \frac{dy(t)}{dt} dt = -\alpha \int_{(n-1)T_s}^{nT_s} y(t)dt + \alpha \int_{(n-1)T_s}^{nT_s} x(t)dt \tag{6.19}$$

We can approximate definite integrals in Eqn. 6.19 by trapezoids to obtain

$$\int_{(n-1)T_s}^{nT_s} y(t)dt = \frac{T_s}{2}(y(nT_s) + y((n-1)T_s)) \tag{6.20}$$

$$\int_{(n-1)T_s}^{nT_s} x(t)dt = \frac{T_s}{2}(x(nT_s) + x((n-1)T_s)) \tag{6.21}$$

The fundamental theorem of calculus yields

$$\int_{(n-1)T_s}^{nT_s} \frac{dy(t)}{dt} dt = y(nT_s) - y((n-1)T_s) \tag{6.22}$$

After normalizing the time index, Eqn. 6.19 becomes

$$y[n] = \frac{1 - \alpha T_s/2}{1 + \alpha T_s/2} y[n-1] + \frac{\alpha T_s/2}{1 + \alpha T_s/2}(x[n] + x[n-1]) \tag{6.23}$$

Eqn. 6.23 represents the difference equation corresponding to the RC filter using bilinear transformation. In general, we can obtain a transformation from continuous-time to discrete-time through bilinear transformation. To do so, let us take the Z-transform of the difference equation in Eqn. 6.23. Rearranging terms, we obtain

$$\frac{Y(z)}{X(z)} = H(z) = \frac{\alpha}{\frac{2(1-z^{-1})}{T_s(1+z^{-1})} + \alpha} \tag{6.24}$$

If we substitute

$$s = \frac{2}{T_s}\frac{1-z^{-1}}{1+z^{-1}} \tag{6.25}$$

into Eqn. 6.8, we can obtain Eqn. 6.24. This conversion works for all LTI continuous-time systems [3]. Therefore, Eqn. 6.25 can be used to find $H(z)$ from $H(s)$ using the bilinear transformation method. In implementation, the only step required is obtaining the difference equation between $y[n]$ and $x[n]$ from $H(z)$. This step is explained in detail in Section 2.5.

The bilinear transformation method also distorts the frequency during transformation. To show how the distortion occurs, let us start with the transformation formula in Eqn. 6.25. As in the backward difference method, we will have $s = j\Omega$ and $z^{-1} = e^{-j\omega}$, which leads to

$$j\Omega = \frac{2}{T_s}\frac{1-e^{-j\omega}}{1+e^{-j\omega}} \tag{6.26}$$

After simplifying Eqn. 6.26, we obtain

$$\omega = 2\ \arctan\left(\frac{\Omega T_s}{2}\right) \tag{6.27}$$

Again, the continuous-time frequency content is transformed to discrete-time frequency by a nonlinear arctangent function as in the backward difference method. In Lab 6.10, we will observe how the nonlinearity affects the filtering operation.

6.4.3 Bilinear Transformation with Prewarping

As indicated in Eqn. 6.27, bilinear transformation warps the continuous-time frequency during operation. We can overcome this problem by prewarping. Let us explain how prewarping can be used on a first-order lowpass filter with transfer function

$$H(s) = \frac{\Omega_C}{s+\Omega_C} \tag{6.28}$$

where Ω_C is the cut-off frequency of the filter.

Assume that we want to convert the filter to discrete time using bilinear transformation. We want to keep the frequency characteristics of the filter around its cut-off frequency during transformation. If we do not use prewarping, the frequency Ω_C will map to discrete time as in Eqn. 6.27. However, we want to have the analog frequency to map to discrete time as in Section 6.10.2. To do so, we prewarp the original frequency value as

$$\Omega_P = \frac{2}{T_s} \tan\left(\frac{\Omega_C T_s}{2}\right) \tag{6.29}$$

where Ω_P is the prewarped frequency. In discrete time, the corresponding frequency value is

$$\omega_C = 2 \arctan\left(\frac{\Omega_P T_s}{2}\right) \tag{6.30}$$

$$= 2 \arctan\left(\frac{T_s}{2}\frac{2}{T_s} \tan\left(\frac{\Omega_C T_s}{2}\right)\right) \tag{6.31}$$

$$= \Omega_C T_s \tag{6.32}$$

As can be seen in Eqn. 6.32, prewarping leads to the desired frequency mapping around ω_C. We can rewrite the transfer function in Eqn. 6.28 using Ω_P as

$$H(s) = \frac{\Omega_P}{s + \Omega_P} \tag{6.33}$$

In order to benefit from prewarping, the transfer function in Eqn. 6.33 must be used in bilinear transformation. We will apply the bilinear transformation method with prewarping to an actual filter in Lab 6.10.

6.4.4 Impulse Invariance

The transformation methods up until now have been based on the differential equation representing the continuous-time system. The impulse invariance method takes a different approach. It uses the impulse response of the system. The aim here is to keep the impulse response of continuous-time and discrete-time systems the same at each sampling instance. This can be achieved by setting $h[n] = T_s h(nT_s)$.

The impulse response of the simple RC filter can be extracted as $h(t) = \alpha e^{-\alpha t} u(t)$ by taking the inverse Laplace transform of Eqn. 6.8. The impulse invariance method leads to

$$h[n] = T_s \alpha e^{-\alpha T_s n} u[n] \tag{6.34}$$

Taking the Z-transform of the discrete-time impulse response (Eqn. 6.34) yields the transfer function

$$H(z) = \frac{Y(z)}{X(z)} = \frac{T_s \alpha}{1 - e^{-\alpha T_s} z^{-1}} \tag{6.35}$$

For the simple RC filter, the corresponding difference equation is

$$y[n] = T_s \alpha x[n] + e^{-\alpha T_s} y[n-1] \tag{6.36}$$

Applying the impulse invariance method does not provide an acceptable result for a continuous-time highpass filter. The main reason for this is aliasing [2]. Therefore, you should follow a different path for obtaining the discrete-time representation of highpass filters. This can be done by transforming a lowpass filter to highpass form with the following equation [2].

$$z^{-1} = -\frac{z^{-1}+a}{1+az^{-1}} \tag{6.37}$$

where

$$a = -\frac{\cos(\frac{\omega_{LC}+\omega_{HC}}{2})}{\cos(\frac{\omega_{LC}-\omega_{HC}}{2})} \tag{6.38}$$

Here, ω_{LC} is the predetermined cut-off frequency of the discrete-time lowpass filter. ω_{HC} is the desired cut-off frequency of the discrete-time highpass filter. We will apply this method to an actual filter in Lab 6.10.

6.5 Choosing the Appropriate Method for Implementation

We have described four methods for representing a continuous-time filter in discrete time. These methods have minor differences, which are summarized as follows. The backward difference method is the simplest. The impulse invariance method is potentially the most complicated but may be effective using tables of Laplace transforms. None of the methods yield a frequency response that is identical to that of the analog filter and each has advantages and disadvantages in that respect. The bilinear transform method is widely used. For example, it is used by MATLAB's Filter Design and Analysis tool.

6.6 Chapter Summary

Digital processing of continuous-time signals allows any analog system to be replaced with its digital counterpart. This leads to digitalization of existing analog systems. The first step here is obtaining the digital version of the analog signal to be processed. The last step is obtaining the analog version of the processed digital signal. We considered these steps through ADC and DAC modules in Chapter 5. The remaining task is to represent the continuous-time system in digital form. We focused on this issue in this chapter. We started with frequency mapping between continuous and discrete-time representations. Then, we introduced four methods to convert a continuous-time system to digital form. These topics give you the necessary knowledge to replace an analog system with its digital version.

6.7 Further Reading

We did not present all methods for representing a continuous-time system in discrete time in this chapter. There are also forward difference and ***step invariance*** methods, which are covered in the literature [3]. We did not consider the forward difference method in this book because it may represent a stable continuous-time system as an unstable discrete-time system. The step invariance method is mostly used in digital control applications. Although we summarized the advantages and disadvantages of the four conversion methods in Section 6.5, this list is not complete. For further coverage, we recommend you consult the wider literature, e.g., [2, 4].

> step invariance
> *A transformation method in which both continuous and discrete-time filters have the same step response at the sampling instants.*

6.8 Exercises

6.8.1 The continuous-time signal $x(t) = \sum_{k=0}^{3} \cos(\pi kt/7)$ is sampled with sampling period $T_s = 0.01$ s. After sampling, what frequency values will be observed?

6.8.2 The discrete-time signal $x[n] = \sum_{k=0}^{3} \cos(\pi kn/7)$ is reconstructed by an appropriate ideal lowpass filter. The sampling period for the overall operation is $T_s = 0.01$ s. What will be the corresponding frequency values after reconstruction?

6.8.3 The impulse response of a continuous-time system is $h(t) = sinc(\pi t/10)$. This system is represented in discrete time.

(a) Choose an appropriate sampling period for this operation.

(b) Obtain the discrete-time representation of the impulse response, $h[n]$, of this system using the backward difference, bilinear transformation, and impulse invariance methods.

6.8.4 Complete Exercise 6.8.3 again for the continuous-time system $h(t) = sinc(\pi t/10)u(t)$. Take $T_s = 10^{-4}$ s as the sampling period.

6.8.5 Suppose we have an ideal lowpass filter with cut-off frequency $\Omega_c = 2\pi \times 500$ rad/s in continuous-time. Assume that the sampling frequency is $f_s = 1100$ Hz. What will the corresponding cut-off frequency value be when the backward difference and bilinear transformation methods are used to represent the ideal lowpass filter in discrete time? Determine a frequency range in continuous-time such that the signal is filtered by one discrete-time representation and not filtered in the other representation?

6.8.6 A continuous-time causal LTI system is represented by the differential equation $\frac{\delta^2 y(t)}{\delta t^2} + 4\frac{\delta y(t)}{\delta t} + 3y(t) = x(t)$. What is the equivalent difference equation for this system when

(a) the backward difference method is used?

(b) the bilinear transformation method is used?

(c) the impulse invariance method is used?

6.8.7 A continuous-time causal LTI system is represented by $H(s) = \frac{1}{s^2+4}$. What is the impulse response of the equivalent discrete-time system when

(a) the backward difference method is used?

(b) the bilinear transformation method is used?

(c) the impulse invariance method is used?

6.8.8 Similar to the backward difference method, there is also a forward difference method for representing a continuous-time system in discrete time [3].

(a) What is the main disadvantage of this method?

(b) Apply this method to obtain the discrete-time equivalent of the RC filter introduced in Figure 6.1. Assume normalized device values, where $\alpha = 1$. Use sampling period $T_s = 0.1$ s for this exercise.

(c) Compare the impulse response of the obtained discrete-time system with the one obtained by the backward difference method.

6.8.9 Similar to the impulse invariance method, there is also a step invariance method for representing a continuous-time system in discrete time [3].

(a) Apply this method to obtain the discrete-time equivalent of the RC filter introduced in Figure 6.1. Assume normalized device values, where $\alpha = 1$. Use sampling period $T_s = 0.1$ s for this exercise.

(b) Compare the impulse response of the obtained discrete-time system with the one obtained by the impulse invariance method.

6.8.10 Implementing the impulse invariance method is described as costly because it will have an exponential term. Why is the exponential term costly in implementation?

6.9 References

[1] Nilsson, J.W. and Riedel, S.A. (2014) *Electric Circuits*, Pearson, Tenth edn.
[2] Oppenheim, A.W. and Schafer, R.W. (2009) *Discrete-Time Signal Processing*, Prentice Hall, Third edn.
[3] Ogata, K. (1987) *Discrete-Time Control Systems*, Prentice Hall.
[4] Proakis, J.G. and Manolakis, D.K. (1995) *Digital Signal Processing: Principles, Algorithms and Applications*, Prentice Hall, Third edn.

6.10 Lab 6

6.10.1 Introduction

This lab introduces the basic concepts in the digital processing of continuous-time signals. As in previous chapters, the main focus is on practical implementation. The first concept to be considered in this chapter is frequency mapping. Then, we will test methods for representing continuous-time filters in digital form. The operations we perform in this chapter are discussed in detail in previous sections.

6.10.2 Frequency Mapping

The aim of this section is to demonstrate how a continuous-time signal can be converted to a discrete-time signal through frequency mapping. We introduced the theoretical background of frequency mapping in Section 6.2. Here, we will focus on its practical implementation.

We will first apply frequency mapping to a continuous-time sinusoidal signal to obtain its discrete-time representation. The continuous-time sinusoidal signal has frequency 250π rad/s with amplitude one. Let us set the sampling frequency to 64000π rad/s. The complete Keil project for this application is given in `Online_Student_Resources\Lab6\frequency_mapping`.

Task 6.1

Run the code in `Online_Student_Resources\Lab6\frequency_mapping`. Plot the corresponding digital signal. Comment on the frequency of the obtained digital signal.

Task 6.2

Now, rearrange the code given in Task 6.1 for the following setup. This time the continuous-time signal is the sum of two sinusoids with frequency values 250π rad/s and 8000π rad/s, respectively. Both sinusoids have amplitude one. The sampling frequency is 64000π rad/s. Plot the obtained digital signal after frequency mapping. Comment on the frequency values of the obtained signal.

To note here, the digital signals obtained in this section correspond to the ones introduced in Section 2.9.3. Here, they are obtained from the corresponding continuous-time signals using frequency mapping.

6.10.3 Continuous-Time Systems

We will use continuous-time systems in the following sections. Therefore, we will use the second-order Sallen–Key lowpass and highpass filters. The circuit diagrams of these filters are shown in Figure 6.2.

Sallen–Key Lowpass Filter

The transfer function of the Sallen–Key lowpass filter is

$$H(s) = \frac{V_{out}(s)}{V_{in}(s)} = \frac{\frac{1}{R_1R_2C_1C_2}}{s^2 + s\left(\frac{1}{R_1C_2} + \frac{1}{R_2C_2}\right) + \frac{1}{R_1R_2C_1C_2}} \tag{6.39}$$

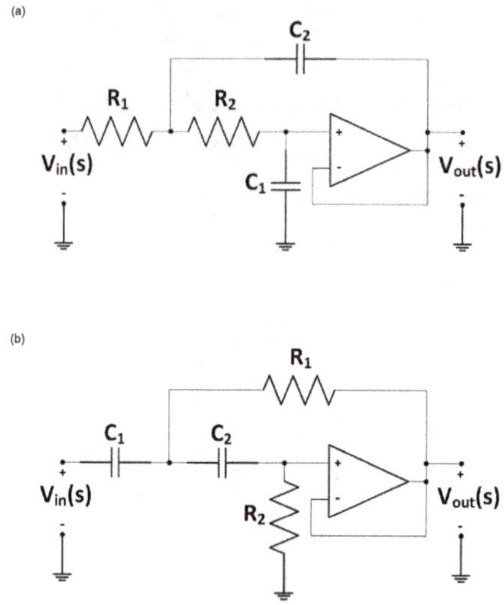

Figure 6.2 Circuit diagram of the second-order Sallen–Key filters. (a) Lowpass filter; (b) Highpass filter.

We select the resistor values R_1=33 kΩ and R_2=33 kΩ and the capacitor values C_1=10 η F and C_2=10 η F. Based on these values, Eqn. 6.39 becomes

$$H(s) = \frac{9182736.455464}{s^2 + 6060.606061s + 9182736.455464} \tag{6.40}$$

A second-order lowpass filter can be formulated as

$$H(s) = \frac{K\Omega_N^2}{s^2 + \frac{\Omega_N}{Q}s + \Omega_N^2} \tag{6.41}$$

The gain for our filter is $K = 1$. The lowpass filter has ***natural frequency*** $\Omega_N =$ 3030.3030305 rad/s. It has a quality factor $Q = 0.5$. The cut-off frequency, Ω_C, for the lowpass filter can be calculated using the equation $|H(j\Omega_C)| = Q|H(0)|$. Based on this, we obtain $\Omega_C = 3030.3030305$ rad/s, which is also its natural frequency.

> natural frequency
> *The frequency at which a system tends to oscillate on its own.*

Sallen–Key Highpass Filter

The transfer function of the Sallen–Key highpass filter is

$$H(s) = \frac{V_{out}(s)}{V_{in}(s)} = \frac{s^2}{s^2 + s\left(\frac{1}{R_2C_1} + \frac{1}{R_2C_2}\right) + \frac{1}{R_1R_2C_1C_2}} \tag{6.42}$$

We select the resistor values R_1=8.2 kΩ and R_2=8.2 kΩ and the capacitor values C_1=10 η F and C_2=10 η F. Based on these values, Eqn. 6.42 becomes

$$H(s) = \frac{s^2}{s^2 + 24390.243902s + 148720999.405116} \tag{6.43}$$

A second-order highpass filter can be formulated as

$$H(s) = \frac{Ks^2}{s^2 + \frac{\Omega_N}{Q}s + \Omega_N^2} \tag{6.44}$$

The gain for our filter is $K = 1$. The highpass filter has natural frequency (also called undamped natural frequency) $\Omega_N = 12195.121951$ rad/s. It has a quality factor of $Q = 0.5$. The quality factor is used to describe the relation between total stored energy and energy loss of the resonator. When the system has a high Q value, damping decreases and the oscillation dies slowly. The cut-off frequency, Ω_C, for the highpass filter can be calculated using the equation $|H(j\Omega_C)| = Q|H(\infty)|$. Based on this, we obtain $\Omega_C = 12195.121951$ rad/s for our filter, which is also its natural frequency.

Simulation of the Sallen–Key Filters

Let us apply the continuous-time signal described in Section 6.10.2 to the Sallen–Key filters. As a reminder, the sinusoids in this signal have frequency values 250π rad/s and 8000π rad/s, respectively, and both have an amplitude of 1. We provide the analog form of this signal in Figure 6.3. As can be seen in this figure, the signal is composed of two sinusoids with slow and fast fluctuations.

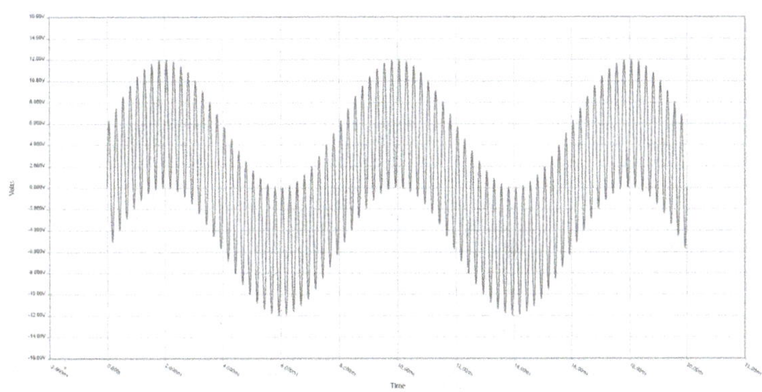

Figure 6.3 Analog signal fed to the Sallen–Key filters.

In order to obtain the output of the Sallen–Key filters, we must construct them with analog components, supply the input signal from a signal generator, and observe the outputs on the oscilloscope. The working principles of the Sallen–Key filters can also be observed in a circuit simulator. There are several online and offline simulators available. You can select the most suitable one for your own needs. We use the online simulator provided by CircuitLab. This simulator can be found at `http://www.circuitlab.com` after creating an account in its student edition, which is free for personal use.

We provide the simulation results in Figure 6.4. As can be seen in this figure, the analog lowpass filter eliminates the sinusoidal signal component with fast fluctuations,

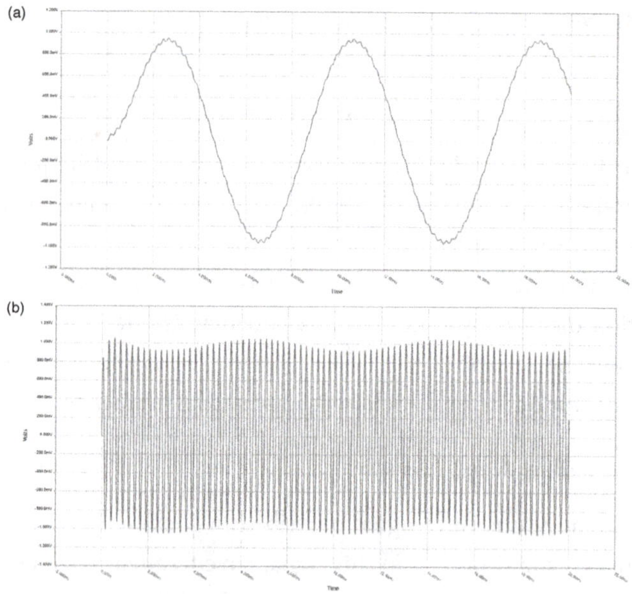

Figure 6.4 Simulation result of the Sallen–Key filters. (a) Output of the lowpass filter; (b) Output of the highpass filter.

and the analog highpass filter eliminates the sinusoidal signal component with slow fluctuations.

6.10.4 Representing a Continuous-Time Filter in Digital Form

The aim in this section is to demonstrate how a continuous-time filter can be represented in digital form. In previous sections, we introduced the four methods as backward difference, bilinear transformation, bilinear transformation with prewarping, and impulse invariance. Here, we will provide practical examples using these methods.

Before starting the conversion process, we must obtain the desired cut-off frequency value of the filters in discrete time. We can use frequency mapping for this purpose.

Task 6.3

Before converting the lowpass and highpass Sallen–Key filters to digital form, we should first obtain their cut-off frequency. Therefore, convert the calculated Ω_C values for the lowpass and highpass Sallen–Key filters from Section 6.10.3 to discrete-time form. While doing this, set the sampling period to $T_s = 1/32000$ s. Hence, the sampling frequency will be $\Omega_s = 64000\pi$ rad/s. These values will be used in the following exercises.

Backward Difference

We introduced the backward difference method in Section 6.4.1. Here, we will use it to convert the continuous-time Sallen–Key filters to digital form. As a reminder, the backward difference method requires replacing s with $(1-z^{-1})/T_s$ in the transfer function. Let us set $T_s = 1/32000$ s. Applying these values to Eqn. 6.40, we obtain the lowpass filter Z-transfer function

$$H(z) = \frac{9182736.455464}{1227122130.408658 - 2241939393.952z^{-1} + 1024000000z^{-2}} \tag{6.45}$$

After simplification, $H(z)$ becomes

$$H(z) = \frac{0.007483z^2}{z^2 - 1.8269896z + 0.834473} \tag{6.46}$$

Task 6.4

Using the backward difference method with $T_s = 1/32000$ s, calculate how the cut-off frequency value of the continuous-time Sallen–Key lowpass filter is mapped to discrete time. To aid with this exercise, Eqn. 6.18 is given below.

$$\omega = \arctan(\Omega T_s) \tag{6.47}$$

Compare the obtained result with the one calculated in Task 6.3.

Task 6.5

Change the T_s from Task 6.4 to both 1/8000 s and 1/128000 s. What are the cut-off frequency values that correspond to these T_s values? Comment on the results with a focus on the arctangent function in Eqn. 6.47.

Task 6.6

Implement the discrete-time filter obtained in Eqn. 6.46 using the tools in Section 2.9.6. Obtain and plot the output of the implemented filter when the input is the signal formed by the sum of two sinusoids given in Section 6.10.2. What do you observe?

Task 6.7

Use the backward difference method to convert the continuous-time Sallen–Key high-pass filter to digital form. Set the sampling period to $T_s = 1/32000$ s. What is the cut-off frequency for this filter?

Task 6.8

Using the tools in Section 2.9.6, implement the discrete-time filter obtained in Task 6.7. Using the signal formed by the sum of two sinusoids given in Section 6.10.2 as the input, obtain and plot the output of the implemented filter. What do you observe?

Bilinear Transformation

We introduced the bilinear transformation method in Section 6.4.2. Here, we will use it to convert the continuous-time Sallen–Key filters to digital form. As a reminder, the bilinear transformation method requires replacing s with $2(1-z^{-1})/(T_s(1+z^{-1}))$ in the transfer function. Let us set $T_s = 1/32000$ s. Applying these values to Eqn. 6.40, we obtain the lowpass filter Z-transfer function

$$H(z) = \frac{9182736.455464(1+2z^{-1}+z^{-2})}{4493061524.360658 - 8173634527.086685z^{-1} + 3717303948.552657z^{-2}} \quad (6.48)$$

After simplification, $H(z)$ becomes

$$H(z) = \frac{0.002044z^2 + 0.004088z + 0.002044}{z^2 - 1.819168z + 0.827343} \quad (6.49)$$

Task 6.9

Using the bilinear transformation method with $T_s = 1/32000$ s, calculate how the cut-off frequency value of the continuous-time Sallen–Key lowpass filter is mapped to discrete time. To aid with this exercise, Eqn. 6.27 is given below.

$$\omega = 2 \arctan\left(\frac{\Omega T_s}{2}\right) \quad (6.50)$$

Compare the obtained result with the one calculated in Task 6.3.

Task 6.10

Change the T_s from Task 6.9 to both 1/8000 s and 1/128000 s. What will be the corresponding cut-off frequency values? Comment on the results with a focus on the arctangent function in Eqn. 6.50.

Task 6.11

Implement the discrete-time filter obtained in Eqn. 6.49 using the tools in Section 2.9.6. Obtain and plot the output of the implemented filter when the input is the signal formed by the sum of two sinusoids given in Section 6.10.2. What do you observe?

Task 6.12

Use the bilinear transformation method to convert the continuous-time Sallen–Key highpass filter to digital form. Set the sampling period to $T_s = 1/32000$ s. What is the cut-off frequency for this filter?

Task 6.13

Using the tools in Section 2.9.6, implement the discrete-time filter obtained in Task 6.12. Using the signal formed by the sum of two sinusoids given in Section 6.10.2 as the input, obtain and plot the output of the implemented filter. What do you observe?

To note here, the discrete-time filters obtained in this section have been covered earlier in this book. The IIR and FIR filter coefficients used in Section 2.9.6 are the coefficients of these filters. In order to understand how they are generated, you should obtain the impulse response of these filters. Then, the desired number of FIR coefficients can be obtained from the impulse response.

Bilinear Transformation with Prewarping

We introduced the bilinear transformation with prewarping method in Section 6.4.3. Here, we will use it to convert the continuous-time Sallen–Key filters to digital form. Therefore, we will start with the bilinear transformation example considered in the previous section. In the previous section, when the transformation was done, an undesired shift in the cut-off frequency occurred. However, we will see in this section that this undesired shift can be avoided by using prewarping. Hence, the analog cut-off frequency can be correctly mapped to digital form.

Let us apply prewarping to a second-order lowpass filter. Assume that we want to convert the filter to discrete time using bilinear transformation. We want the analog cut-off frequency, Ω_C, to map to discrete time as in Section 6.10.2.

From the derivation given in Section 6.4.3, we can rewrite the transfer function in Eqn. 6.41 as

$$H(s) = \frac{\Omega_P^2}{s^2 + 2\Omega_P s + \Omega_P^2} \tag{6.51}$$

where $\Omega_P = (2/T_s)\tan(\Omega_C T_s/2)$. After bilinear transformation, we obtain

$$H(z) = \frac{\left(\frac{2}{T_s}\tan\left(\frac{\Omega_C T_s}{2}\right)\right)^2}{\left(\frac{2}{T_s}\frac{1-z^{-1}}{1+z^{-1}}\right)^2 - 2\left(\frac{2}{T_s}\frac{1-z^{-1}}{1+z^{-1}}\frac{2}{T_s}\tan\left(\frac{\Omega_C T_s}{2}\right)\right) + \left(\frac{2}{T_s}\tan\left(\frac{\Omega_C T_s}{2}\right)\right)^2} \tag{6.52}$$

If we replace T_s with 1/32000 s and Ω_C with 3030.3030305 rad/s, Eqn. 6.52 becomes

$$H(z) = \frac{9196478.301942(1 + 2z^{-1} + z^{-2})}{4493365385.478987 - 8173607043.396115z^{-1} + 3717027571.124898z^{-2}} \tag{6.53}$$

After simplification, $H(z)$ becomes

$$H(z) = \frac{0.002047z^2 + 0.004094z + 0.002047}{z^2 - 1.819039z + 0.827226} \tag{6.54}$$

For this transfer function, the prewarped analog cut-off frequency is calculated as Ω_C=3032.569587 rad/s. Then, using Eqn. 6.50, the cut-off frequency will be $\omega_C =$ 0.094697 rad/sample. Therefore, prewarping worked as expected.

Task 6.14

Implement the discrete-time filter obtained in Eqn. 6.54 using the tools in Section 2.9.6. Obtain and plot the output of the implemented filter when the input is the signal formed by the sum of two sinusoids given in Section 6.10.2. What do you observe?

Task 6.15

Use the bilinear transformation method with prewarping to convert the continuous-time Sallen–Key highpass filter to digital form. Apply the prewarping such that the cut-off frequency value of the continuous-time filter is mapped correctly to digital form. Set the sampling period to $T_s = 1/32000$ s. How does the new digital filter differ from the one obtained through the bilinear transform alone? Does prewarping work as expected?

Task 6.16

Using the tools in Section 2.9.6, implement the discrete-time filter obtained in Task 6.15. Obtain and plot the output of the implemented filter when the input is the signal formed by the sum of the two sinusoids given in Section 6.10.2. What do you observe?

Impulse Invariance

We introduced the impulse invariance method in Section 6.4.4. Here, we will use it to convert the continuous-time Sallen–Key filters to digital form. As opposed to the previous methods, the user must have the impulse response of the continuous-time filter at hand.

The continuous-time Sallen–Key lowpass filter impulse response can be obtained by applying the inverse Laplace transform to Eqn. 6.40 to obtain

$$h(t) = 9182736.4566575te^{-3030.3030305t}u(t) \tag{6.55}$$

As a reminder, the impulse invariance method has the transformation $h[n] = T_s h(nT_s)$. Applying it to Eqn. 6.55, we obtain

$$h[n] = 9182736.4566575nT_s^2 e^{-3030.3030305nT_s}u[n] \tag{6.56}$$

Applying the Z-transform to the impulse response in Eqn. 6.56, we obtain the discrete-time transfer function

$$H(z) = \frac{9182736.4566575T_s^2 e^{-3030.3030305T_s}z^{-1}}{(1 - e^{-3030.3030305T_s}z^{-1})^2} \tag{6.57}$$

Setting the sampling period to $T_s = 1/32000$ s, Eqn. 6.57 becomes

$$H(z) = \frac{0.0081573z^{-1}}{(1 - 0.9096485z^{-1})^2} \tag{6.58}$$

After simplification, $H(z)$ becomes

$$H(z) = \frac{0.0081573z}{z^2 - 1.819297z + 0.8274604} \tag{6.59}$$

The cut-off frequency value for the impulse invariance method can be calculated as

$$\omega_C = \Omega_C T_s \text{ when } |\omega_C| < \pi \tag{6.60}$$

The real discrete-time cut-off frequency is calculated as 0.094697 rad/sample. However, Eqn. 6.60 is only accurate when the continuous-time filter is bandlimited. In practice, a filter cannot be exactly bandlimited. Hence, the real discrete-time frequency will be slightly different to the result obtained in Eqn. 6.60.

Task 6.17

Using the tools in Section 2.9.6, implement the obtained discrete-time filter. Obtain and plot the output of the implemented filter when the input is the signal formed by the sum of two sinusoids given in Section 6.10.2. What do you observe?

Task 6.18

Use the impulse invariance method to convert the continuous-time Sallen–Key highpass filter to digital form. Set the sampling period to $T_s = 1/32000$ s. Implement the obtained discrete-time filter using the tools in Section 2.9.6. Obtain and plot the output of the implemented filter when the input is the signal formed by the sum of two sinusoids given in Section 6.10.2. What do you observe?

As mentioned in Section 6.4.4, applying the impulse invariance method does not provide an acceptable result for a continuous-time highpass filter. Let us take the lowpass filter in Eqn. 6.59 as an example. This filter has a cut-off frequency of $\omega_{LC} = 0.094697$ rad/sample. Using Eqn. 6.60, the desired cut-off frequency for the highpass filter, ω_{HC}, can be calculated as 0.38109756 rad/sample. We can obtain the highpass filter from the lowpass filter in Eqn. 6.59. To do so, we can use Eqns. 6.37 and 6.38. The resulting discrete-time transfer function for the highpass filter is

$$z^{-1} = -\frac{z^{-1} + a}{1 + az^{-1}} \tag{6.61}$$

where

$$a = -\frac{\cos\left(\frac{\omega_{LC}+\omega_{HC}}{2}\right)}{\cos\left(\frac{\omega_{LC}-\omega_{HC}}{2}\right)} \tag{6.62}$$

The resulting discrete-time transfer function for the highpass filter will be

$$H(z) = \frac{0.701825452z^2 - 1.4038854z + 0.70182545}{z^2 - 1.3523936z + 0.457242} \tag{6.63}$$

Task 6.19

Implement the obtained discrete-time filter using the tools in Section 2.9.6. Obtain and plot the output of the implemented filter when the input is the signal formed by the sum of two sinusoids given in Section 6.10.2. What do you observe?

7

Structures for Discrete-Time LTI Systems

Contents

7.1 Introduction

A discrete-time LTI system can be implemented in more than one way. Although each different implementation will have the same input–output relation, its structure may be different. One structure might require less memory or be more robust to parameter quantization effects than another, especially when fixed-point implementation (to be introduced in Chapter 10) is considered. This chapter is about structures for LTI systems. We start with a brief review of LTI systems. Then, we consider direct form representations. Then, we explore cascade, transposed, and lattice forms, which are extensively used in practical applications.

7.2 LTI Systems Revisited

Before describing alternative structures for discrete-time LTI systems, we revisit the key properties of LTI systems.

7.2.1 Basic Definitions

Linear and time-invariant (LTI) systems are widely studied in literature because they can be represented and analyzed by well-defined mathematical tools. Hence, they can be designed, modified, and implemented easily. Before exploring the properties of LTI systems, let us first recall the representation of a discrete-time LTI system. In the digital signal processing literature, the terms system and filter are used interchangeably.

A generic, single-input, single-output discrete-time LTI system can be represented as in Figure 7.1.

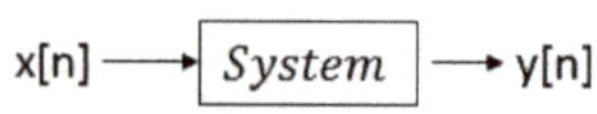

Figure 7.1 Representation of an LTI discrete-time system. $x[n]$ and $y[n]$ are the input and output signals, respectively.

The relationship between the input and output signals can be represented by constant-coefficient difference equations in an LTI system. Hence, for an LTI system the relationship between $x[n]$ and $y[n]$ is

$$y[n] + \alpha_1 y[n-1] + \cdots + \alpha_{L-1} y[n-L+1] = \beta_0 x[n] + \cdots + \beta_{K-1} x[n-K+1] \quad (7.1)$$

where α and β are constant coefficients.

Rearranging Eqn. 7.1 to make $y[n]$ the subject, we obtain

$$y[n] = \sum_{k=0}^{K-1} \beta_k x[n-k] - \sum_{l=1}^{L-1} \alpha_l y[n-l] \quad (7.2)$$

The input–output relationship can also be represented by the convolution sum operation as

$$y[n] = h[n] * x[n] = \sum_{k=-\infty}^{\infty} h[k]x[n-k] \quad (7.3)$$

where $h[n]$ is the impulse response of the system. Although Eqn. 7.3 may be of limited practical use, it states that, as long as we know $h[n]$, we can find the output of the LTI system in response to any input signal.

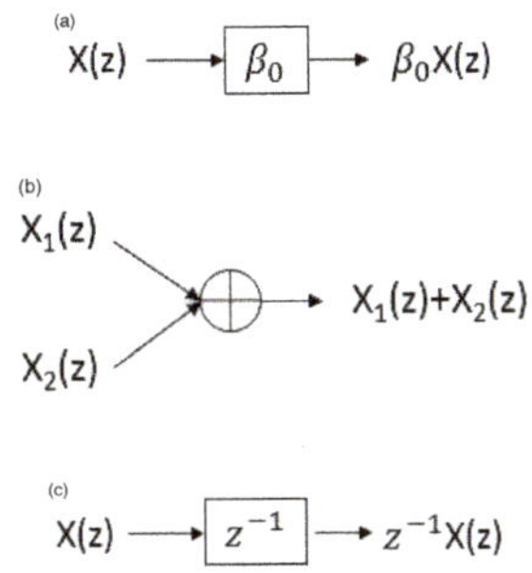

Figure 7.2 Schematic representation of multiplication by a constant, addition, and a delay element. (a) Multiplication; (b) Addition; (c) Delay.

We can represent the input–output relationship of Eqn. 7.2 in the z-domain as

$$Y(z) = X(z)\sum_{k=0}^{K-1}\beta_k z^{-k} - Y(z)\sum_{l=1}^{L-1}\alpha_l z^{-l} \tag{7.4}$$

As can be seen here, the difference equation in the time domain becomes an algebraic equation in the complex domain.

We can represent the Z-transfer function of the system as

$$H(z) = \frac{Y(z)}{X(z)} = \frac{\sum_{k=0}^{K-1}\beta_k z^{-k}}{\sum_{l=0}^{L-1}\alpha_l z^{-l}} \tag{7.5}$$

where $\alpha_0 = 1$. Eqn. 7.5 indicates that $H(z)$ is the ratio of two polynomials for an LTI system. In the constant-coefficient difference equation, the numerator polynomial holds the coefficients of the $x[n]$ terms, and the denominator polynomial holds the coefficients of the $y[n]$ terms.

7.2.2 Block Diagram Representation of LTI Systems

As can be seen in Eqn. 7.4, there are three basic operations in this representation, namely multiplication by a constant, addition, and multiplication by z^{-1}. From the shift in time property of the Z-transform, we know that multiplying a signal by z^{-1} in the complex domain corresponds to shifting it by one sample in the time domain. This is called the delay operation. Thus, any LTI system can be represented using only delay elements, multipliers, and adders. Each element can be represented schematically as in Figure 7.2.

Figure 7.3 shows a block diagram representation of Eqn. 7.4. This representation allows us to analyze the discrete-time system visually. It also allows us to construct different structural forms of the same system, which we will cover in later sections.

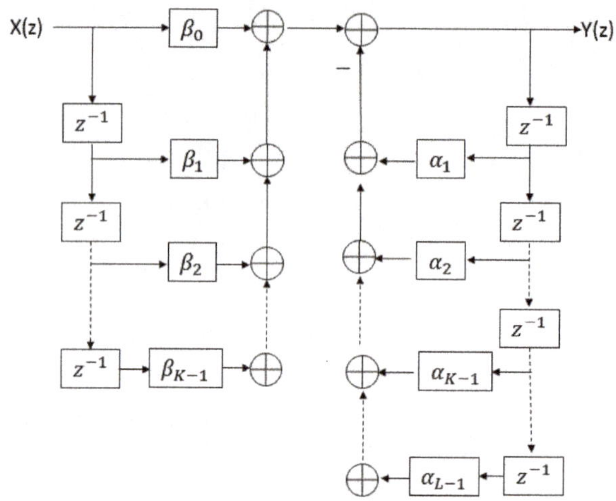

Figure 7.3 Block diagram representation of the general LTI system.

7.2.3 Characteristics of LTI Systems

There are two summation terms in Eqn. 7.4. The first one only takes current and past input signal values. This sum is called the feedforward term. The second sum only takes past output signal values. This sum is called the feedback term.

The feedback term does not exist in an FIR filter. This is the fundamental difference between FIR and IIR filters. Because an FIR filter has only feedforward terms in its difference equation, the denominator term in its Z-transfer function is equal to one.

An IIR filter has both feedforward and feedback terms in its difference equation. Correspondingly, both the numerator and denominator of its Z-transfer function are polynomials in z.

7.3 Direct Forms

The first structure type to be considered is direct form, which is based on the difference equation representing a general LTI system given in Eqn. 7.4. There are two versions of direct form. Let us start with direct form I.

7.3.1 Direct Form I

The difference equation representing a general LTI system given in Eqn. 7.4 can be directly implemented. The structure obtained will be as in Figure 7.4. This structure is called direct form I. Let us provide an example of this structure.

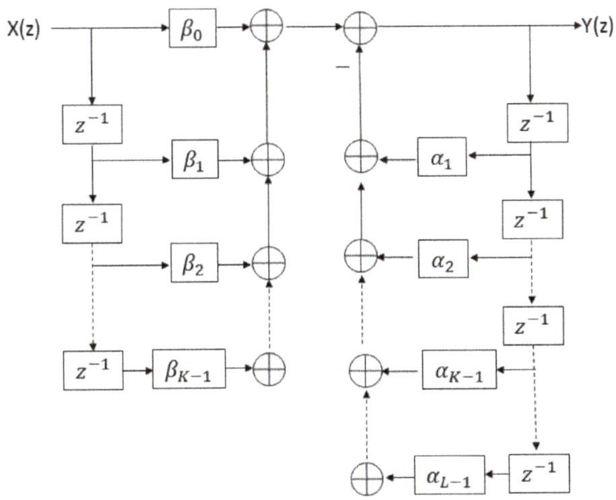

Figure 7.4 Direct form I structure of a generic IIR filter.

Example 7.1 Direct form I structure of an IIR filter

Let us reconsider the IIR filter represented by the difference equation

$$\begin{aligned} Y(z) = {} & X(z)(0.002044 + 0.004088z^{-1} + 0.002044z^{-2}) \\ & - Y(z)(-1.819168z^{-1} + 0.827343z^{-2}) \end{aligned} \tag{7.6}$$

which was considered in Task 8. As a reminder, this difference equation was obtained by applying bilinear transformation to the Sallen–Key lowpass filter in Task 8. The direct form structure of this difference equation will be as in Figure 7.5.

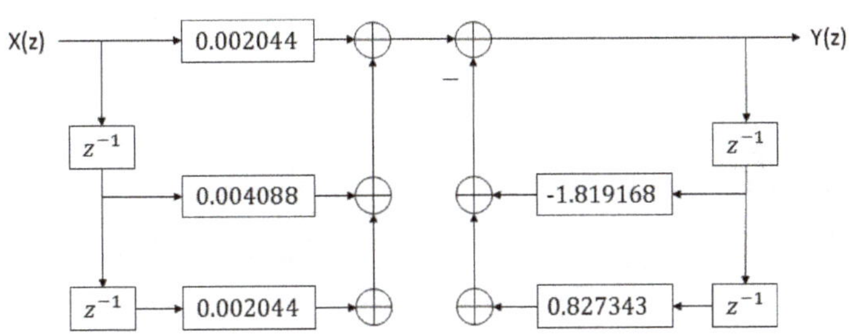

Figure 7.5 Direct form I structure of the IIR filter in Example 7.1.

7.3.2 Direct Form II

The direct form I representation can be simplified further. To do so, we must rewrite Eqn. 7.5 as

$$H(z) = \frac{Y(z)}{X(z)} = H_1(z)H_2(z) \tag{7.7}$$

where

$$H_1(z) = \sum_{k=0}^{K-1} \beta_k z^{-k} \tag{7.8}$$

and

$$H_2(z) = \frac{1}{\sum_{l=0}^{L-1} \alpha_l z^{-l}} \tag{7.9}$$

Using the linearity property, we can rewrite Eqn. 7.7 as

$$H(z) = \frac{Y(z)}{X(z)} = H_2(z)H_1(z) \tag{7.10}$$

Eqn. 7.10 allows us to combine the delay elements used in the system. As a result, we will have fewer delay elements in implementation. Figure 7.6 shows the direct form II structure for an IIR filter. Now, let us obtain the direct form II structure of the IIR filter in Example 7.1.

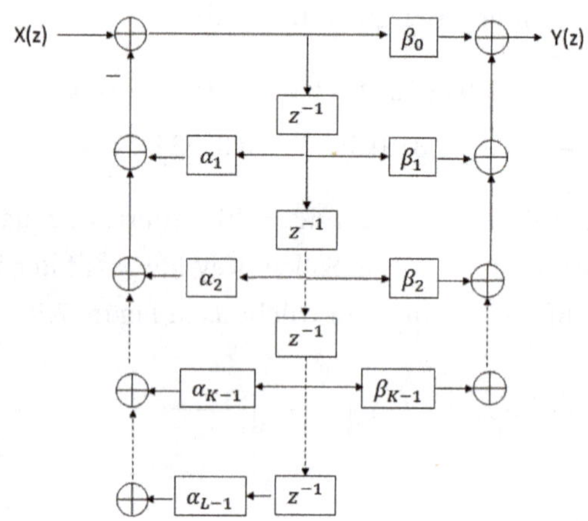

Figure 7.6 Direct form II representation of the generic IIR filter.

Example 7.2 Direct form II structure of the IIR filter in Example 7.1

Figure 7.7 shows the direct form II structure of the IIR filter considered in Example 7.1. As can be seen in this figure, the number of delay elements used in this structure is half of that in direct form I.

Although direct forms are straightforward to construct and easy to implement, they have a major shortcoming. They are sensitive to parameter quantization effects in implementation. Therefore, direct forms are not recommended for real-life applications [1]. For use in real-life applications, we now introduce cascade form.

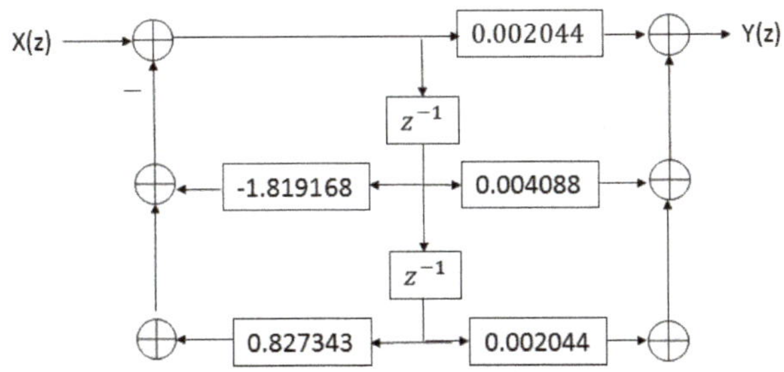

Figure 7.7 Direct form II structure of the IIR filter in Example 7.1.

7.4 Cascade Form

The second structure type to be considered is cascade form. In this structure, the transfer function of the IIR filter is represented in terms of simpler blocks. Thus, Eqn. 7.5 can be written as

$$H(z) = \prod_{c=1}^{C} H_c(z) \tag{7.11}$$

where each $H_c(z)$ is a simple filter. More specifically, if $H_c(z)$ is taken as the second-order filter

$$H_c(z) = \frac{\beta_{c0} + \beta_{c1}z^{-1} + \beta_{c2}z^{-2}}{1 + \alpha_{c1}z^{-1} + \alpha_{c2}z^{-2}} \tag{7.12}$$

The representation in Eqn. 7.11 is called the biquad cascade form. This form is robust to filter coefficient quantization effects [2]. Therefore, it is generally preferred for implementation.

Each second-order term in the biquad cascade form can be represented in direct form II. This will also decrease the total number of delay elements in the system. We provide the block diagram of such a filter in Figure 7.8.

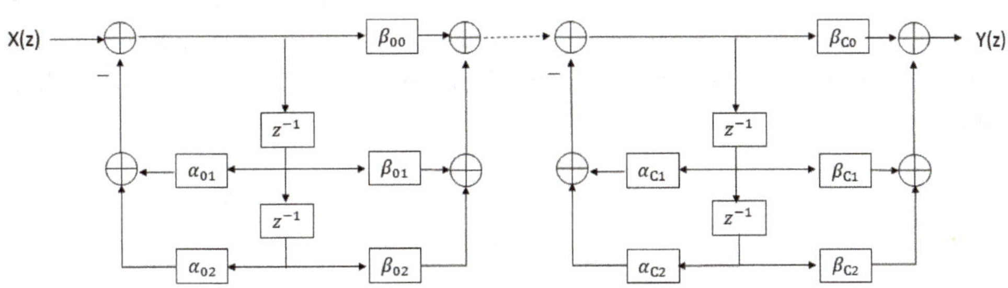

Figure 7.8 Biquad cascade form representation of a generic IIR filter.

7.5 Transposed Form

The third structure type to be considered is the transposed form. Transposed form of a structure can be obtained by reversing the directions in it as well as switching the input

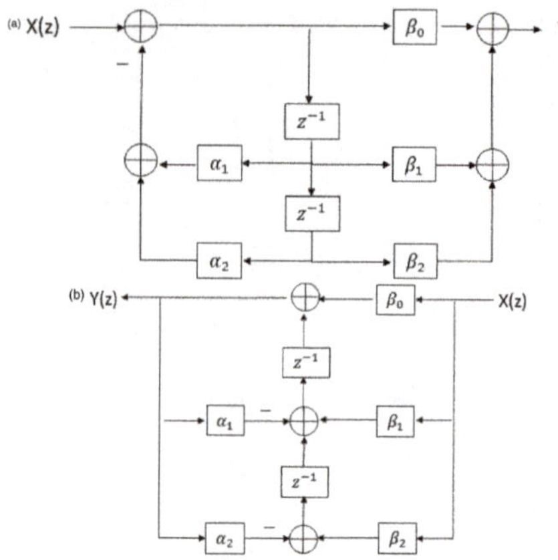

Figure 7.9 Original and transposed form of the biquad direct form II IIR filter. (a) Original form; (b) Transposed form.

and output locations. In doing this, the system characteristics remain unchanged. The main difference in the transposed form is in the implementation of poles and zeros. In the original direct form, poles are implemented first. In transposed form, zeros are implemented first, which may affect the filter characteristic's robustness to coefficient quantization and noise [2].

We provide the original and transposed forms of a biquad IIR filter in Figure 7.9. As can be seen in this figure, the filter coefficients and the number of delay elements used are the same in both structures. However, the input signal is processed in a different order. Let us now consider the transposed form of the IIR filter in Example 7.2.

Example 7.3 Transposed form of the IIR filter in Example 7.2

By reversing directions and switching the input and output locations, we can obtain the transposed form of the IIR filter in Example 7.2. This is shown in Figure 7.10. As can be seen in this figure, the filter coefficients and the number of delay elements used are the same in the original and transposed forms. However, the input signal is processed in a different order.

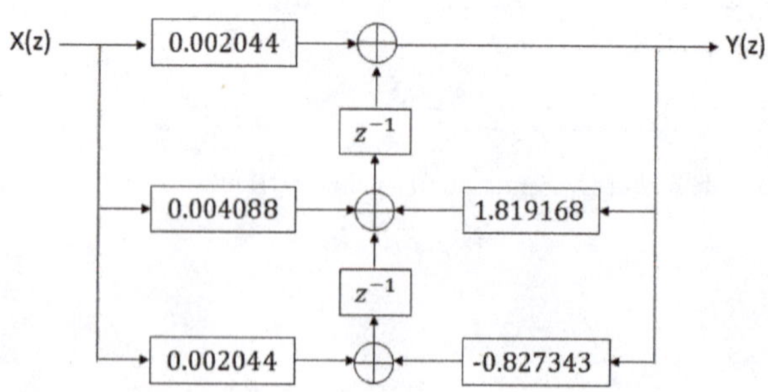

Figure 7.10 Transposed form of the IIR filter in Example 7.2.

7.6 Lattice Filters

The fourth structure type to be considered is the lattice filter, which provides yet another cascade structure for implementation. We will now consider FIR and IIR lattice filter structures.

7.6.1 FIR Lattice Filter

FIR lattice filters can be constructed using intermediate signals $f_m[n]$ and $g_m[n]$ as

$$f_0[n] = g_0[n] = x[n] \tag{7.13}$$
$$f_m[n] = f_{m-1}[n] + k_m g_{m-1}[n-1] \text{ for } m = 1,2,\cdots,M \tag{7.14}$$
$$g_m[n] = k_m f_{m-1}[n] + g_{m-1}[n-1] \text{ for } m = 1,2,\cdots,M \tag{7.15}$$
$$y[n] = f_M[n] \tag{7.16}$$

where k_m $(m = 1,2,\cdots,M)$ are called reflection coefficients. You can refer to the literature to see how these signals are obtained and for the algorithms used to obtain lattice filter coefficients [1, 2]. Throughout this book, we will use MATLAB's predefined functions for this purpose.

Figure 7.11 shows the block diagram of an FIR lattice filter based on the definitions in Eqns. 7.13 to 7.16. As can be seen in this figure, the FIR lattice filter is composed of a cascade connection of basic butterfly-like structures.

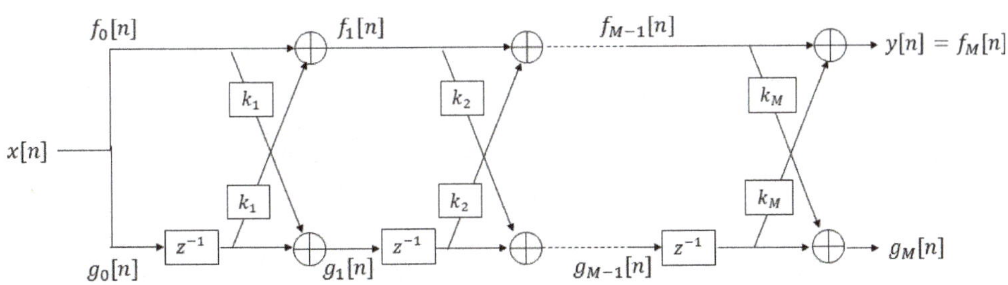

Figure 7.11 Block diagram of an FIR lattice filter.

Example 7.4 Lattice structure of an FIR filter

We demonstrate how to obtain the lattice structure of an FIR filter in Section 2.9. As the FIR filter we produce has 64 coefficients, it is not feasible to show the overall lattice structure. Let us consider the first four filter coefficients as 0.002044, 0.007806, 0.014554, and 0.020018. We can apply MATLAB's `tf2latc` function to obtain the corresponding FIR lattice filter reflection coefficients, which are 0.5265, 0.3190, and 9.7935. The FIR lattice structure obtained is shown in Figure 7.12.

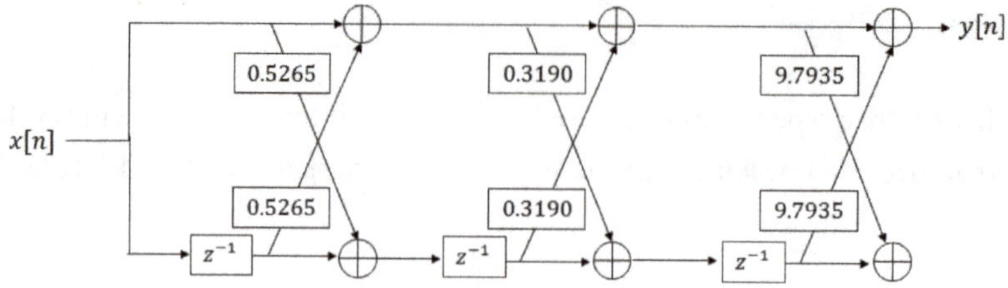

Figure 7.12 Lattice structure of the FIR filter considered in Example 7.4.

7.6.2 IIR Lattice Filter

An IIR filter can also be implemented in lattice form. Here, there will be two intermediate signals, $f_m[n]$ and $g_m[n]$. Based on these, the IIR lattice filter structure can be constructed as

$$f_N[n] = x[n] \tag{7.17}$$

$$f_{m-1}[n] = f_m[n] + k_m g_{m-1}[n-1] \text{ for } m = N, N-1, \cdots, 1 \tag{7.18}$$

$$g_m[n] = k_m f_{m-1}[n] + g_{m-1}[n-1] \text{ for } m = N, N-1, \cdots, 1 \tag{7.19}$$

$$y[n] = \sum_{m=0}^{N} v_m g_m[n] \tag{7.20}$$

where k_m $(m = 1, 2, \cdots, M)$ are called the reflection coefficients. v_m $(m = 1, 2, \cdots, N)$ are called the ladder coefficients. As in the FIR lattice filter, you can refer to the literature for how these signals are obtained [1, 2].

Figure 7.13 shows the block diagram of an IIR lattice filter based on the definitions in Eqns. 7.17 to 7.20. As with the FIR lattice filter, the IIR lattice filter is composed of a cascade connection of basic butterfly-like structures with extra loops. Let us represent the IIR filter in Example 7.1 in lattice form.

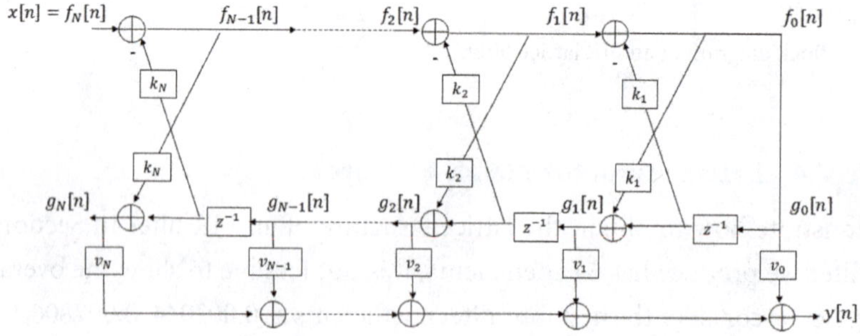

Figure 7.13 Block diagram of an IIR lattice filter.

Example 7.5 Lattice structure of the IIR filter in Example 7.1

We can obtain the lattice structure of the IIR filter in Example 7.1 using MATLAB's `tf2latc` function. This gives us the IIR lattice filter reflection and ladder coefficients $k = -0.9955, 0.8273$ and $v = 0.0081, 0.0078, 0.0020$, respectively. The parameters k and v are computed iteratively. Because this iteration becomes long for high-degree filters, it is best to use predefined MATLAB functions. The IIR lattice structure obtained is shown in Figure 7.14.

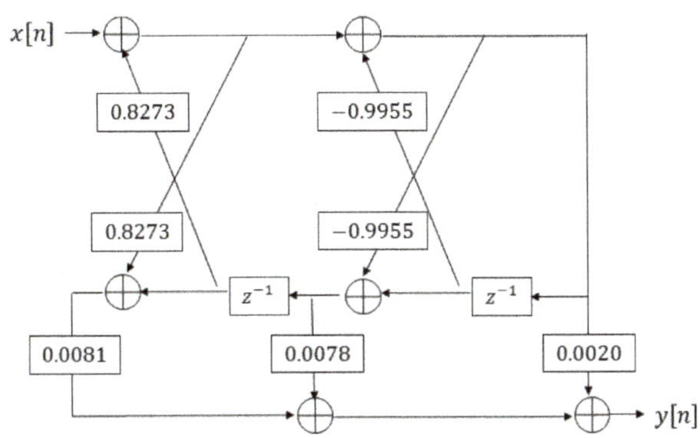

Figure 7.14 Lattice structure of the IIR filter in Example 7.1.

7.7 Chapter Summary

This chapter dealt with different structures to implement an FIR or IIR filter. Although the input–output relationship of these structures is the same, their characteristics will differ. Hence, each has its own advantages and disadvantages. Here, we specifically considered structures available in the CMSIS-DSP library. These structures are the same ones we consider in Lab 8.11.

7.8 Further Reading

Using different LTI system structures leads to a difference in implementation. When fixed-point arithmetic is used while realizing a filter, quantization on filter coefficients may cause undesired effects. Books by Proakis and Oppenheim analyze these effects from a theoretical perspective in detail [1, 2]. We suggest that you consult these books for more information on the structures for LTI systems.

7.9 Exercises

7.9.1 An LTI system is represented as $H(z) = \frac{(z+1)(z+2)}{(z^2-0.01)(z^2+0.5)}$. Obtain

(a) the direct form I representation of this system.

(b) the direct form II representation of this system.

7.9.2 Obtain the biquad cascade form of the system given in Exercise 7.9.1.

7.9.3 Obtain the transposed form of the structures given in Exercise 7.9.1 and Exercise 7.9.2.

7.9.4 Find the IIR lattice form of the filter $H(z)$ given in Exercise 7.9.1.

7.9.5 An FIR filter has impulse response $h[n] = \sum_{k=0}^{5}(k+1)\delta[n-k]$. Find the lattice form of this filter.

7.10 References

[1] Proakis, J.G. and Manolakis, D.K. (1995) *Digital Signal Processing: Principles, Algorithms and Applications*, Prentice Hall, Third edn.

[2] Oppenheim, A.W. and Schafer, R.W. (2009) *Discrete-Time Signal Processing*, Prentice Hall, Third edn.

Note

This chapter does not include an accompanying lab.

8

Digital Filter Design

Contents

8.1 Introduction

This chapter is about digital filter design. First, the definition of an ideal filter must be fully understood. Then comes filter design specifications. Based on these, we can introduce IIR and FIR filter design methods. Here, we will briefly explain filter design methodology from a theoretical perspective. More detail on these aspects can be found in the literature [1, 2, 3]. In this book, we will focus on digital filter design using software (MATLAB and Octave). We will also consider implementation details of the designed digital filter.

8.2 Ideal Filters

In theory, one can define a filter to pass only certain frequency components of an input signal. Such a filter is called ideal because it shows the most desirable characteristics in the frequency domain. However, in practice, it is not possible to construct such a filter because ideal filters are noncausal. However, ideal filters provide valuable insight into the filtering operation in the frequency domain. In this section, we will begin with ideal lowpass, bandpass, and highpass filters.

8.2.1 Ideal Lowpass Filter

An ideal lowpass filter (LPF) allows low-frequency components of an input signal to pass, but blocks high-frequency components. The frequency response of an ideal discrete-time LPF can be represented as

$$H_{ILPF}(e^{j\omega}) = \begin{cases} 1 & |\omega| \leq \omega_s \\ 0 & |\omega| > \omega_s \end{cases} \text{ where } -\pi < \omega < \pi \tag{8.1}$$

where ω_s is the cut-off frequency of the filter. Eqn. 8.1 indicates that the ideal LPF passes input signal components with frequency value less than ω_s and stops all other signal components. In other words, the pass-band of the ideal LPF is $|\omega| \leq \omega_s$. The stop-band of this filter is $\omega_s < |\omega| \leq \pi$. We show the magnitude of $H_{ILPF}(e^{j\omega})$ in Figure 8.1. Here, we only show the frequency range $\omega \in [0, \pi]$ because $|H_{ILPF}(e^{j\omega})|$ is symmetrical around $\omega = 0$. We will apply the same reasoning in the following sections.

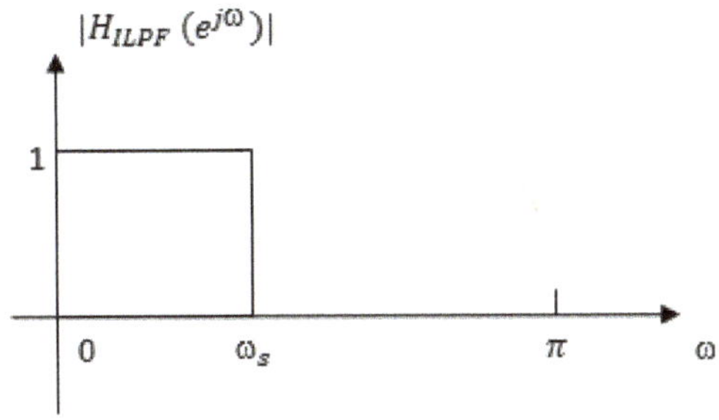

Figure 8.1 Magnitude component of $H_{ILPF}(e^{j\omega})$.

The inverse DTFT of $H_{ILPF}(e^{j\omega})$ is

$$h_{ILPF}[n] = \frac{1}{2\pi}\int_{-\pi}^{\pi} H_{ILPF}(e^{j\omega})e^{jn\omega}d\omega \tag{8.2}$$

$$= \frac{1}{2\pi}\int_{-\omega_s}^{\omega_s} e^{jn\omega}d\omega \tag{8.3}$$

$$= \frac{1}{2\pi}\left[\frac{e^{jn\omega}}{jn}\right]_{-\omega_s}^{\omega_s} \tag{8.4}$$

$$= \frac{1}{n\pi}\left[\frac{e^{jn\omega_s} - e^{-jn\omega_s}}{2j}\right] \tag{8.5}$$

$$= \frac{\sin(n\omega_s)}{n\pi} \tag{8.6}$$

This is not valid for $n = 0$, so we have

$$h_{ILPF}[n] = \begin{cases} \frac{\sin(\omega_s n)}{n\pi} & n \neq 0 \\ \frac{\omega_s}{\pi} & n = 0 \end{cases} \tag{8.7}$$

As the impulse response is non-zero for negative n, this filter is noncausal and unrealizable in practice.

8.2.2 Ideal Bandpass Filter

An ideal bandpass filter (BPF) passes input signal components with a frequency within a predefined range. Hence, the frequency response of an ideal discrete-time BPF can be represented as

$$H_{IBPF}(e^{j\omega}) = \begin{cases} 0 & |\omega| < \omega_{s1} \\ 1 & \omega_{s1} \leq |\omega| \leq \omega_{s2} \\ 0 & \omega_{s2} < |\omega| \leq \pi \end{cases} \tag{8.8}$$

where ω_{s1} and ω_{s2} are called the stop-band frequencies of the filter. Eqn. 8.8 indicates that the ideal BPF passes input signal components with a frequency value between ω_{s1} and ω_{s2} and it stops all other signal components. In other words, the pass-band of the ideal BPF is $\omega_{s1} \leq |\omega| \leq \omega_{s2}$. The magnitude of $H_{IBPF}(e^{j\omega})$ is shown in Figure 8.2.

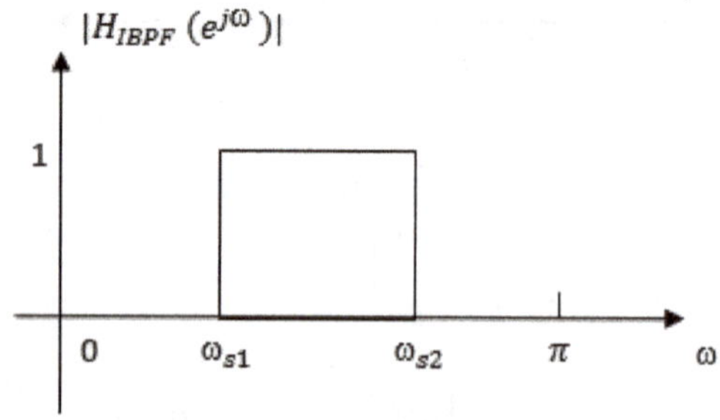

Figure 8.2 Magnitude component of $H_{IBPF}(e^{j\omega})$.

Taking the inverse DTFT of $H_{IBPF}(e^{j\omega})$ yields

$$h_{IBPF}[n] = \begin{cases} \frac{\sin(\omega_{s2}n) - \sin(\omega_{s1}n)}{n\pi} & n \neq 0 \\ \frac{\omega_{s2} - \omega_{s1}}{\pi} & n = 0 \end{cases} \tag{8.9}$$

The ideal bandpass filter is noncausal and unrealizable.

8.2.3 Ideal Highpass Filter

An ideal highpass filter (HPF) passes high-frequency components of an input signal. Hence, the frequency response of an ideal discrete-time HPF can be represented as

$$H_{IHPF}(e^{j\omega}) = \begin{cases} 0 & |\omega| < \omega_s \\ 1 & \omega_s \leq |\omega| \leq \pi \end{cases} \tag{8.10}$$

where ω_s is called the cut-off frequency of the filter. Eqn. 8.10 indicates that the ideal HPF passes input signal components with frequency values higher than ω_s and it stops all other signal components. In other words, the pass-band of the ideal HPF filter is described by $\omega_s \leq |\omega| \leq \pi$. The stop-band of the filter is described by $|\omega| < \omega_s$. The magnitude of $H_{IHPF}(e^{j\omega})$ is shown in Figure 8.3.

Taking the inverse DTFT of $H_{IHPF}(e^{j\omega})$ yields

$$h_{IBPF}[n] = \begin{cases} \frac{-\sin(\omega_s n)}{n\pi} & n \neq 0 \\ 1 - \frac{\omega_s}{\pi} & n = 0 \end{cases} \tag{8.11}$$

The ideal highpass filter is noncausal and unrealizable.

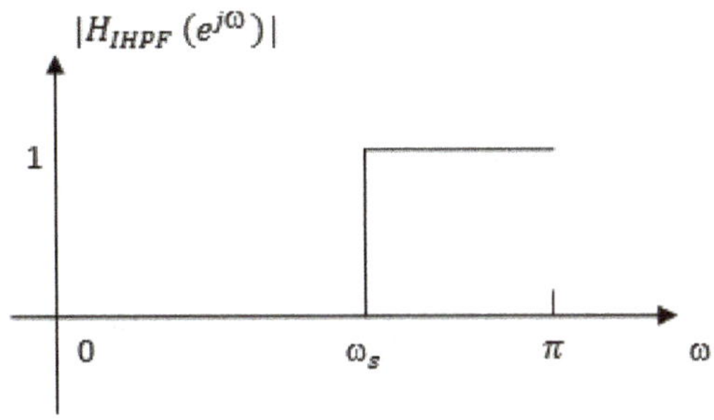

Figure 8.3 Magnitude component of $H_{IHPF}(e^{j\omega})$.

8.3 Filter Design Specifications

As mentioned in the previous section, ideal filters cannot be realized because they are noncausal. In order to have a realizable filter corresponding to its ideal form, we must relax the filter design criteria to obtain the corresponding approximate causal filter, which can be performed in three steps. First, a realizable filter need not have the characteristics of a flat pass-band. There can be a tolerance in the pass-band. This is represented as $1 \pm \delta_p$. Second, the realizable filter cannot totally stop the stop-band frequency components. Instead, there can be a tolerance added for the stop-band. This is represented as δ_s. Third, the transition from pass-band to stop-band cannot be abrupt in a realizable filter. Hence, there should be a transition band between these two regions. As a result, the filter will be designed to have both pass-band (ω_p) and stop-band (ω_s) frequencies representing the transition band.

We provide the aforementioned filter specifications in Figure 8.4. Here, we use a lowpass filter as an example and only focus on the magnitude of the frequency response. We will benefit from Figure 8.4 throughout this chapter when designing realizable filters.

8.4 IIR Filter Design Techniques

IIR filters are extensively studied in continuous-time, so most of their design methods are in this domain. The usual practice in designing a discrete-time IIR filter is as follows. First, design specifications in discrete time are projected to continuous-time. Here, we can benefit from the methods introduced in Chapter 6. Second, the IIR filter is designed

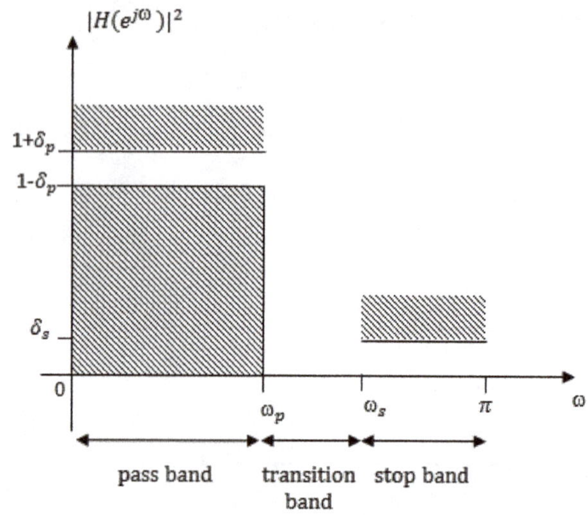

Figure 8.4 Filter design specifications.

in continuous-time. Third, the designed filter is converted to discrete-time using the methods introduced in Chapter 6. Specifically, we will use the bilinear transformation method for converting a continuous-time IIR filter to a discrete-time one.

Designing an IIR filter in continuous-time can be taken as approximating the frequency response of the corresponding ideal filter. In other words, we would ideally like to have $\delta_p = 0$, $\delta_s = 0$, and $\omega_p = \omega_s$ in Figure 8.4. As we have seen in Section 8.2, this is not possible. Therefore, we approximate this ideal case by adding tolerance values and using predefined filter structures such as Butterworth, Chebyshev, and Elliptic.

We will not explore the theory behind IIR filter design. However, you should be familiar with the basics of this process, so will now briefly explain IIR filter design using Butterworth, Chebyshev, and Elliptic filters in continuous-time.

8.4.1 Butterworth Filters

Butterworth filters have maximally flat pass-band and stop-band characteristics. In continuous-time, the magnitude (in squared form) of a Butterworth filter is

$$|H_B(j\Omega)|^2 = \frac{1}{1+(\Omega/\Omega_c)^{2N}} \tag{8.12}$$

where Ω_c is the cut-off frequency and N is the filter order.

While designing an IIR filter, the user should set either the parameter N or tolerance values. After the filter is designed, the transfer function of the Butterworth filter will have all its poles equally spaced on the unit circle in the complex plane. For more information on how to design the Butterworth filter based on a given criterion, please see [1, 2, 3].

8.4.2 Chebyshev Filters

Although the Butterworth filter provides maximally flat pass-band and stop-band characteristics, it produces a high-order filter. In case this is not required, one can allow ripples in either the pass or stop-band. As a result, the filter order decreases. This can be achieved by Chebyshev filters.

There are two Chebyshev filter types. The type I Chebyshev filter has equiripple characteristics in the pass-band. It has the form

$$|H_{CI}(j\Omega)|^2 = \frac{1}{1+\epsilon^2 V_N^2(\Omega/\Omega_c)} \tag{8.13}$$

where $V(\cdot)$ is the Nth order Chebyshev polynomial and $\epsilon^2 \ll 1$.

The Nth order Chebyshev polynomial can be defined as $V_N(x) = \cos(N\cos^{-1}(x))$. This form can be described in an iterative way as $V_{N+1}(x) = 2xV_N(x) - V_{N-1}(x)$, where $V_0(x) = 1$ and $V_1(x) = x$.

The type II Chebyshev filter has equiripple characteristics in the stop-band. It has the form

$$|H_{CII}(j\Omega)|^2 = 1 - \frac{1}{1+\epsilon^2 V_N^2(\Omega_c/\Omega)} \tag{8.14}$$

As the Chebyshev filter is designed, its poles will lie on an ellipse in the complex plane. For more information on how to design the Chebyshev filter based on a given criterion, please see [1, 2, 3].

8.4.3 Elliptic Filters

Finally, we consider elliptic filters. Here, the equiripple characteristics are evenly distributed in the stop-band. Therefore, elliptic filters have similar characteristics to Chebyshev filters. The elliptic filter has the form

$$|H_E(j\Omega)|^2 = \frac{1}{1+\epsilon^2 R_N^2(\Omega/\Omega_c)} \tag{8.15}$$

where $R_N(x)$ is the rational function defined as (for N even)

$$R_N(x) = r_0 \prod_{n=1}^{N} \frac{(x-a_n)}{(x-b_n)} \tag{8.16}$$

Here, r_0 is normalizing constant such that $R_N(1) = 1$. The constants a_n and b_n are zeros and poles of the function.

8.5 FIR Filter Design Techniques

FIR filters can be directly designed in discrete time. Therefore, FIR filter design methods are different than the ones introduced in the previous section. In this section, we focus on two main FIR filter techniques. For more information on the remaining FIR filter design methods, please see [1, 2, 3].

8.5.1 Design by Windowing

FIR filter design by the windowing method is based on the idea of modifying an ideal filter impulse response to make it realizable. In other words, we select an ideal filter impulse response first, and then we multiply it by a window function such that the final form becomes causal and finite.

Assume that the ideal filter of interest has impulse response $h_d[n]$. As mentioned in Section 8.2, this impulse response can be defined for $n \in [-\infty, \infty]$. We can obtain a realizable FIR filter by multiplying $h_d[n]$ by a window function $w[n] = u[n] - u[n-M]$. The resulting impulse response becomes

$$h[n] = h_d[n]w[n] \tag{8.17}$$

where the impulse response of the realizable FIR filter, $h[n]$, will have length M+1.

The window function used in Eqn. 8.17 defines the characteristics of the final impulse response. Therefore, there are several predefined window functions used in the literature. Some of these can be categorized as rectangular, Bartlett (triangular), Hann, Hamming, Blackman, or Keiser. For more information on these window functions and detailed analysis of their effects in the frequency domain, please see [1, 2, 3].

8.5.2 Least-Squares or Optimal Filter Design

Although FIR filter design by windowing is straightforward to implement, it does not provide complete control of the filter pass and stop-band characteristics. Least-squares or optimal filter design methods handle this shortcoming.

In general, least-squares filter design methods aim to minimize the integral of the square of the error between the desired and actual frequency responses. Hence, a prototype and parametric filtering function can be selected. The least-squares method, which we will explain in detail in Chapter 10, can be used to adjust the parameters such that the defined error decreases gradually.

One specific algorithm for optimal filter design was proposed by Parks and McClellan. This method uses the Chebyshev approximation theory in connection with the minimization of the error as in the least-squares case. The optimization of this filter design method is based on minimizing the maximum error between the desired and actual frequency responses. The theory behind this method is given in detail in [1, 2, 3].

8.6 Filter Design using Software

In this book, we use software (MATLAB and Octave) to design IIR and FIR filters. MATLAB is a commercial software used in most engineering disciplines. Octave is a free software supported by its community. Octave shares most, but not all, of MATLAB's functionality. Therefore, we use MATLAB as the main software for filter design in this book and explain concepts accordingly. We provide the corresponding Octave functions wherever possible.

MATLAB and Octave have predefined functions for filter design in the command window. MATLAB also has a filter design and analysis tool (FDATool) with an aesthetic graphical user interface (GUI). Through it, FIR and IIR filters can be designed by setting performance criteria. The filter design method can also be set in the GUI. FDATool also provides options for analyzing the designed filter. This includes the magnitude and phase response, pole-zero, step, and impulse response characteristics. As the filter is designed in FDATool, its coefficients can be exported from the same GUI. Therefore, we suggest that you use FDATool if possible.

8.6.1 FDATool Graphical User Interface

FDATool can be activated by the `fdatool` command in MATLAB. Then, the "Design Filter" panel will appear as in Figure 8.5. This GUI provides options and boxes to set all filter parameters.

FDATool provides several options to design a filter. The user can select the filter type (lowpass, highpass, bandpass, stop-band, etc.) by checking the radio button in the lower-left rectangle. The user can select the filter type (FIR or IIR) from the same location. Each filter type has its design methods in the accompanying drop down menu. The central region of FDATool's GUI has a dynamic setup. It changes with respect to the filter type being designed. The basic parameters that can be set are as follows. Filter order can be selected or it can be left as *minimum order*. If this option is selected, the filter order will be set automatically according to the remaining design parameters. F_s stands for the sampling frequency. *wpass* and *wstop* indicate the pass and stop-band frequencies, respectively. These and other parameters are also shown graphically in the upper region of FDATool's GUI. The user can select the magnitude specifications of the filter being designed on the right side of the GUI. There are two options in the "Units" drop down list of the magnitude specifications section. The first option is *dB*, and the second option is *Linear*. The design specifications given in Section 8.3 are based on this selection. The user can click the *Design Filter* button after all parameters are set, and the filter will be designed automatically based on the design criteria provided.

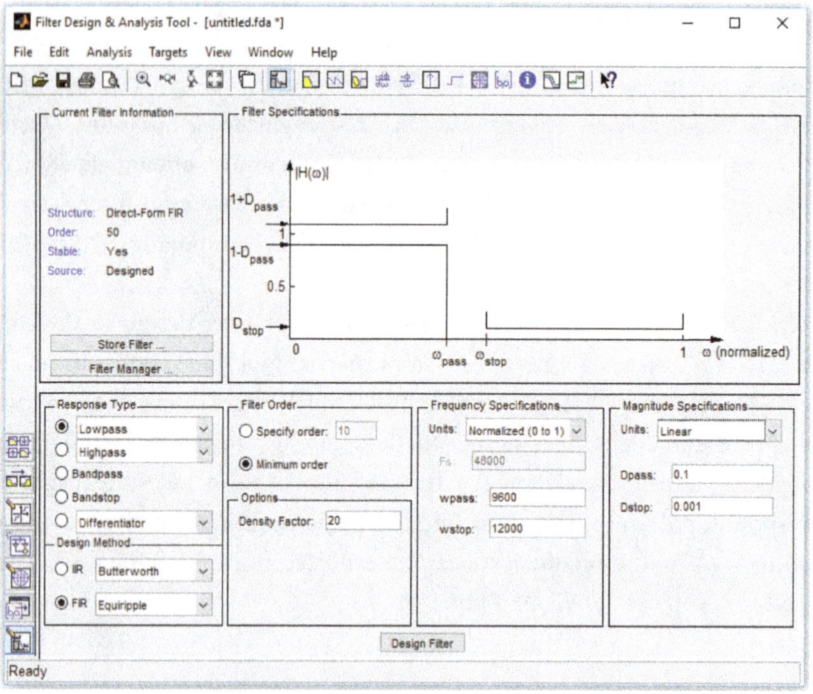

Figure 8.5 FDATool graphical user interface.

8.6.2 FIR Filter Design

FDATool has several options for FIR filter design. These are equiripple, least-squares, window, constrained least-squares, complex equiripple, maximally flat, least Pth-norm, constrained equiripple, generalized equiripple, constrained band equiripple, and interpolated FIR. More information on these design methods can be found in the MATLAB help files. The user can select the most appropriate option based on their filter design constraints.

Let us consider the discrete-time signal introduced in Lab 6. This signal is the sum of two sinusoids with angular frequencies $\pi/128$ rad/sample and $\pi/4$ rad/sample, respectively. We can rewrite this signal as

$$x[n] = \sin(\pi n/128) + \sin(\pi n/4) \tag{8.18}$$

We can design FIR lowpass and highpass filters to separate the two sinusoids in the signal. Let us start with a lowpass filter design using FDATool.

Example 8.1 FIR lowpass filter design using FDATool

Because the two sinusoids have distinct angular frequencies in Eqn. 8.18, we can pick suitable pass-band and stop-band frequency values. Let us take $wpass = 0.02\pi$ rad/sample and $wstop = 0.1\pi$ rad/sample. Let the order of the filter be 32. Finally, let us set the FIR filter type as `Equiripple`. The *magnitude response* of the designed FIR filter will be as in Figure 8.6. To note here, the magnitude is given in terms of dB in this figure. We will use this filter for filtering the signal in Eqn. 8.18.

> **magnitude response**
> *The measure of the output magnitude of a system or device in response to an input stimulus.*

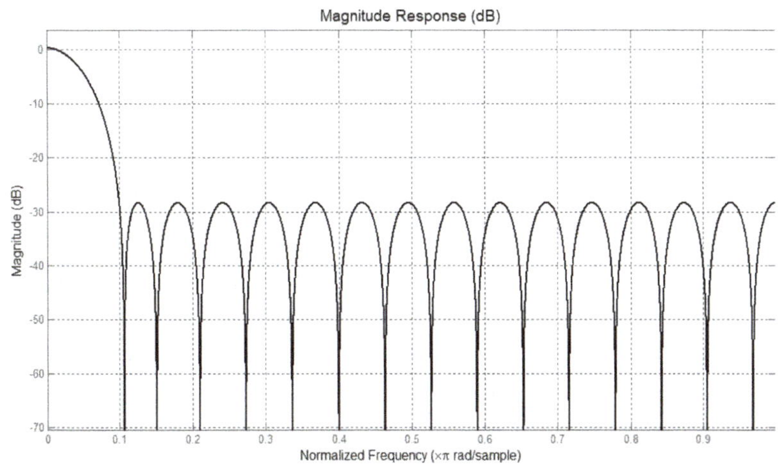

Figure 8.6 Magnitude response of the designed FIR filter in MATLAB.

Octave has the following FIR filter design functions: `cl2bp`, `fir1`, `fir2`, `firls`, `qp_kaiser`, `remez`, and `sgolay`. These can be used to design FIR filters based on different methodologies. Octave has an active community, which means these functions may be altered or expanded over time. Let us now consider an FIR lowpass filter design in Octave.

Example 8.2 FIR lowpass filter design using Octave

We can use Octave to design an FIR filter with similar specifications as in Example 8.1. Here, we set the cut-off frequency of the filter to 0.02 rad/sample. We also set the order of the filter to 32. By using Octave's `fir1` function, we can obtain the magnitude response of the designed FIR filter as in Figure 8.7. Again, the magnitude is given in terms of dB in this figure. We will use this filter on the signal in Eqn. 8.18.

We should note a shortcoming of Octave. It has only one function, `fir1`, for an FIR filter. The workings of this function are not explained in any resources. Therefore, it may never be possible to generate the same filter in both MATLAB and Octave. This is the reason that the outputs in Figures 8.6 and 8.7 are not the same.

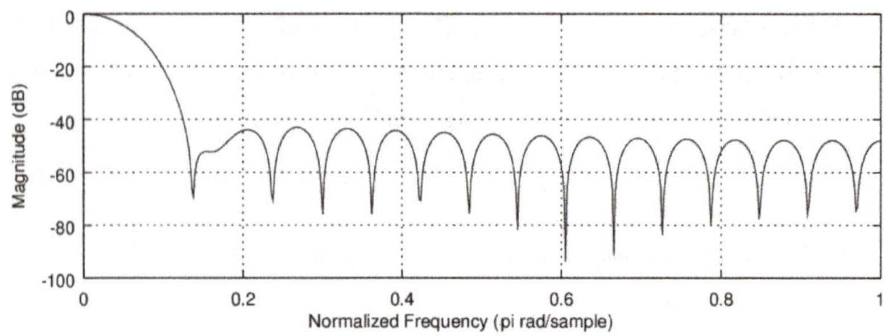

Figure 8.7 Magnitude response of the designed FIR filter in Octave.

8.6.3 IIR Filter Design

FDATool has several options for IIR filter design. These are Butterworth, Chebyshev, elliptic, maximally flat, least Pth-norm, and constrained least Pth-norm. More information on these design methods can be found in the MATLAB help files. The user can select the most appropriate option based on their filter design constraints.

As in the previous section, we can design IIR lowpass and highpass filters to separate the two sinusoids in Eqn. 8.18. Let us start with a lowpass filter design using FDATool.

Example 8.3 IIR lowpass filter design using FDATool

We can design an IIR filter to extract the sinusoid with angular frequency $\pi/128$ rad/sample in Eqn. 8.18. To do so, let us take *wstop* $= 0.02$ rad/sample. Let the order of the filter be two. Finally, let us set the IIR filter type as `Butterworth`. The magnitude response of the designed IIR filter will be as in Figure 8.8. We will use this filter on the signal in Eqn. 8.18.

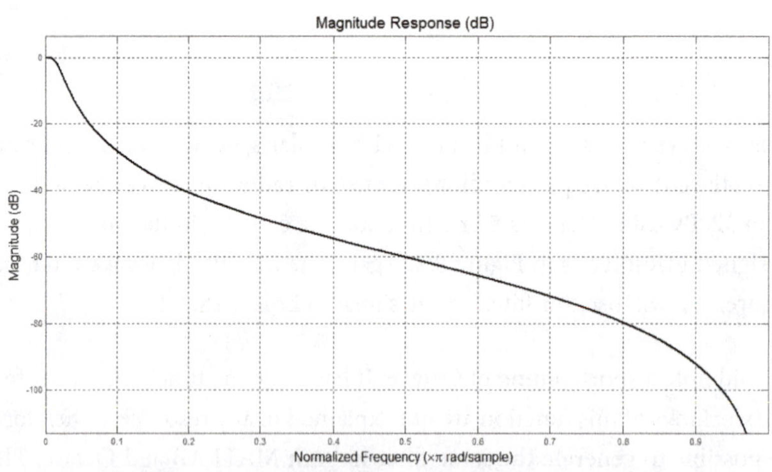

Figure 8.8 Magnitude response of the designed IIR filter in MATLAB.

Octave has the following IIR filter design functions: `besselap`, `besself`, `bilinear`, `buttap`, `butter`, `buttord`, `cheb`, `cheb1ap`, `cheb1ord`, `cheb2ap`, `cheb2ord`, `cheby1`, `cheby2`, `ellip`, `ellipap`, `ellipord`, `iirlp2mb`, `impinvar`, `invimpinvar`, `ncauer`, `pei_tseng_notch`, and `sftrans`. As with FIR filters, these functions can be used to design IIR filters with any given design criterion. Let us now consider an IIR lowpass filter design in Octave.

Example 8.4 IIR lowpass filter design using Octave

We can use Octave to design an IIR filter with similar specifications as in Example 8.3. Here, we set the cut-off frequency of the filter to 0.02 rad/sample. We also set the order of the filter to two. By using Octave's `butter` function, we can obtain the magnitude response of the designed IIR filter as in Figure 8.9. To note here, the magnitude is given in terms of dB in this figure. We will use this filter on the signal in Eqn. 8.18.

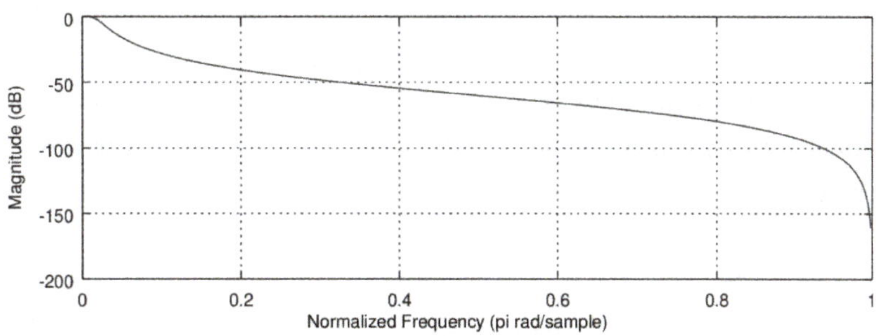

Figure 8.9 Magnitude response of the designed IIR filter in Octave.

8.6.4 FIR Filter Structure Conversions

As mentioned in Chapter 7, FIR and IIR filters have different structures to be implemented. Let us first consider FIR filter structures. When the *Design Filter* button is pressed, FDATool automatically creates an FIR filter with direct form structure. In order to convert the filter structure, the FIR filter must be designed as minimum-phase or maximum-phase. These options can be found under the generalized equiripple or constrained band equiripple design methods. After the FIR filter is designed using one of these methods, the user should click on *Edit* $\rightarrow$ *Convert Structure*. The "Convert Structure" window will open as in Figure 8.10. Here, *Lattice Moving-Average Minimum Phase* (or *Maximum Phase*) corresponds to the lattice structure.

Before converting the FIR filter to lattice structure, save the first coefficient of the designed FIR filter with direct form structure. The output of the lattice filter must be scaled down by this coefficient.

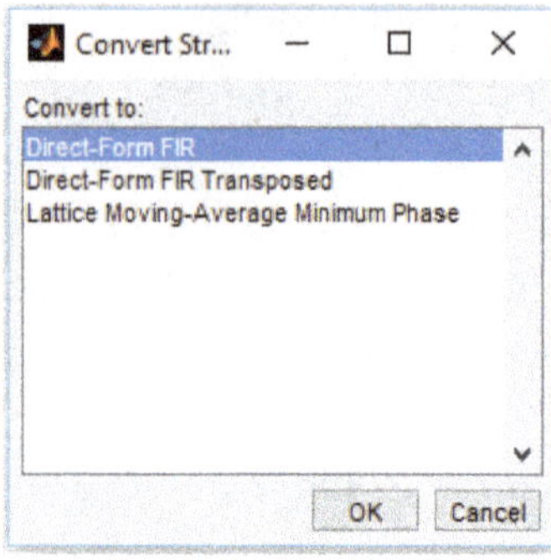

Figure 8.10 Convert Structure window for FIR filters.

8.6.5 IIR Filter Structure Conversions

Let us now consider IIR filter structures. When the *Design Filter* button is pressed, FDATool automatically creates an IIR filter with direct form II structure. In order to convert the filter structure, the user must click on *Edit* → *Convert Structure*. The "Convert Structure" window will open as in Figure 8.11. Here, *Direct-Form I, SOS* corresponds to direct form I, *Direct-Form 2 Transposed, SOS* corresponds to transposed direct form II, and *Lattice Autoregressive Moving Average (ARMA)* corresponds to the lattice structure.

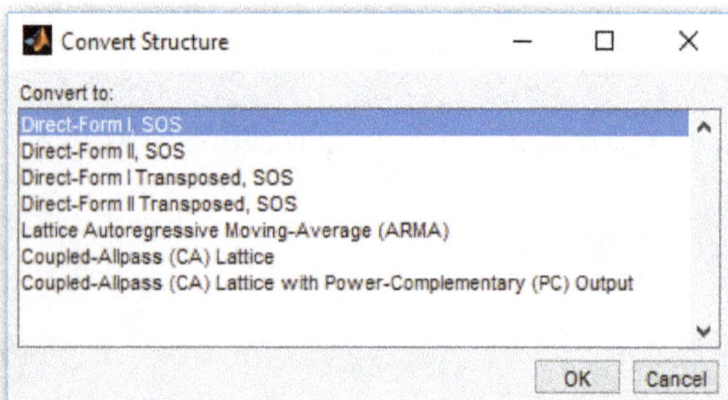

Figure 8.11 Convert Structure window for IIR filters.

Octave has the following functions for IIR filter structure conversions.

- `sos2tf`: Converts series second-order sections to direct form.
- `tf2sos`: Converts direct form filter coefficients to series second-order sections.
- `tf2zp`: Converts a transfer function to a poles and zeros representation.
- `zp2sos`: Converts a poles and zeros filter to second-order sections.
- `zp2tf`: Converts a poles and zeros filter to a transfer function.

These functions can be used to convert a given structure to another form. Octave does not have as many filter structure conversion functions. You can consult online resources for more information on filter structure conversion in Octave.

8.6.6 Exporting Filter Coefficients

FDATool creates ***double precision*** floating point (64-bit) filter coefficients by default. However, most microcontrollers use ***single precision*** floating point (32-bit) format. Therefore, you should convert the filter coefficients from double to single precision. To do so, click on the *Set quantization parameters* button. This is the third button from the top at the bottom left of the GUI. Change "Filter arithmetic" to "Single-precision floating-point" in the opened window as in Figure 8.12.

> **double precision**
> *A computer number format that occupies 8 bytes (64 bits) in computer memory and represents a wide, dynamic range of values by using a floating point.*

> **single precision**
> *A computer number format that occupies 4 bytes (32 bits) in computer memory and represents a wide dynamic range of values by using a floating point.*

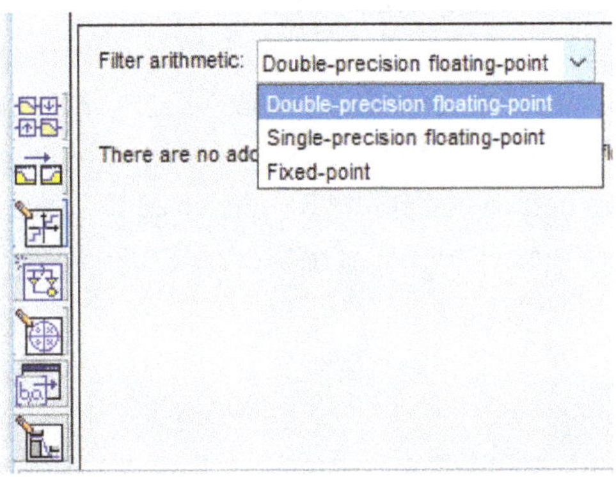

Figure 8.12 Changing the precision of filter coefficients.

As mentioned in the previous section, while designing an IIR filter, its structure must be selected. Assume that we use biquad cascade (SOS) direct form-I representation for implementation. After converting the filter structure, we should click on the *Edit → Reorder and Scale Second-Order Sections*. Then, we should select the "Scale" check box and click *Apply* as in Figure 8.13. Now, IIR filter coefficients are ready to be exported.

We require designed filter coefficients in implementation. We can obtain these in a C header file by first exporting them to the MATLAB Workspace and then converting to a CMSIS compliant form. To do so, click the "File" tab and then "export" under FDATool. The "Export To" and "Export As" fields should be "Workspace" and "Coefficients," respectively. The "Variable Names" option changes according to filter and structure type.

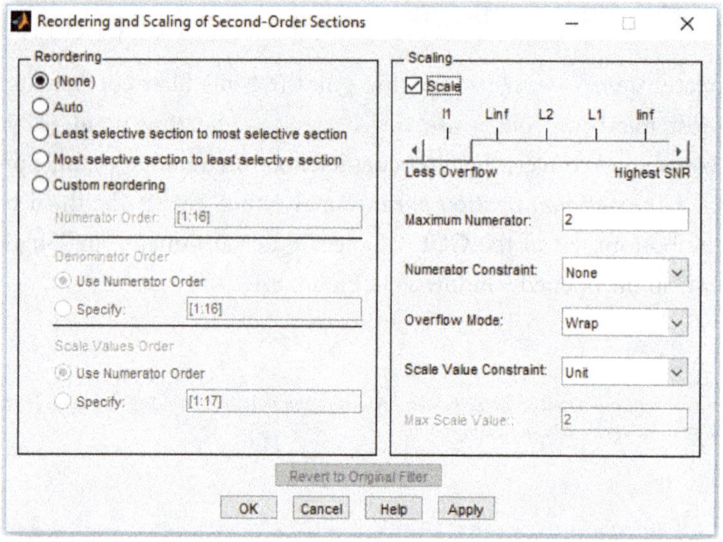

Figure 8.13 Reordering and scaling second-order sections of an IIR filter.

Exported filter coefficients will not be compatible with CMSIS-DSP library filtering functions. Therefore, they should be transformed using the m-files given in `Online_Student_Resources\Lab8\filter_coefficient_formatting`. These m-files generate CMSIS compliant header files.

As there is no GUI in Octave, we cannot export filter coefficients directly. We can use the m-files in `Online_Student_Resources\Lab8\filter_coefficient_formatting` to generate the coefficient header file after designing the filter in Octave. To note here, IIR filter coefficients should be converted to series second-order sections (SOS) before generating the header file as in Section 8.6.5.

8.7 Chapter Summary

Designing a filter is one of the most important aspects of digital signal processing. We considered the topic from a practical viewpoint in this chapter. Therefore, we did not focus on theoretical filter design methods. Instead, we took them for granted and benefited from MATLAB and Octave in designing a filter with given design criteria. We believe that this approach is more useful in practical digital signal processing applications.

8.8 Further Reading

We did not focus on the theoretical aspects of digital filter design in this chapter; the literature on this topic is vast. If you would like to go deep into the theory of digital filter design, you can refer to a number of books in the literature, e.g., [1, 2, 3].

8.9 Exercises

8.9.1 What is a stop-band filter? How can it be constructed from a bandpass filter?

8.9.2 What is a notch filter? How is it different from the stop-band filter?

8.9.3 Research the comb filter structure and its application areas.

8.9.4 A discrete-time signal is represented as $x[n] = \sum_{k=0}^{5} x_k[n]$, where $x_k[n] = \cos(k\pi n/8)/(k+1)$. Design an ideal filter to extract the first two components ($x_0[n]$ and $x_1[n]$) of $x[n]$.

8.9.5 Repeat Exercise 8.9.4, but extract the middle two components ($x_2[n]$ and $x_3[n]$) of $x[n]$.

8.9.6 Repeat Exercise 8.9.4, but extract the last two components ($x_4[n]$ and $x_5[n]$) of $x[n]$.

8.9.7 Repeat Exercise 8.9.4, but extract $x_3[n]$ from $x[n]$.

8.9.8 Repeat Exercise 8.9.4, but extract $x_2[n]$ and $x_4[n]$ from $x[n]$.

8.9.9 Design an FIR filter using FDATool as in Example 8.1 to extract the low-frequency component of the signal in Eqn. 8.18.

8.9.10 Repeat Exercise 8.9.9 using Octave.

8.9.11 Design an IIR filter using FDATool as in Example 8.3 to extract the low-frequency component of the signal in Eqn. 8.18.

8.9.12 Repeat Exercise 8.9.11 using Octave.

8.10 References

[1] Oppenheim, A.W. and Schafer, R.W. (2009) *Discrete-Time Signal Processing*, Prentice Hall, Third edn.
[2] Mitra, S.K. (2010) *Digital Signal Processing*, McGraw-Hill, Fourth edn.
[3] Proakis, J.G. and Manolakis, D.K. (1995) *Digital Signal Processing: Principles, Algorithms and Applications*, Prentice Hall, Third edn.

8.11 Lab 8

8.11.1 Introduction

We introduced general methods for designing digital filters in MATLAB and Octave in previous sections. This lab is on digital filter design and the associated filter structures. Therefore, we will start by introducing the structures available in the CMSIS-DSP library. Then, we will focus on converting filter coefficients to the CMSIS-DSP library format. Next, we will implement the filters designed in previous sections. Finally, we will design a three-band audio equalizer using FIR and IIR filters. We provide equivalent MATLAB and Octave exercise throughout this chapter because each platform has its own nuances. This does not mean that you should attempt exercises using

```
void arm_biquad_cascade_df1_init_f32(arm_biquad_casd_df1_inst_f32 *S,
uint8_t numStages, float32_t *pCoeffs, float32_t *pState)
/*
S: points to an instance of the floating-point Biquad cascade structure.
numStages: number of 2nd-order stages in the filter.
pCoeffs: points to the filter coefficients array.
pState: points to the state array.
*/

void arm_biquad_cascade_df1_f32(const arm_biquad_casd_df1_inst_f32 *S,
float32_t *pSrc, float32_t *pDst, uint32_t blockSize)
/*
S: points to an instance of the floating-point Biquad cascade structure.
pSrc: points to the block of input data.
pDst: points to the block of output data.
blockSize: number of samples to process per call.
*/

void arm_scale_f32(float32_t *pSrc, float32_t scale, float32_t *pDst,
uint32_t blockSize);
/*
pSrc: is the pointer for the input data buffer.
scale: is the scaling factor.
pDst: is the pointer for the output data buffer.
blockSize: is the number of samples in the vector.
*/
```

Listing 8.1 Predefined functions to implement biquad cascade filters.

both MATLAB and Octave. You should select one platform and complete its relevant exercises.

8.11.2 Filter Structures in the CMSIS-DSP Library

The CMSIS-DSP library has predefined functions for LTI system structures, which we introduced in Chapter 7. Here, we provide their prototypes. For more information on these functions, please see the CMSIS-DSP library documentation. To note here, we introduced the direct form FIR functions in the CMSIS-DSP library in Section 2.9.6. You can consult that section for more information.

Biquad Cascade Direct Form I Structure

The CMSIS-DSP library has predefined functions for implementing biquad cascade filters in the direct form I representation. We provide these in Listing 8.1. Here, the first function performs the initialization operation for the filter. The second function performs the filtering operation. There is an additional function, arm_scale_f32(), to scale the output according to filter gain, which will be obtained from the MATLAB exercises as the parameter G.

Biquad Cascade Direct Form II Transposed Structure

The CMSIS-DSP library also has predefined functions for implementing biquad cascade filters in the direct form II transposed structure. These are given in Listing 8.2. Here, the first and second functions perform the initialization and filtering operations,

```
void arm_biquad_cascade_df2T_init_f32(arm_biquad_cascade_df2T_instance_f32 *S,
uint8_t numStages, float32_t *pCoeffs, float32_t *pState)
/*
S: points to an instance of the filter data structure.
numStages: number of 2nd-order stages in the filter.
pCoeffs: points to the filter coefficients.
pState: points to the state buffer.
*/

void arm_biquad_cascade_df2T_f32(const arm_biquad_cascade_df2T_instance_f32 *S,
float32_t *pSrc, float32_t *pDst, uint32_t blockSize)
/*
S: points to an instance of the filter data structure.
pSrc: points to the block of input data.
pDst: points to the block of output data
blockSize: number of samples to process.
*/
```

Listing 8.2 Predefined functions to implement biquad cascade filters in direct form II transposed structure.

```
void arm_fir_lattice_init_f32(arm_fir_lattice_instance_f32 *S, uint16_t numStages,
float32_t *pCoeffs, float32_t *pState)
/*
S: points to an instance of the floating-point FIR lattice structure.
numStages: number of filter stages.
pCoeffs: points to the coefficient buffer. The array is of length numStages.
pState: points to the state buffer. The array is of length numStages.
*/

void arm_fir_lattice_f32(const arm_fir_lattice_instance_f32 *S, float32_t *pSrc,
float32_t *pDst, uint32_t blockSize)
/*
S: points to an instance of the floating-point FIR lattice structure.
pSrc: points to the block of input data.
pDst: points to the block of output data.
blockSize: number of samples to process.
*/
```

Listing 8.3 Predefined functions to implement an FIR lattice filter.

respectively. The scaling function `arm_scale_f32` given in Listing 8.1 is also used here.

FIR Lattice Filter

FIR lattice filters are also available in the CMSIS-DSP library. Predefined functions for these are given in Listing 8.3. As in previous structures, the first function performs the initialization operation for the FIR filter. The second function performs the filtering operation.

IIR Lattice Filter

The CMSIS-DSP library also has predefined functions to implement IIR lattice filters as given in Listing 8.4. Here, the first and second functions perform initialization and filtering operations, respectively.

```
void arm_iir_lattice_init_f32(arm_iir_lattice_instance_f32 *S, uint16_t numStages,
float32_t *pkCoeffs, float32_t *pvCoeffs, float32_t *pState, uint32_t blockSize)
/*
S: points to an instance of the floating-point IIR lattice structure.
numStages: number of stages in the filter.
pkCoeffs: points to the reflection coefficient buffer. The array is of length
numStages.
pvCoeffs: points to the ladder coefficient buffer. The array is of length
numStages+1.
pState: points to the state buffer. The array is of length numStages+blockSize.
blockSize: number of samples to process.
*/

void arm_iir_lattice_f32(const arm_iir_lattice_instance_f32 *S, float32_t *pSrc,
float32_t *pDst, uint32_t blockSize)
/*
S: points to an instance of the floating-point IIR lattice structure.
pSrc: points to the block of input data.
pDst: points to the block of output data.
blockSize: number of samples to process.
*/
```

Listing 8.4 Predefined functions to implement an IIR lattice filter.

8.11.3 Implementing a Filter using Different Structures

As explained in the previous section, the CMSIS-DSP library has predefined functions for implementing different structures. We have used these functions to implement the

IIR lowpass filter

$$H(z) = \frac{0.002044z^2 + 0.004088z + 0.002044}{z^2 - 1.819168z + 0.827343} \tag{8.19}$$

in `Online_Student_Resources\Lab8\filtering_examples`. You will benefit from these projects when implementing the filters considered here. To note here, the IIR filter of interest was initially introduced in Eqn. 6.1.

In this section, we will use the predefined CMSIS-DSP library functions to implement the FIR and IIR filters designed in Section 8.4. The aim here is to become familiar with filter structure implementation issues.

We will also apply the input signal

$$x[n] = \sin(\pi n/128) + \sin(\pi n/4) \tag{8.20}$$

to the implemented filter to observe the result of filtering. This input signal is specifically chosen because it has one component with low-frequency characteristics and another component with high-frequency characteristics. Therefore, the performance of the implemented filter can be observed using this signal.

Finally, we will compare the implementation details of the different structures considered in this section, so we will focus on the computation load required by each structure. This will aid in selecting an appropriate structure for implementation.

Please remember that the output must be scaled by the scaling factor G after all the filtering operations in this section. This scaling factor is handled in the filter coefficient export scripts in `Online_Student_Resources\Lab8\filter_coefficient_formatting`.

Direct Form I Structure

For storing filter coefficients in the direct form I structure, FDATool has a single multidimensional array and single one-dimensional array. The first array stores the series second-order sections (SOS), and the second array stores the scaling factors. However, the coefficient array used in the biquad cascade direct form I initialization function in the CMSIS-DSP library is one-dimensional. Transformation between these two forms can be done using the m-file given in `Online_Student_Resources\Lab8\filter_coefficient_formatting`. As this m-file is run, a CMSIS-DSP library compliant header will be generated in the MATLAB working directory.

We will first consider converting the designed filter coefficients to direct form I structure. This exercise will aid us in understanding how this conversion is done.

Task 8.1

To fully grasp the filter coefficient conversion procedure, use the lowpass filter coefficients designed in Example 8.3. Convert them to direct form I structure. The procedure for this is given in Section 8.4. We must apply two more steps to implement the obtained filter structure. First, export the coefficients to a MATLAB workspace as explained in Section 8.4. Second, run the m-file in `Online_Student_Resources\Lab8\filter_coefficient_formatting` to transform the coefficients from SOS form to CMSIS-DSP library format.

Now, the obtained direct form I structure is ready to be implemented. Write a C code to implement the given structure. Plot the filtered signal after running the C code. Repeat the same procedure to implement the highpass filter designed in Exercise 8.9.11. Comment on the results.

Unlike FDATool in MATLAB, there is no GUI in Octave for designing a filter. Hence, we must use command line functions for filter design as explained in Section 8.4. However, the filter design functions discussed there do not generate CMSIS-DSP library compliant coefficients. To transform the coefficients, we will again benefit from the m-file given in `Online_Student_Resources\Lab8\filter_coefficient_formatting`. Before transformation, the coefficients need to be converted to series SOS form. IIR filter structure conversions for Octave are given in Section 8.4. We should mention a shortcoming of Octave, which is its limited number of functions for filter implementation. The workings of these functions are not explained in any available resources. You should be aware of this shortcoming while using Octave.

The aim of the following exercise is to show how the designed filter coefficients can be converted to direct form I structure in Octave. If you have already completed Task 8.1, then skip Task 8.2. Completing one of them is sufficient to understand the working principles of the structure conversion mechanism.

Task 8.2

We can also use Octave to understand the filter conversion procedure. Therefore, use the designed lowpass filter coefficients from Example 8.4. Convert them to direct form I structure. Write a C code to implement the given structure. Plot the filtered signal after running the C code. Repeat the same procedure to implement the highpass filter designed in Exercise 8.9.12. Comment on the results.

Direct Form II Transposed Structure

As in the direct form I structure, FDATool's direct form II transposed structure has one multidimensional array and one one-dimensional array for storing filter coefficients. We should follow the same steps as in the previous section to obtain CMSIS-DSP library compliant filter coefficients.

To observe the direct form II transposed structure conversion steps, let us repeat Task 8.1 using the direct form II transposed structure.

Task 8.3

In MATLAB, repeat Task 8.1 using the direct form II transposed structure. What do you observe?

As with Task 8.2, we ask that you only complete Task 8.4 if you did not complete Task 8.3.

Task 8.4

In Octave, repeat Task 8.2 using the direct form II transposed structure. To note here, designing an IIR lowpass filter in Octave was discussed in Example 8.4. This example may help with implementation.

Filter design functions in Octave do not generate CMSIS-DSP library compliant coefficients. Again, we will apply the same steps as in the previous section for this purpose.

FIR Lattice Filter

FDATool's Lattice Moving-Average Minimum Phase (or Maximum Phase) structure has one one-dimensional array for storing coefficients. This array is compliant with the FIR lattice filter initialization function in the CMSIS-DSP library. To export coefficients to the header file, we will benefit from the m-file given in `Online_Student_Resources\Lab8\filter_coefficient_formatting`.

We considered FIR filter design in Example 8.1. This function will be helpful when implementing an FIR filter in the FIR lattice structure.

Task 8.5

Repeat Task 8.1 with the FIR lattice structure. Use the FIR filter design from Example 8.1. What is the important point here?

IIR Lattice Filter

Finally, we discuss the implementation of the IIR lattice. FDATool's IIR lattice filter structure has two one-dimensional arrays for storing filter coefficients, namely `K` and `V`, and they store reflection and ladder coefficients, respectively. The same naming convention is used in the CMSIS-DSP library. However, `K` and `V` should be placed in reversed order to be used by the corresponding CMSIS-DSP library function. This can be done using the m-file given in `Online_Student_Resources\Lab8\filter_coefficient_formatting`.

Task 8.6

We would like to observe the result of converting a structure to IIR lattice form in filtering. Repeat Task 8.1 using the IIR lattice structure. How does this implementation differ from the one considered in Task 8.1?

Comparing the Implemented Filters

In the previous sections, we have applied different structures to implement the same filter. You should understand the advantages and disadvantages of these structures, and as a result, be able to select a suitable structure for any given application. In this section, we compare filter structures.

Task 8.7

Compare the outputs obtained from the implementations in Tasks 8.1, 8.3, 8.5, and 8.6 (or Tasks 8.2 and 8.4 if using Octave). Further, compare the computation load by tabulating clock cycles (just for filtering), memory usage in flash, and RAM for each implementation.

This exercise should clearly indicate the advantages and disadvantages of each considered structure. This will help you to select the most suitable structure for your design specifications.

8.11.4 Three-Band Audio Equalizer Design

Filter design can be explained in several ways. One way is to select a real-life problem and design suitable filter(s) for it. Here, we will use three-band audio equalizer implementation as the real-life problem to be solved. An equalizer is a special hardware or software that adjusts the gain of different frequency components of a signal. An equalizer uses filters for this purpose. Here, each filter is used to pass a selected frequency band. Then, the selected frequency band can be boosted or suppressed using an adjustable gain constant attached to the output of the corresponding filter.

We will design the three-band audio equalizer described by the block diagram in Figure 8.14. In this figure, `LPF`, `BPF`, and `HPF` stand for lowpass, bandpass, and highpass filters, respectively. The constants k_1, k_2, and k_3 represent the gain factor for each filter output. We will use both FIR and IIR filters in this design.

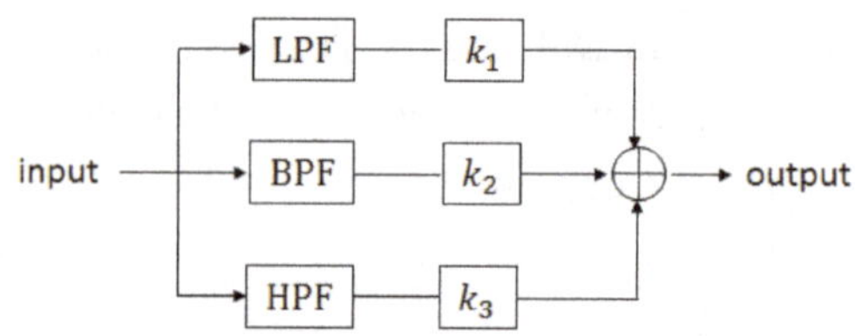

Figure 8.14 Block diagram of a three-band audio equalizer system.

Because the chirp signal sweeps a given frequency range, we will use it to test the designed filters. The total number of elements in the chirp signal is set to 2048 and its frequency range is set between 0 rad/sample and π rad/sample. We have provided the C code of such a chirp signal in `Online_Student_Resources\Lab4\STFT_calculations`.

FIR Filter Design

The equalizer can be implemented using only FIR filters. To do so, we will first design the filters. Then, we will test the operation of the equalizer by applying different gain factors.

As the filter order increases, its operation characteristics (such as filtering quality) improve. Therefore, we will start with designing high-order filters for our equalizer.

Task 8.8

For the equalizer shown in Figure 8.14, design 128th order FIR filters for LPF, BPF, and HPF. Use the Hamming window in this design. Set the normalized angular cut-off frequency for the lowpass filter to $\pi/24$ rad/sample. For the bandpass filter, set the normalized angular cut-off frequency values to $\pi/24$ rad/sample and $\pi/6$ rad/sample. Finally, set the normalized angular cut-off frequency value of the highpass filter to $\pi/6$

rad/sample. Use the FIR direct form structure for implementation. Plot the magnitude response of the designed filters. Based on these plots, comment on whether they satisfy the design constraints.

Now, we will implement the designed filters to realize the equalizer.

Task 8.9

Implement the filters designed in Task 8.8. Test the implementation with the three gain value sets $\{k_1 = 1, k_2 = 0, k_3 = 0\}$, $\{k_1 = 0, k_2 = 1, k_3 = 0\}$, and $\{k_1 = 0, k_2 = 0, k_3 = 1\}$. Apply the chirp signal as the equalizer input. Comment on the results.

Although a high-order filter has desirable characteristics, it has two fundamental shortcomings. First, as the filter order increases, the delay in the filter also increases. Second, the memory and the CPU time required to implement the filter increase. Therefore, we now focus on designing low-order filters for our equalizer.

Task 8.10

To observe the effect of filter order on equalizer design, repeat Task 8.8, but use 32nd-order FIR filters. In the same line, repeat Task 8.9 such that the equalizer is used by changing its coefficients. Do the outputs change when the filter order is changed from 128 to 32? If the outputs do change, analyze how they change.

Here, you should have learned how to design FIR filters of different types and orders. Moreover, you should have observed the filtering operation. Further, you should have observed that as filter order increases, attenuation in the stop band increases and the quality of filtering improves.

IIR Filter Design

An equalizer can also be implemented using IIR filters. As in the previous section, we will first design the filters. Then, we will test the operation of the designed equalizer by applying different gain factors.

Filter coefficients decrease as we start using IIR filters. We will focus on this issue first.

Task 8.11

For the equalizer in Figure 8.14, design 16th order IIR filters for LPF, BPF, and HPF. Use the Butterworth design method. Set the cut-off frequency values as in Task 8.8. Use the biquad cascade direct form II transposed filter structure. Plot the magnitude response of the designed filters. Based on these plots, comment on whether they satisfy design constraints.

Now, we will implement the designed filters to realize the equalizer.

Task 8.12

Test the implementation on the three gain value sets $\{k_1 = 1, k_2 = 0, k_3 = 0\}$, $\{k_1 = 0, k_2 = 1, k_3 = 0\}$, and $\{k_1 = 0, k_2 = 0, k_3 = 1\}$. Apply the chirp signal as input to the equalizer. Comment on the results.

9

Adaptive Signal Processing

Contents

9.1 Introduction

Adaptive signal processing is an alternative approach for solving signal processing problems. The main property of these systems is that they can learn from data, so system parameters can be updated by a change in the properties of the processed data at hand. Adaptive signal processing can be explained using random processes and probability theory. However, these two topics require heavy mathematical derivations. Therefore, in this chapter, we will only review the basics of adaptive signal processing and focus mainly on its applications.

To explain adaptive signal processing concepts, we will begin with the definition of an adaptive filter. Then, we will introduce the steepest descent method. Based on this, we will consider the least mean squares and normalized least mean squares methods. Finally, we will provide adaptive filter examples based on their application areas.

9.2 What is an Adaptive Filter?

The usual practice in signal processing is to design a filter for a specific application. An adaptive filter is a special type of digital filter because its characteristics (or parameters more specifically) can be modified based on the target objective or the incoming signal properties. As a result, adaptive filters perform better compared to their static counterparts.

For a static filter, design parameters are set to suite the assumed working condition of the filter. Then, a computer tool can be used to design the filter as discussed in Chapter 8. The result will be a filter structure with fixed coefficients. If the assumed working condition of the filter changes, then it will not work properly or perhaps not at all. In such cases, adaptive filters are the solution.

In contrast to digital filter design, as explored in Chapter 8, there is no design phase when forming an adaptive filter. Although the filter structure type and order must be set by the user, the filter parameters are adjusted by an algorithm depending on the given data. The algorithm's goal is to adapt the filter to meet a performance criterion. An adaptation (learning) rule is used to calculate filter parameters based on a given training set. This is a semi-intelligent mechanism (i.e., the adaptive filters can be changed on the fly), which can adapt itself to changing working conditions.

Adaptation is not an easy task in continuous-time signal processing, but in a digital system (such as a microcontroller) it is straightforward because the filtering operation is performed by a code fragment. The user can define the filter parameter adjustment rule on the digital system as another code fragment. Therefore, filter parameters can be changed easily by modifying the relevant code fragment.

Adaptive filters can be used in a wide variety of applications. Their usage is necessary in some applications, such as system identification and noise cancellation. In system identification, an unknown system needs to be identified (or replicated) by a known system. Here, the identification process corresponds to adjusting parameters of the known system using the data at hand. In noise cancellation, a system is formed that adapts to the noise characteristics of the incoming signal in order to cancel it out. It is not possible to solve both applications using standard filter design methodologies. On the contrary, adaptive filters are time-varying. They are also considered nonlinear [1]. Therefore, they do not obey LTI conditions, which makes them difficult to analyze. Further, coefficient convergence of an adaptive filter is not always guaranteed to satisfy the best possible performance criteria. You should be aware of these shortcomings when implementing an adaptive filter.

9.3 Steepest Descent Method

There are several parameter adaptation methods in literature [1]. Some of these can only be explained through the theory on random signal processing. Random signal processing is based on statistical methods of analyzing signals with nonstationary characteristics. Some of these have elegant theory behind them, but their implementation on a resource constrained digital system (such as a microcontroller) is not feasible. Therefore, in this book, we only focus on the steepest (or gradient) descent method.

The theory behind the steepest descent method is elegant. This method uses a performance criterion (or associated cost function to be minimized) to adjust the given system parameters based on a target objective. The cost function can be tailored to suit the specific problem at hand. Let us assume that we wish to construct a cost function for parameter adaptation. This function should be continuously differentiable with respect to its parameter values. The steepest descent method is an iterative process. It begins with predefined parameter values, and at each iteration, the parameter values are updated to meet the desired performance criterion. If this criterion is set in terms of a minimal value, then the gradient of the cost function with respect to parameters shows the increment direction of the cost function. Therefore, updating parameters by the negative of the gradient allows the performance criterion to be met at each iteration. Here, we should also include a positive step size parameter, which affects the applied update rate. As a result, we can implement an updating rule for the steepest descent method.

Let us assume that parameter vector $\underline{\beta} = [\beta_0, \beta_1 \cdots, \beta_{N-1}]$ is updated. Based on the steepest descent rule, updating can be formulated as

$$\underline{\beta}(n+1) = \underline{\beta}(n) - \mu \left. \frac{\partial J(\underline{\beta})}{\partial \underline{\beta}} \right|_{\underline{\beta}=\underline{\beta}(n)} \tag{9.1}$$

where J is the performance criterion to be minimized with respect to $\underline{\beta}$, $\underline{\beta}(n)$ represents the old parameter vector, which is to be updated, $\underline{\beta}(n+1)$ is the new parameter vector calculated after updating, and μ is the positive step size parameter. Next, we will introduce the least mean squares and the normalized least mean squares methods, which are based on the steepest descent rule.

9.4 Least Mean Squares Method

This method uses the mean squared error between the desired output, $d[n]$, and the adaptive filter's output, $y[n]$, as the cost function, J, to be minimized. Error is defined as $e[n] = y[n] - d[n]$, which means $J = E\left[e^2[n]\right]$.

$$y[n] = \sum_{k=0}^{N-1} \beta_k(n)x[n-k] \tag{9.2}$$

or in vector form,

$$y[n] = \underline{\beta}^T \underline{x}[n] \tag{9.3}$$

where $\underline{x}[n] = [x[n], x[n-1], \cdots, x[n-N+1]]^T$.

Then, the error between the desired output and the adaptive filter's output becomes

$$e[n] = d[n] - \underline{\beta}^T \underline{x}[n] \tag{9.4}$$

We can formulate the parameter updating by the steepest descent rule (using Eqn. 9.1) as follows.

$$\underline{\beta}(n+1) = \underline{\beta} - \mu \frac{\partial e^2[n]}{\partial \underline{\beta}} \tag{9.5}$$

$$= \underline{\beta}(n) - \mu \frac{\partial (d[n] - \underline{\beta}^T \underline{x}[n])^2}{\partial \underline{\beta}} \tag{9.6}$$

$$= \underline{\beta}(n) + 2\mu(d[n] - \underline{\beta}^T x[n])x[n] \tag{9.7}$$

or in simplified form,

$$\underline{\beta}(n+1) = \underline{\beta}(n) + \mu e[n]x[n] \tag{9.8}$$

where the positive step size parameter μ takes into account the multiplier of two in Eqn. 9.7.

The iterative process in Eqn. 9.8 starts with an arbitrary initial value $\underline{\beta}(0)$ and ends when the minimum of the cost function is reached. To note here, the FIR filter structure is selected in the derivation. Selecting the FIR filter structure ensures that the cost function is a quadratic function of $\underline{\beta}$, which means it has a single global minimum. Therefore, the steepest descent approach is justified.

Convergence analysis of the parameters in this iterative process requires a stochastic approach, which is not covered in this book. We strongly suggest Haykin's book [1] for the details of convergence analysis.

9.5 Normalized Least Mean Squares Method

The parameter update rule in Eqn. 9.8 is straightforward to implement. It also provides good results in adjusting parameters. However, the parameter convergence rate may be slow for some applications. To overcome this problem, the normalized least mean squares (normalized LMS) method can be used. This method differs from LMS in that it has an additional normalization coefficient in step size. As a result, the parameter update rule in Eqn. 9.8 becomes

$$\underline{\beta}(n+1) = \underline{\beta}(n) + \frac{\mu}{P[n]} e[n]\underline{x}[n] \tag{9.9}$$

where $P[n]$ is the normalization coefficient introduced to speed up the parameter convergence rate. It is calculated from the input signal as

$$P[n] = \sum_{k=0}^{N-1} x^2[n] \tag{9.10}$$

9.6 Adaptive Filter Applications

The application areas of adaptive filters can be categorized as one of system identification, equalization, prediction, or noise cancellation. We now consider each application category in detail. All of these share a similar adaptive filter block as in Figure 9.1.

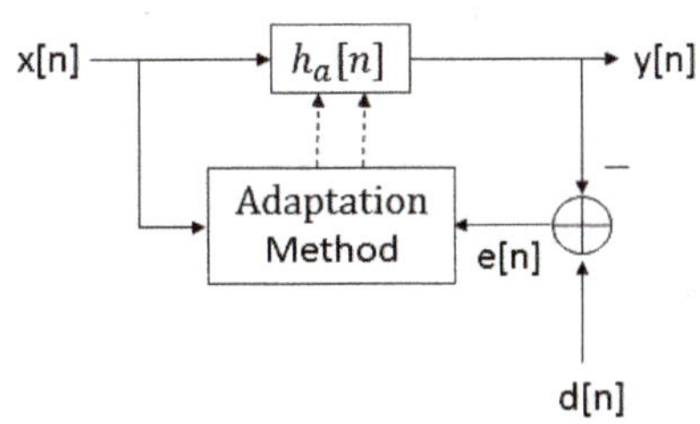

Figure 9.1 Basic adaptive filter block. Dotted connections represent the parameter adjustment mechanism between the adaptation method and the adaptive filter.

In Figure 9.1, $h_a[n] = \sum_{k=0}^{N-1} \beta_k \delta[n-k]$. To note here, we chose an FIR filter for adaptation, but an IIR filter could also have been used. The adaptation method used here is either LMS or normalized LMS.

9.6.1 System Identification

This application category aims to represent an unknown system from its input–output relationship, which is done as follows. The unknown system and adaptive filter are connected in parallel. A known input signal is fed to both systems. The desired output, $d[n]$, is obtained from the unknown system. The adaptive filter adjusts its parameters to mimic the input–output relationship of the unknown system. The block diagram for system identification is shown in Figure 9.2. We now provide an example of system identification.

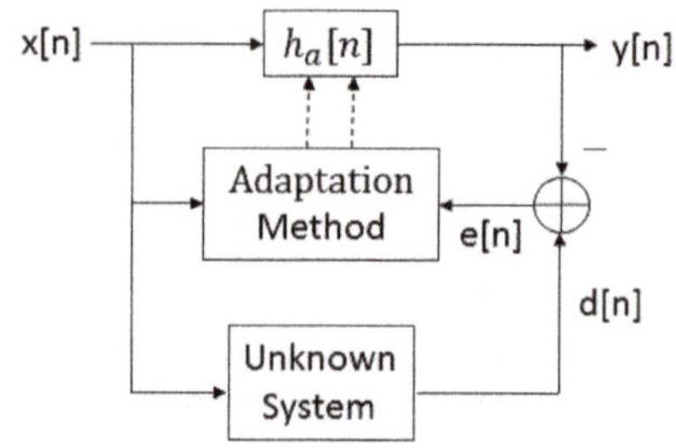

Figure 9.2 System identification block diagram.

Example 9.1 System identification

To show how system identification works, we use a third-order IIR filter as the unknown system. The transfer function of this filter is

$$H(z) = \frac{0.018099z^3 + 0.054297z^2 + 0.054297z + 0.018099}{z^3 - 1.760042z^2 + 1.182893z - 0.278059} \tag{9.11}$$

The impulse response of the IIR filter of interest, $h[n]$, is shown in Figure 9.3 (b). As can be seen in this figure, the impulse response approximately stabilizes after the 30th sample. Therefore, we select 32 as the order of the adaptive filter to be used in system identification.

We feed a pseudorandom signal to both the unknown system and the adaptive filter. Using the LMS method, we aim to identify the unknown system (third-order IIR filter) by adjusting the adaptive filter (32nd-order FIR filter) parameters.

We provide the error in terms of iteration number in Figure 9.3 (a). As can be seen in this figure, the error gradually decreases to zero as iteration number increases. Therefore, characteristics of the adaptive filter can be taken after some number of iterations.

We provide the overlapped impulse responses of the unknown system and the adaptive filter in Figure 9.3 (b). As can be seen in this figure, the impulse responses are almost the same. This indicates that the adaptive filter has correctly identified the input–output relationship of the unknown system, so the obtained filter can be used instead of the unknown system.

We provide the complete C code related to this example in Lab 9.12.3, where we solve the problem using both LMS and normalized LMS methods.

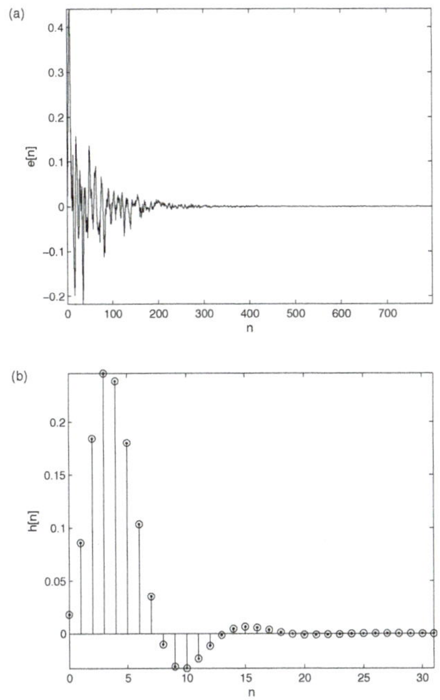

Figure 9.3 System identification example. (a) Error in terms of iteration number; (b) Impulse response of the unknown IIR filter (filled circles) and that of the adaptive filter (empty circles).

9.6.2 Equalization

In some cases, we may wish to obtain the original signal processed by an unknown system. In other words, we may want to obtain the inverse model of the original system. As a result, we expect to obtain the delayed version of input signal fed to the unknown system. This is called equalization. Here, the unknown system and adaptive filter are connected in cascade form. The block diagram for equalization is shown in Figure 9.4.

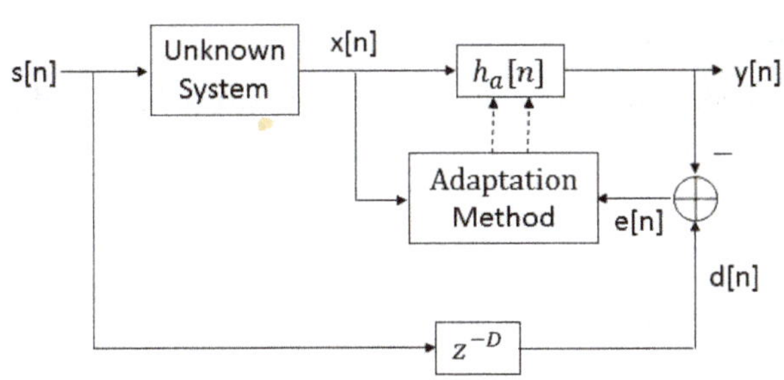

Figure 9.4 Equalization block diagram.

As can be seen in Figure 9.4, the input of the adaptive filter is the output of the unknown system. We want to find $y[n] = \hat{s}[n]$ such that the original signal is obtained (at least approximately) as a result of equalization.

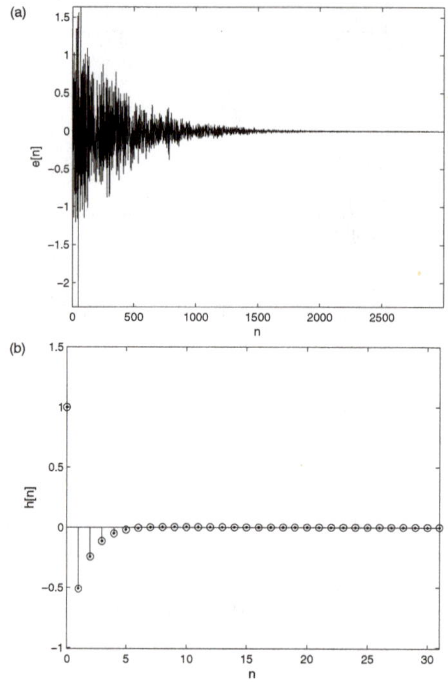

Figure 9.5 Equalization example. (a) Error in terms of iteration number; (b) Impulse response of the inverse of the filter to be equalized (filled circles) and that of the adaptive filter (empty circles).

We must add a delay (by D samples) to the desired output in equalization. This is to compensate for the lag induced through the filtering operations of the unknown system and the adaptive filter. To note here, the parameter adaptation process is performed on a predefined signal before the system is used for equalization. Otherwise, it would not be possible to obtain the desired output.

We now provide an example of equalization.

Example 9.2 Equalization

To demonstrate how equalization works, we use a second-order IIR filter as the unknown system. The transfer function of this filter is as follows. The impulse response of the inverse system ($1/H(z)$) is shown in Figure 9.5 (b).

$$H(z) = \frac{z^2 - 1.3z + 0.4}{z^2 - 1.809524z + 0.818594} \tag{9.12}$$

We use a 32nd-order FIR filter as the equalizer. We feed the pseudorandom signal to the system. We aim to obtain the inverse of the unknown system using the LMS method.

We provide the error in terms of iteration number in Figure 9.5 (a). As can be seen in this figure, error gradually decreases to zero as the iteration number increases, so the adaptive filter starts mimicking the inverse of the unknown system.

We provide the overlapped impulse responses of the inverse system and the adaptive filter in Figure 9.5 (b). As can be seen in this figure, both impulse responses are almost the same. This indicates that the adaptive filter has identified the inverse of the unknown system, so it can be used as an equalizer.

9.6.3 Prediction

An adaptive filter can also be used to predict a signal's future values from its past ones. For prediction, a delayed version of the input signal is fed to the adaptive filter in the parameter adjustment phase. The desired signal is the input itself. The adaptation mechanism aims to predict present input signal values from the past input samples. Then, after the parameter adjustment phase, the adaptive filter can be used to predict future values of input signal. The block diagram for this process is shown in Figure 9.6. As an example, we aim to predict future values of a sinusoidal signal.

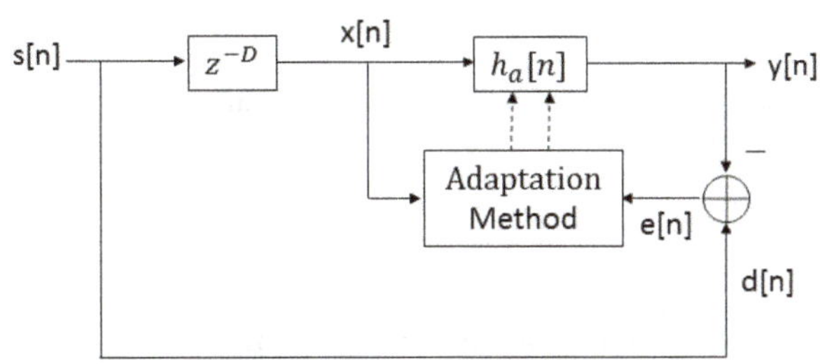

Figure 9.6 Prediction setup.

Example 9.3 Prediction

Assume we have a sinusoidal signal and its past values, and we want to predict its future values based on these. We experimentally observe that an 8th-order FIR filter is sufficient for predicting the sinusoidal signal. We can apply prediction using the setup in Figure 9.6. Here, the adaptation method used is LMS. The error in terms of iteration number is shown in Figure 9.7 (a). As can be seen in this figure, error gradually decreases to zero as the iteration number increases, so we know the adaptive filter starts predicting the sinusoidal signal from its past samples. We also provide the predicted sinusoidal signal in Figure 9.7 (b). This figure indicates that the predictor works as expected in obtaining the sinusoidal signal from its initial samples.

The complete C code for this example is available in Lab 9.12.3, where we solve the problem using both the LMS and normalized LMS methods.

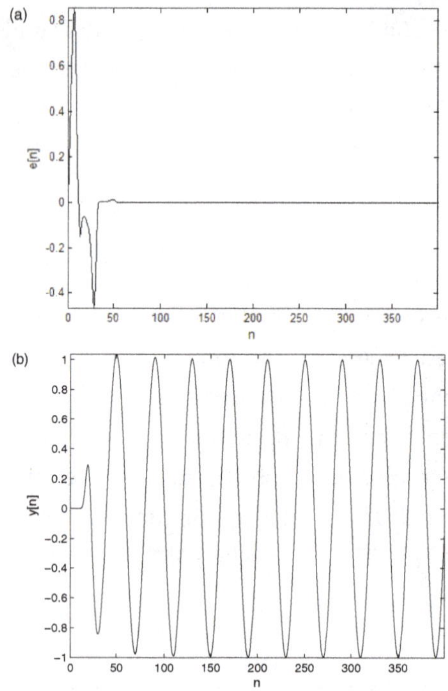

Figure 9.7 Prediction example. (a) Error in terms of iteration number; (b) $y[n]$ in terms of iteration number.

9.6.4 Noise Cancellation

An adaptive filter can also be used for noise cancellation. Here, the aim is to filter out the noise term in a given signal. Noise is not a deterministic signal, so it is modeled as random. This means we cannot subtract the noise term from the original signal to eliminate its effects. Therefore, we assume that the noise in the signal and observed noise signals are correlated. Using this correlation, we aim to estimate noise in the signal via adaptive filtering. After estimation, we subtract this component from the signal to obtain its noise cancelled form. The block diagram for noise cancellation is shown in Figure 9.8.

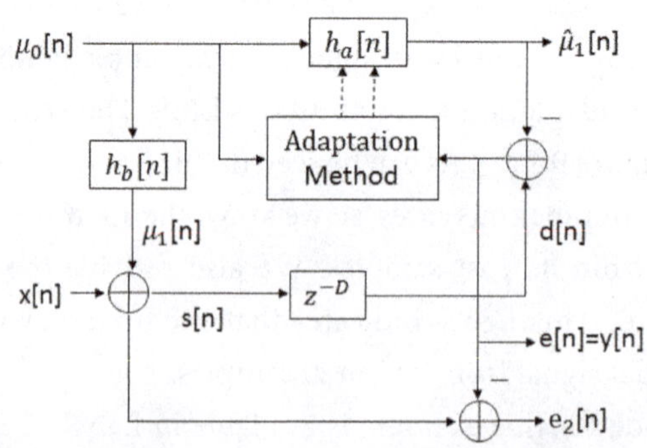

Figure 9.8 Noise cancellation block diagram.

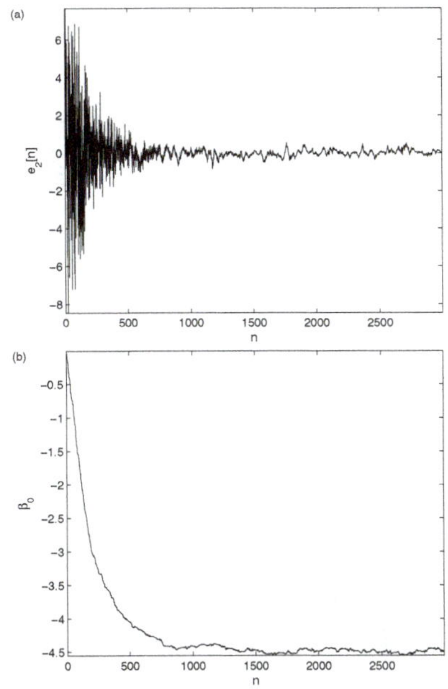

Figure 9.9 Noise cancellation example: part I. (a) e_2 in terms of iteration number; (b) β_0 in terms of iteration number.

In Figure 9.8, $\mu_0[n]$ is the noise signal. The noise signal applied to the original signal, $x[n]$, is $\mu_1[n]$. This signal is obtained by filtering $\mu_0[n]$ through the lowpass filter, $h_b[n]$. Different from our previous examples, the error signal is also used as the output after noise cancellation. Here, we add the signal component $e_2[n]$ to indicate the error minimization property of the adaptive filter. Let us now consider a noise cancellation example.

Example 9.4 Noise cancellation

In this example, we want to cancel out the additive noise signal applied to a sinusoidal signal. To do so, we use a 32nd-order FIR filter (as the adaptive filter) and the LMS method. We provide the error ($e_2[n]$) in terms of iteration number in Figure 9.9 (a). As can be seen in this figure, the error gradually decreases to zero as the iteration number increases. We also provide the convergence of the FIR filter coefficient, β_0, in terms of iteration number in Figure 9.9 (b). This figure indicates that the selected coefficient converges to a value as the iteration number increases.

We provide the noise contaminated signal, $s[n]$, in Figure 9.10 (a). In fact, this is a sinusoidal signal with additive Gaussian noise. We provide the signal after noise cancellation in Figure 9.10 (b). As can be seen in this figure, the noise cancellation setup

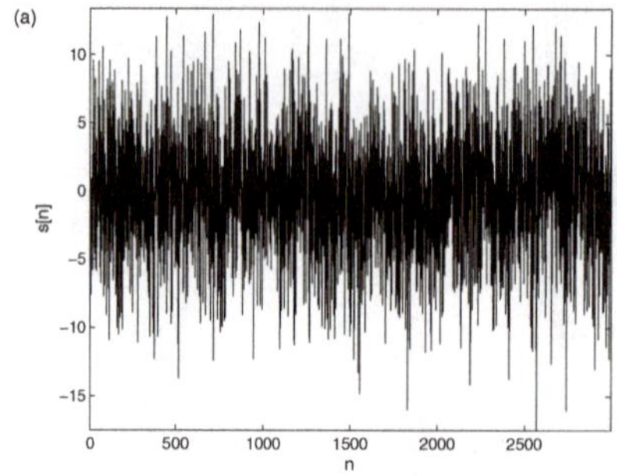

(a) Original signal with noise.

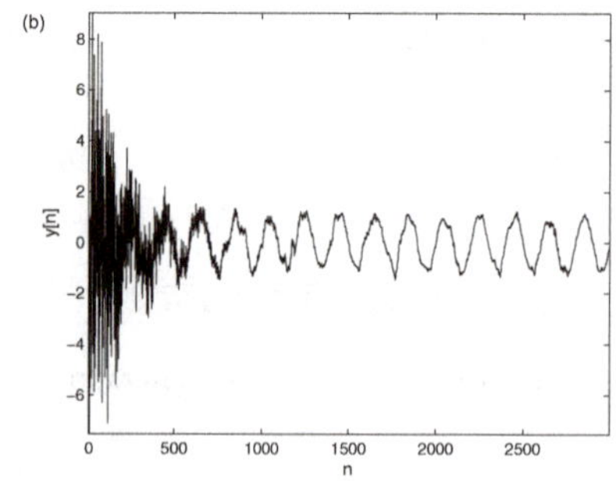

(b) Output signal after noise cancellation.

Figure 9.10 Noise cancellation example: part II.

works as expected. Hence, the sinusoidal signal can be observed at output after a certain number of iterations.

As with the previous examples, we provide the complete C code in Lab 9.12.3, where we solve the problem using both the LMS and normalized LMS methods.

9.7 Performance Analysis of an Adaptive Filter

The topics introduced in this section require detailed analysis and explanation using random signal processing with stochastic terms. We strongly suggest Haykin's book [1] for such a derivation. In this section, we only provide the useful results derived for practical applications. As mentioned in the previous section, it is not easy to analyze an adaptive filter. However, there are some benchmark measures that can be used for this purpose, and we introduce them in this section.

9.7.1 Stability

Step size directly affects the adaptation in the parameter adjustment step. A large step size means that adaptation uses crude parameter adjustment. Hence, setting an

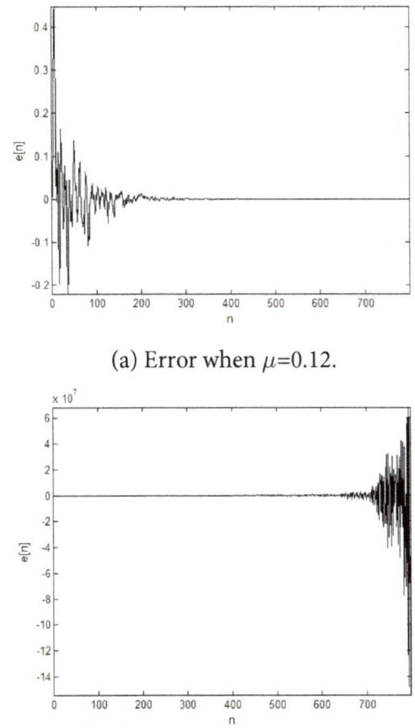

(a) Error when μ=0.12.

(b) Error when μ=0.24.

Figure 9.11 Stability of LMS filter for different step sizes.

inappropriate step size leads to false parameter values, which may lead to iterative instability. For LMS-based adaptive filters, the stability condition is ensured when the step size is

$$0 < \mu < \frac{2}{LP_x} \tag{9.13}$$

where L is the length of the adaptive filter and P_x is the power of the input signal [2]. Hence, in order to ensure stability, μ should be decreased when L is increased. We now provide an example on the stability of an LMS filter in terms of step size.

Example 9.5 LMS filter stability: effect of step size

Let us reconsider the system identification LMS filter in Example 9.1, where $L = 32$ and $P_x = 0.334$. The step size should be between 0 and 0.187. Given $\mu = 0.12$ and $\mu = 0.24$, the resulting error signals will be as in Figure 9.11 (a) and (b), respectively. As can be seen in these figures, the system starts to diverge when the step size exceeds 0.187.

Example 9.6 LMS filter stability: effect of filter length

Let us take the previous example and increase the filter length to 32 for $\mu = 0.12$. The resulting error signals for $L = 32$ and $L = 128$ are shown in Figures 9.12 (a) and (b),

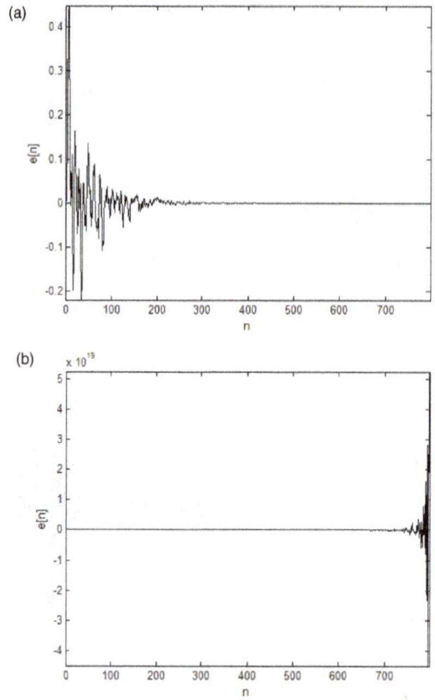

Figure 9.12 Stability of the LMS filter for different filter lengths. (a) Error when $L = 32$; (b) Error when $L = 128$.

respectively. As a reminder, the maximum value of μ is 0.05 for $L = 128$. Therefore, with this setup, the system becomes unstable for $\mu = 0.12$.

Example 9.7 Normalized LMS filter stability: effect of step size

Let us reconsider the system identification in Example 9.1 using the normalized LMS. Setting $\mu = 1.5$ and $\mu = 2.5$, the resulting error signals are shown in Figures 9.13 (a) and (b), respectively. As can be seen in these figures, the system starts to oscillate when the step size exceeds two.

9.7.2 Convergence Time

The convergence time of an adaptive filter can be formulated as

$$\tau \cong \frac{1}{\mu \lambda_{min}} \tag{9.14}$$

where λ_{min} is the smallest eigenvalue of the input correlation matrix [2]. By looking at this equation, we can say that increasing μ results in a lower convergence time. However, the stability condition (as discussed in Section 9.7.1) should also be taken into account at this step. Furthermore, the difference between the minimum and maximum eigenvalues

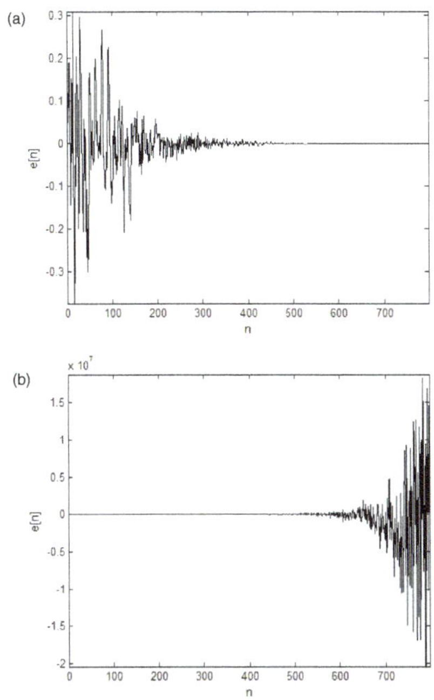

Figure 9.13 Stability of normalized LMS filter for different step sizes. (a) Error when μ=1.5; (b) Error when μ=2.5.

(eigenvalue spread) of the input correlation matrix has a direct effect on convergence time. For further information on these issues, please see [2]. Let us now reconsider the convergence time of the adaptive filter in Example 9.1.

Example 9.8 LMS filter: convergence time

Let us focus on the convergence time of the adaptive filter (using the LMS method) in Example 9.1. Given $\mu = 0.024$ and $\mu = 0.12$, the resulting error signals are shown in Figures 9.14 (a) and (b). As can be seen in these figures, the system converges faster when μ is increased. The same result can be observed when the normalized LMS method is used.

9.7.3 Input with Noise

In real life, there is always noise affecting the input signal. In order to analyze the effect of this noise, we will reconsider the system identification in Example 9.1.

Example 9.9 LMS filter: input with noise

We can add Gaussian white noise to the input signal in Example 9.1. The resulting error signal will be as in Figure 9.15. As can be seen in this figure, noise at the input directly affects the performance of the adaptive filter.

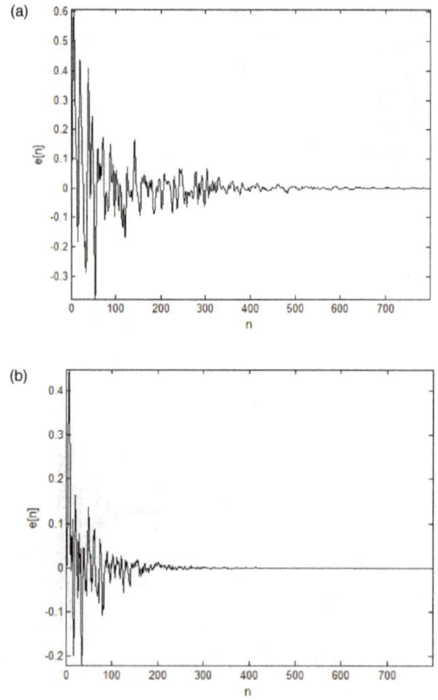

Figure 9.14 Comparison of convergence time for different step size values. (a) Error when μ=0.024; (b) Error when μ=0.12.

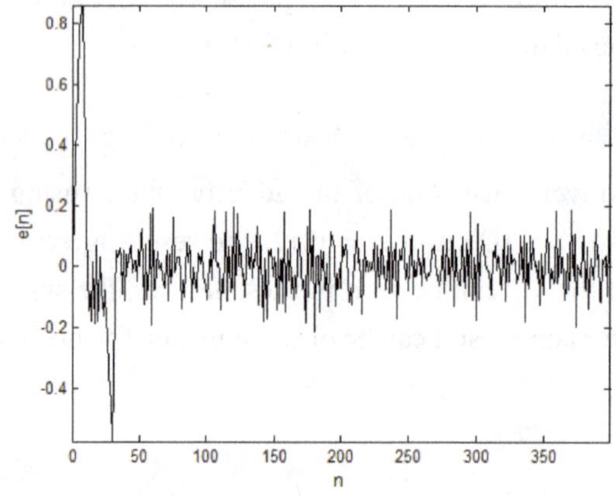

Figure 9.15 The error signal for noisy input.

Example 9.10 LMS filter: input with noise multiple trials

We can apply multiple trials and average the results from the previous example. Then, the error signal will be as shown in Figure 9.16. As can be seen in this figure, averaging smooths the result.

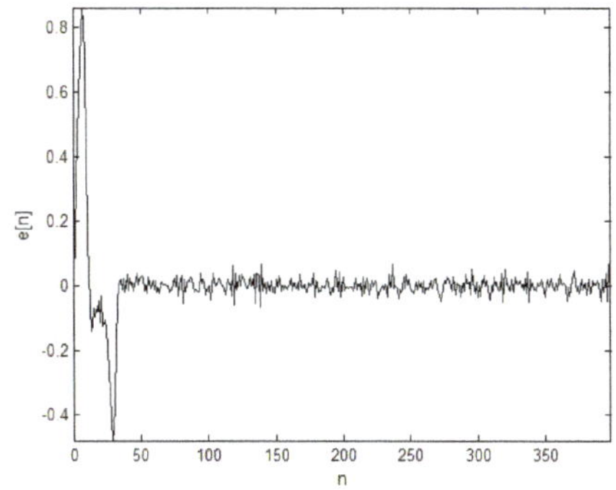

Figure 9.16 The error signal for the noisy input when multiple trials are realized and results are averaged.

9.8 Chapter Summary

Adaptive signal processing is a valuable approach for certain applications. In this chapter, we explored properties of adaptive filtering from a practical viewpoint. Further, we briefly introduced the theoretical background. Then, we applied adaptive filtering to four application areas. You will benefit from these as a starting point for your own adaptive filtering applications.

9.9 Further Reading

Adaptive filters and related concepts are considered in several books [1, 3, 4, 5, 6, 7]. From our perspective, the book by Haykin is the most comprehensive of these [1], and we suggest that you consult it for more details on the theory of adaptive filters.

9.10 Exercises

9.10.1 Use the LMS-based system identification method to estimate the system characteristics of the highpass filter obtained in Lab 15. Use the pseudorandom signal given in the `Online_Student_Resources\Lab9\system_identification` project as the input signal.

(a) Change the step size in the estimation step to observe its effects.

(b) Change the filter length in the estimation step to observe its effects.

9.10.2 Repeat Exercise 9.10.1 using the normalized LMS method. Comment on the differences between the LMS and normalized LMS methods for this problem.

9.10.3 Use the LMS-based system identification method to equalize the input signal $\sin(\pi n/48)$ when it is applied to the system with transfer function

$$H(z) = \frac{z^2 - 1.809524z + 0.818594}{z^2 - 1.3z + 0.4}$$

First, train the adaptive filter with pseudorandom input until absolute error is below 0.005 for 1000 sequential iterations. Then, stop the adaptation process and use the resulting adaptive filter to equalize the signal $\sin(\pi n/48)$. The `float float_rand(float min, float max)` function given in the `Online_Student_Resources\Lab9\pseudorandom_input` project can be used to create continuous pseudorandom data.

(a) Change the step size in the equalization step to observe its effects.

(b) Change the filter length in the estimation step to observe its effects.

9.10.4 Repeat Exercise 9.10.3 using the normalized LMS method. Comment on the differences between using the LMS and normalized LMS methods in this problem.

9.10.5 Say that the first 500 samples of a signal are given as $s[n] = 0.5\sin(\pi n/4) + 0.25\sin(\pi n/40) + 0.25\sin(\pi n/10)$.

(a) Use an LMS-based adaptive filter to predict the remaining 200 samples of $s[n]$.

(b) Change the step size in the prediction step to observe its effects.

(c) Change the filter length in the estimation step to observe its effects.

9.10.6 Repeat Exercise 9.10.5 using normalized LMS. Comment on the differences between using the LMS and normalized LMS methods in this problem.

9.10.7 Let there be a system with an input signal of $x[n] = \sin(\pi n/48)$. Let there also be a noise source in the system, which affects the input signal. This noise source is mathematically described as $\mu_0[n] = \sin(\pi n/4)$.

(a) Use the LMS-based noise cancellation method to obtain the original signal from the noisy input signal.

(b) Change the step size in the noise cancellation step to observe its effects.

(c) Change the filter length in the estimation step to observe its effects.

9.10.8 Repeat Exercise 9.10.7 using the normalized LMS method. Comment on the differences between using the LMS and normalized LMS methods in this problem.

9.11 References

[1] Haykin, S. (2014) *Adaptive Filter Theory*, Pearson, Fifth edn.
[2] Kuo, S.M., Lee, B.H., and Tian, W. (2013) *Real-Time Digital Signal Processing: Fundamentals, Implementations and Applications*, Wiley, Third edn.
[3] Widrow, B. and Stearns, S.D. (1985) *Adaptive Signal Processing*, Prentice-Hall.
[4] Sayed, A.H. (2008) *Adaptive Filters*, Wiley.
[5] Diniz, P.S.R. (2013) *Adaptive Filtering: Algorithms and Practical Implementation*, Springer, Fourth edn.
[6] Boroujeny, B.F. (2013) *Adaptive Filters: Theory and Applications Second Edition*, Wiley, Second edn.
[7] Poularikas, A.D. (2014) *Adaptive Filtering: Fundamentals of Least Mean Squares with MATLAB*, CRC Press.

9.12 Lab 9

9.12.1 Introduction

This lab is about the practical implementation of adaptive signal processing. We first consider the examples introduced in this chapter. We provide complete implementation details here. Then, we consider the discrete system considered in Section 6.10.4. We apply system identification tools to obtain an FIR filter with similar characteristics. We compare the original and identified systems in terms of their impulse and frequency responses.

9.12.2 CMSIS Implementation of the LMS and Normalized LMS methods

The CMSIS-DSP library has predefined functions for the LMS and normalized LMS methods. We provide these in Listing 9.1. Here, the first two functions are for LMS and the last two are for normalized LMS.

```

void arm_lms_init_f32(arm_lms_instance_f32 *S, uint16_t numTaps,
float32_t *pCoeffs, float32_t *pState, float32_t mu, uint32_t blockSize)
/*
S: points to an instance of the floating-point LMS filter structure.
numTaps: number of filter coefficients.
pCoeffs: points to the coefficient buffer.
pState: points to state buffer.
mu: step size that controls filter coefficient updates.
blockSize: number of samples to process.
*/

void arm_lms_f32(const arm_lms_instance_f32 *S, float32_t *pSrc, float32_t *pRef,
float32_t *pOut, float32_t *pErr, uint32_t blockSize)
/*
S: points to an instance of the floating-point LMS filter structure.
pSrc: points to the block of input data.
pRef: points to the block of reference data.
pOut: points to the block of output data.
pErr: points to the block of error data.
blockSize: number of samples to process.
*/


void arm_lms_norm_init_f32(arm_lms_norm_instance_f32 *S, uint16_t numTaps,
float32_t *pCoeffs, float32_t *pState, float32_t mu, uint32_t blockSize)
/*
S: points to an instance of the floating-point LMS filter structure.
numTaps: number of filter coefficients.
pCoeffs: points to coefficient buffer.
pState: points to state buffer.
mu: step size that controls filter coefficient updates.
blockSize: number of samples to process.
*/

void arm_lms_norm_f32(arm_lms_norm_instance_f32 *S, float32_t *pSrc,
float32_t *pRef, float32_t *pOut, float32_t *pErr, uint32_t blockSize)
/*
S: points to an instance of the floating-point normalized LMS filter structure.
pSrc: points to the block of input data.
pRef: points to the block of reference data.
```

```
pOut: points to the block of output data.
pErr: points to the block of error data.
blockSize: number of samples to process.
*/
```

Listing 9.1 CMSIS-DSP library LMS and normalized LMS functions.

The functions in Listing 9.1 can be applied to blocks of data. Therefore, they can process `blockSize` samples through the filter. However, parameter adjustment is done for each input sample. In these functions, the length of the state buffer exceeds the length of the coefficient array by `blockSize`-1 samples. This allows for circular buffering, which will be introduced in Chapter 11. State buffers are updated after each block of data is processed.

9.12.3 Adaptive Filter Applications

We considered four different adaptive filter applications in Section 9.6. We provided an example for each application area. In this section, we explore the implementation details of each of these examples.

System Identification

The first adaptive filter application introduced in the book was system identification. The complete project for this example can be found in `Online_Student_Resources\Lab9\system_identification`. Here, we provide LMS and normalized LMS realizations with and without the CMSIS-DSP library.

Task 9.1

In our first exercise, we become familiar with the implementation details of system identification. Run the LMS code (without using CMSIS-DSP library functions) in `Online_Student_Resources\Lab9\system_identification` and plot error with respect to iteration number. You should obtain similar results to those given in Figure 9.3. Here, the input is a pseudorandom signal generated by the project given in `Online_Student_Resources\Lab9\pseudorandom_input`.

Repeat this exercise with the LMS code using CMSIS-DSP library functions. Do you observe any changes? Under normal conditions, you should not.

We can apply system identification to the discrete-time filter obtained through the bilinear transformation with prewarping method in Section 6.10.4. Let us first obtain the impulse and frequency response of this system. These will be used for comparison purposes in the following sections.

Task 9.2

In order to obtain the impulse and frequency response of this system, we must first obtain the time domain difference equation of the IIR filter given in Section 6.4.3. Then, using the information given in Chapter 3, we must implement this difference equation in C to obtain the impulse response of the system. Based on this, plot the obtained impulse response of the system. Further, obtain the frequency response of the system using the information given in Chapter 4. Plot the obtained frequency response of the system. We will use these results in the following exercises.

Task 9.3

We can use the LMS method to identify the IIR filter in Task 9.2. The input for this system is the pseudorandom signal used in Task 9.1. The size of this input signal is $N = 800$. The power of the input signal can be calculated as

$$P_x = \frac{1}{800}\sum_{n=0}^{799} x^2[n] = 0.334 \tag{9.15}$$

We can select an FIR filter of length $L = 128$ as an adaptive filter. As a reminder, to ensure stability, the step size interval of the adaptive filter should satisfy

$$0 < \mu < \frac{2}{LP_x} \tag{9.16}$$

In this case, $0 < \mu < 0.047$.

Use direct implementation of the LMS method (without CMSIS-DSP library functions) to identify the FIR filter coefficients (used as an adaptive filter).

Obtain the output of the adaptive filter to the given input $x[n]$. Visually compare this result with the output of the original IIR filter. How did the adaptive filter perform?

Obtain the impulse and frequency responses of the adaptive filter. Visually compare the results with the ones obtained in Task 9.2. Did the adaptive filter mimic the actual system sufficiently?

Task 9.4

Although we can expect no difference, repeat Task 9.3 using CMSIS-DSP library functions. The idea here is to use the predefined functions for system identification. After running the code, did you observe any differences?

Task 9.5

This exercise involves using the normalized LMS method for system identification. Therefore, we can use the normalized LMS method instead of the LMS method used in Tasks 9.3 and 9.4. Comment on the differences observed when the normalized LMS method is used instead of the LMS method.

Equalization

The second adaptive filter application introduced in the book was equalization. The complete project for this example can be found in `Online_Student_Resources\Lab9\equalization`. Here, we provide the LMS and normalized LMS realizations with and without CMSIS-DSP library usage.

Task 9.6

We will become familiar with the implementation details of equalization in this exercise. Run the LMS code (without using CMSIS-DSP library functions) in the mentioned directory and plot error with respect to iteration number, the impulse response of the inverse of the unknown system, and the adaptive filter. You should obtain results that are comparable with Figure 9.5. Repeat the exercise with the LMS code using CMSIS-DSP library functions. Do you observe any changes? Under normal conditions, you should not.

Prediction

The third adaptive filter application introduced in the book was prediction. The complete project for this example can be found in `Online_Student_Resources\Lab9\prediction`. Here, we provide LMS and normalized LMS realizations with and without CMSIS-DSP library usage.

Task 9.7

In order to become familiar with the implementation details of prediction, run the LMS code (without using CMSIS-DSP library functions) in the mentioned directory. Plot error and output with respect to iteration number. You should obtain similar results to those in Figure 9.7.

Again, repeat this exercise with the LMS code using CMSIS-DSP library functions. Do you observe any changes? Under normal conditions, you should not.

Noise Cancellation

The fourth and final adaptive filter application introduced in the book was noise cancellation. The complete project for this example can be found in `Online_Student_`

`Resources\Lab9\noise_cancellation`. Here, we provide LMS and normalized LMS realizations with and without CMSIS-DSP library usage.

Task 9.8

In order to become familiar with the implementation details of noise cancellation, run the LMS code (without using CMSIS-DSP library functions) in the mentioned directory. Plot error (e_2) and β_0 with respect to iteration number, the original signal with noise, and the output signal after noise cancellation. You should obtain similar results to those in Figures 9.9 and 9.10.

Repeat this exercise with the LMS code using CMSIS-DSP library functions. Do you observe any changes? Under normal conditions, you should not.

9.12.4 Performance Analysis of an Adaptive Filter

We considered the performance analysis of an adaptive filter in Section 9.7. There, we provided examples to explain the performance analysis steps. We explore the implementation details of these examples in this section.

Task 9.9

Example 9.5 was about the stability of LMS-based adaptive filters with respect to step size. Implement this example. Plot error with respect to iteration number. Compare your results with Figure 9.11.

Task 9.10

We considered the stability of LMS-based adaptive filters with respect to filter length in Example 9.6. Implement this example. Plot error with respect to iteration number. Compare your results with Figure 9.12.

Task 9.11

In Example 9.7, as well as in Task 9.5 of this lab, we considered the stability of normalized LMS-based adaptive filters with respect to step size. Implement this example. Plot error with respect to iteration number. Compare your results with Figure 9.13.

Task 9.12

Example 9.8 was about the convergence time of LMS-based adaptive filters. Implement this example. Plot error curves for $\mu = 0.024$ and $\mu = 0.12$. Compare your results with Figure 9.14.

Task 9.13

We considered input with noise for LMS-based adaptive filters in Example 9.9. Implement this example. Use the C function `0.1*rand_normal(0.0,0.5)` as the noise signal. This function will generate random Gaussian white noise with mean 0 and variance 0.5. Plot the error curve. Compare your result with Figure 9.15.

Task 9.14

We considered input with noise for LMS-based adaptive filters when multiple trials are realized in Example 9.10. Implement this example for 15 trials. Use the C function `0.1*rand_normal(0.0,0.5)` as the noise signal. This function will generate random Gaussian white noise with mean 0 and variance 0.5. Plot the error curve. Compare your result with Figure 9.16.

10

Fixed-Point Implementation

Contents

10.1 Introduction

Rational numbers can be represented in a digital system as either fixed-point or floating-point. Up to now, we have used floating-point numbers in all DSP algorithms. These algorithms can also be implemented using fixed-point numbers. This may be advantageous in terms of memory usage and speed, but their use in DSP algorithms is not straightforward. Thus, we must not make the decision to use fixed-point numbers lightly. This chapter aims to clarify the characteristics of fixed-point numbers so they can be used effectively. We first introduce the floating-point and fixed-point number representations. Then, we consider conversion between the two. Finally, we explore MATLAB's fixed-point designer toolbox for filter design. All of these concepts can be understood better by implementing them. This is covered in the accompanying lab.

10.2 Floating-Point Number Representation

The number of bits assigned to the integer and decimal parts of a number are not predefined in the floating-point representation. Instead, the number of bits assigned differs for each number depending on its number of significant digits. Therefore, a much wider range of values can be handled using the floating-point number representation.

In the floating-point representation, a binary decimal number is defined as $N = (-1)^S \times 2^E \times F$, where S is the sign bit, E represents the exponent value, and F represents the parts of the number after the decimal separator. The floating-point number N is stored in memory as *SEF*. To represent the floating-point number as $N = (-1)^S \times 2^E \times F$, the number should be normalized so that the integer part has only one digit. The exponent will be biased by $2^{(e-1)} - 1$, where e is the number of bits used for E in a given format. Finally, a certain number of bits are assigned to S, E, and F depending on the standard format used for representation. Most digital systems use the IEEE 754 standard for floating-point numbers. This standard is summarized in Table 10.1.

The C language has `float` and `double` data types to represent floating-point numbers, which correspond to the `single` and `double` formats, respectively, in Table 10.1. The corresponding CMSIS-DSP library-type definitions are `float_32` and `float_64`, respectively.

Table 10.1 The IEEE 754 standard for floating-point number representation.

Format	Exponent bias	# bits for S	# bits for E	# bits for F	# total bits
half	15	1	5	10	16
single	127	1	8	23	32
double	1023	1	11	52	64
quad	16383	1	15	112	128

10.3 Fixed-Point Number Representation

In this representation, the number of bits assigned to the integer and decimal parts are fixed. A fixed-point number is represented as Qp.q, where Q is the sign, and p and q are the number's integer and decimal parts, respectively. We provide the signed bit formats used in the CMSIS-DSP library for the fixed-point representation in Table 10.2. As can be seen in this table, fixed-point numbers are assumed to be normalized between −1 and 1. We will explore this normalization in Section 10.4.

Table 10.2 Fixed-point signed number representations used in CMSIS-DSP library.

Format	Min.	Max.	Resolution	# p bits	# q bits	# total bits
Q1.7	−1	$1-2^{-7}$	2^{-7}	1	7	8
Q1.15	−1	$1-2^{-15}$	2^{-15}	1	15	16
Q1.31	−1	$1-2^{-31}$	2^{-31}	1	31	32
Q1.63	−1	$1-2^{-63}$	2^{-63}	1	63	64

The C language does not have basic data types to store the fixed-point numbers summarized in Table 10.2. Instead, we can represent a fixed-point number using the `signed char`, `short`, `int`, or `long long` basic C data types. The `stdint.h` header file redefines these for ease of use. The CMSIS-DSP library also has predefined fixed-point type definitions. We tabulate basic C, `stdint.h`, and CMSIS-DSP library definitions for signed fixed-point numbers in Table 10.3. In this table, we deliberately added shorthand notation for the numbers so that they are consistent with CMSIS-DSP library usage.

Table 10.3 Fixed-point signed number representations in C, the `stdint.h` header file, and CMSIS-DSP library.

Format	Shorthand	C data type	`stdint.h`	CMSIS
Q1.7	Q7	signed char	int8_t	q7_t
Q1.15	Q15	short	int16_t	q15_t
Q1.31	Q31	int	int32_t	q31_t
Q1.63	Q63	long long	int64_t	q63_t

Table 10.3 summarizes only fixed-point signed numbers. The `stdint.h` header file has unsigned fixed-point representations as well. We provide them in Table 10.4 along with the corresponding C data types.

Table 10.4 Fixed-point unsigned number representations in C and the `stdint.h` header file.

Format	C data type	stdint.h
UQ1.7	unsigned char	uint8_t
UQ1.15	unsigned short	uint16_t
UQ1.31	unsigned int	uint32_t
UQ1.63	unsigned long long	uint64_t

10.4 Conversion between Fixed-Point and Floating-Point Numbers

Conversion between fixed-point and floating-point numbers is a necessity during DSP operations. We discuss the conversion herein. We also consider these conversions using CMSIS-DSP library functions in Lab 10.11.2.

10.4.1 Floating-Point to Fixed-Point Number Conversion

We can convert a floating-point number to fixed-point form using the following equations.

$$Q7 = (\texttt{q7\_t})(\mathit{float} \times 2^{7}) \tag{10.1}$$

$$Q15 = (\texttt{q15\_t})(\mathit{float} \times 2^{15}) \tag{10.2}$$

$$Q31 = (\texttt{q31\_t})(\mathit{float} \times 2^{31}) \tag{10.3}$$

In these equations, a floating-point number (represented as `float`) is converted to the corresponding fixed-point number. To note here, the floating-point number should be scaled to be between -1 and 1 before applying any conversion. Next, it is multiplied by a coefficient that is a power of 2. Then, casting is applied, where (`q7_t`), (`q15_t`), and (`q31_t`) represent the casting operation. Hence, the scaled float value is converted to the corresponding fixed-point number. All of these conversion operations use saturating arithmetic, which means that results outside the selected fixed-point range will be saturated to the maximum allowable value.

Example 10.1 Floating-point to fixed-point number conversion

In this example, we analyze how a floating-point number can be represented in fixed-point form. Therefore, we convert the four numbers 0.2, 1/3, PI/4, and 8. Please note that PI is the constant corresponding to the irrational number π defined in the `arm_math.h` header file. We will convert these numbers to the Q1.7, Q1.15, and Q1.31 fixed-point formats.

Conversion operations are only applicable to numbers in the range $[-1,1]$. Hence, the number 8 must be scaled into this interval. Therefore, we divide it by 10 before converting it to fixed-point form. The results of the conversions are given in Table 10.5.

Table 10.5 Floating-point numbers and their representation in the fixed-point Q1.7, Q1.15, and Q1.31 formats.

Number	Q1.7 value	Q1.15 value	Q1.31 value
0.2	25	6553	429496736
1/3	42	10922	715827904
PI/4	100	25735	1686629760
8/10	102	26214	1717986944

10.4.2 Fixed-Point to Floating-Point Number Conversion

We can convert a fixed-point number to a floating-point one using the following equations.

$$\mathit{float} = (\texttt{float32\_t})(Q7 \times 2^{-7}) \tag{10.4}$$

$$\mathit{float} = (\texttt{float32\_t})(Q15 \times 2^{-15}) \tag{10.5}$$

$$\mathit{float} = (\texttt{float32\_t})(Q31 \times 2^{-31}) \tag{10.6}$$

Here, Q7, Q15, and Q31 represent the fixed-point number to be converted. We first multiply the number with the coefficient as the negated power of 2 as in Eqns. 10.4, 10.5, and 10.6. Then, we apply casting, where (`float32_t`) represents the casting operation.

Example 10.2 Fixed-point to floating-point number conversion

In this example, we analyze how a fixed-point number can be represented in floating-point form. We start with the representations in Table 10.5. We will convert these fixed-point numbers back to floating-point ones. The results of these conversions are given in Table 10.6.

We also provide the error obtained after these conversions in terms of percentages in Table 10.7. As can be seen in this table, the Q1.31 number format gives almost no error

Table 10.6 Fixed-point numbers and their floating-point equivalents.

Number	Q1.7 to float	Q1.15 to float	Q1.31 to float
0.2	0.1953125	0.199981689	0.200000003
1/3	0.328125	0.333312988	0.333333343
PI/4	0.78125	0.785369873	0.785398185
8	7.96875	7.99987793	8.0

when a floating-point number is converted to this representation and then back again. As expected, the worst-case scenario occurs for the Q1.7 number format.

Table 10.7 Percentage error values after conversion opearations.

Number	Q1.7 Error (%)	Q1.15 Error (%)	Q1.31 Error (%)
0.2	2.34375143	0.00915676355	0.0
1/3	1.56250298	0.00610649586	0.0
PI/4	0.52816331	0.00360482186	0.0
8	0.39062500	0.00152587891	0.0

10.4.3 Conversion between Different Fixed-Point Number Representations

Finally, we can apply conversion between different fixed-point number representations. The equations to carry out these conversions are given below.

$$Q15 = (\texttt{q15\_t})(Q7 << 8) \tag{10.7}$$

$$Q31 = (\texttt{q31\_t})(Q7 << 24) \tag{10.8}$$

$$Q7 = (\texttt{q7\_t})(Q15 >> 8) \tag{10.9}$$

$$Q31 = (\texttt{q31\_t})(Q15 << 16) \tag{10.10}$$

$$Q7 = (\texttt{q7\_t})(Q31 >> 24) \tag{10.11}$$

$$Q15 = (\texttt{q15\_t})(Q31 >> 16) \tag{10.12}$$

where (`q7_t`), (`q15_t`), and (`q31_t`) represent casting operations. Here, it is important to note that binary multiplication and division operations are carried out by shifting operations, which are represented by << and >>, respectively. Then, casting leads to the final fixed-point number representation.

10.5 Fixed-Point Operations

Up to now, we have introduced several digital signal processing methods, including convolution, FFT, downsampling and interpolation, filter design, and adaptive filtering. Without mentioning it, we have used floating-point representation for all methods via the `float` data type. All these methods can be implemented using fixed-point number representations as well. Here, CMSIS-DSP library functions will be very useful. We cover the full implementation details of fixed-point operations in Lab 10.11. We only explore filter design using fixed-point representations in this chapter.

10.6 MATLAB Fixed-Point Designer Toolbox

Designing a fixed-point filter requires extensive testing and adjustment. Therefore, it is not easy to design a filter with fixed-point coefficients. In Lab 8.11, we construct an equalizer using lowpass, bandpass, and highpass filters. In this section, we will focus on the fixed-point version of the designed lowpass filter. To do so, we will extensively use MATLAB's fixed-point designer toolbox, which can convert floating-point filter coefficients to fixed-point ones. It also allows the user to adjust the word length for filter coefficients. The user can check any possible overflow based on the given input after conversion. The user can also obtain filter coefficients in integer form.

10.6.1 Filter Coefficient Conversion

MATLAB's fixed-point designer toolbox can be used within FDATool. Let us see a simple example on its usage. The FIR filter to be converted is given in Figure 10.1.

In order to design a fixed-point filter with the same specifications as a floating-point one, click on the *Set quantization parameters* button, which is the third button from the top at the bottom left of the FDATool GUI. Then, change "Filter arithmetic" to "Fixed-point." The three windows to adjust fixed-point configurations after the filter arithmetic is changed to fixed-point are "Coefficients," "Input/Output," and "Filter Internals." The user can change the numerator word length in the "Coefficients" window as shown in Figure 10.2. The numerator fraction length is automatically adjusted for the best precision. If the user wants to adjust the numerator fraction length manually, *Filter precision* must be changed from "Full" to "Specify all." The user can also enable dynamic range analysis by clicking on the *Scale the numerator coefficient to fully utilize the entire dynamic range* check-box. Dynamic range analysis provides better scaled coefficients for the input data range.

The user can change the input word and fraction lengths in the "Input/Output" window (shown in Figure 10.3) by changing *Filter precision* from "Full" to "Specify all."

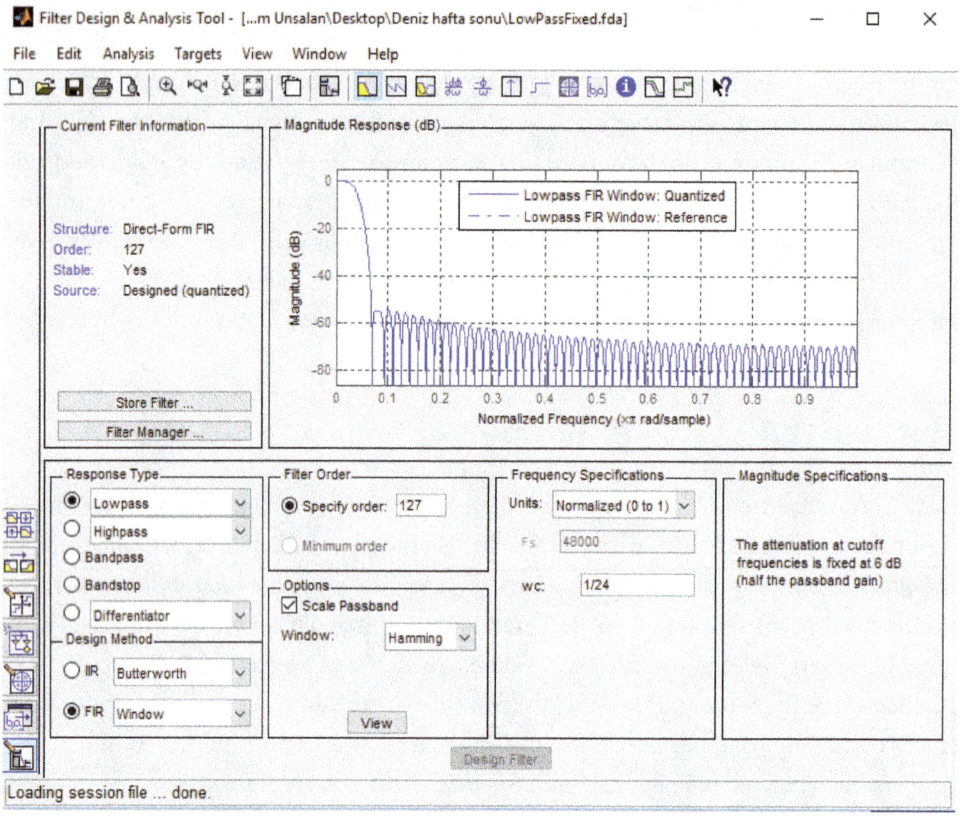

Figure 10.1 FDATool design of the FIR filter to be converted.

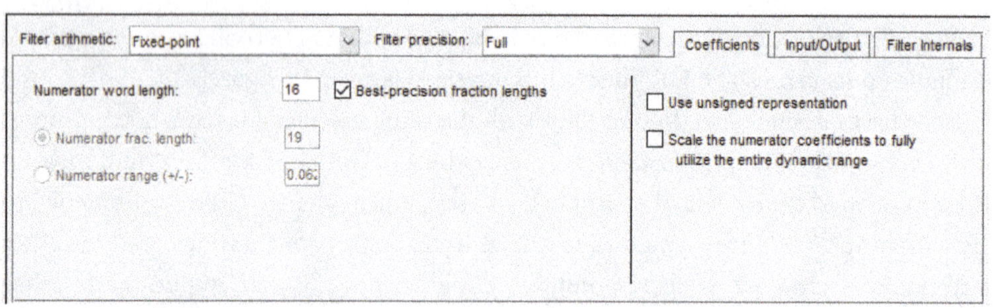

Figure 10.2 Coefficients window.

The user can adjust the lengths of the product word, fraction, accumulator word, and accumulator fraction. The rounding and overflow modes can also be adjusted. This can be done in the "Filter Internals" window (shown in Figure 10.4) by changing *Filter precision* from "Full" to "Specify all."

The user can generate the quantized coefficients by clicking on the *Apply* button after making all these configuration changes.

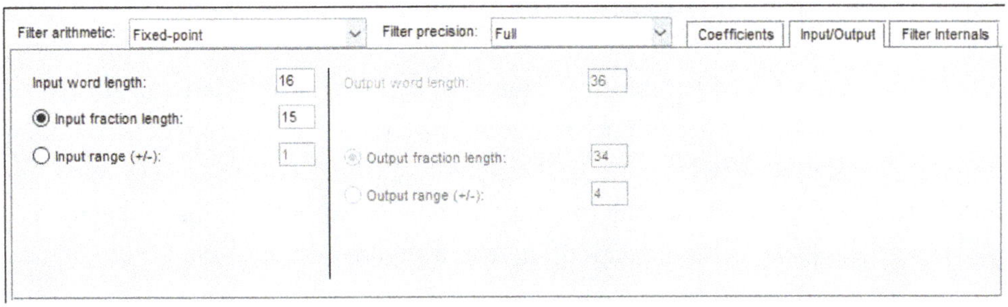

Figure 10.3 Input/Output window.

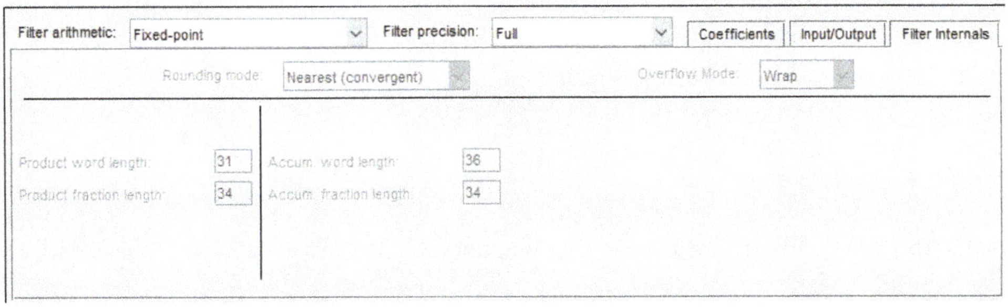

Figure 10.4 Filter Internals window.

10.6.2 Filter Coefficient Conversion Problems

You should be extremely cautious while obtaining the fixed-point filter coefficients from a floating-point design. Although the designed filter in floating-point form is stable, it can become unstable after converting its coefficients to fixed-point form. One specific example is as follows.

Example 10.3 Elliptical IIR filter: fixed-point conversion

Let us design an elliptical IIR lowpass filter. Let the sampling frequency of the filter be 48000 Hz. Let the pass-band frequency be 480 Hz. Although the designed filter in floating-point form is stable, it becomes unstable when filter coefficients are converted to fixed-point Q1.7 format. This is the worst-case scenario that can happen.

Elliptic filters are known to cause such problems when implemented in fixed-point form [1]. The designed filter with floating-point coefficients may still be stable after converting its coefficients to fixed-point. However, the filter response may deviate from the original floating-point form. Therefore, you should check the characteristics of the filter after converting its coefficients to fixed-point form.

10.7 Chapter Summary

Fixed-point numbers can be used in all the DSP algorithms considered in this book. This chapter covered the techniques required to perform operations with fixed-point numbers. We started with the ways floating-point and fixed-point numbers are represented. Then, we revisited the concepts that we introduced in the earlier chapters. We also provided implementation details of these in the accompanying lab.

10.8 Further Reading

Fixed-point operations heavily depend on implementation. For more in-depth coverage of this topic, you can consult [1, 2, 3, 4]. We also suggest that you revisit the CMSIS-DSP library user manual in relation to this issue.

10.9 Exercises

10.9.1 The aim in this exercise is observing the precision loss when a float value is converted to fixed-point form.

(a) Write some basic C code declaring a float variable `p` to store π in floating-point form.

(b) Convert `p` to the fixed-point Q7, Q15, and Q31 forms. Then, convert the results back to floating-point form. What are the differences between the original and converted forms?

10.9.2 A discrete-time system is represented as $y[n] = y[n-1] + \cos(x[n+1])$. The system is initially at rest. What will be the output of this system given the below input signals? Plot the output signals for $n = 0, \cdots, 20$. Use the fixed-point Q7 number format while solving the problem.

(a) $x[n] = \delta[n]$.

(b) $x[n] = u[n]$.

10.9.3 Determine the inverse DTFT of the below signals. Use the fixed-point Q15 number format while solving the problem.

(a) $X(e^{j\omega}) = \sin(4\omega\pi/8)$.

(b) $X(e^{j\omega}) = \cos(-2\omega + 1/2)$.

10.9.4 We want to reconstruct the signal $x(t) = \cos(\pi t/10)$ with sampling frequency $f_s = 10$ Hz. Plot the reconstructed signal for the below cases. Use the fixed-point Q31 number format while solving the problem.

(a) An ideal LPF with an appropriate cut-off frequency.

(b) A ZOH circuit with an appropriate T_d value.

10.9.5 An FIR filter has impulse response $h[n] = \sum_{k=0}^{5}(k+1)\delta[n-k]$. Find the lattice form of this filter. Use the fixed-point Q7 number format while solving the problem.

10.9.6 A discrete-time signal is represented as $x[n] = \sum_{k=0}^{5} x_k[n]$, where $x_k[n] = \cos(k\pi n/8)/(k+1)$. Design an ideal filter to extract the first two components ($x_0[n]$ and $x_1[n]$) of $x[n]$. Use the fixed-point Q15 number format while solving the problem.

10.9.7 Repeat Exercise 9.10.1 using the fixed-point Q31 number format.

10.9.8 Design a filter such that it will be stable in floating-point representation. Will it become unstable when represented in fixed-point Q7 number format?

10.10 References

[1] Welch, T.B., Wright, C.H.G., and Morrow, M.G. (2011) *Real-Time Digital Signal Processing from MATLAB to C with the TMS320C6x DSPs*, CRC Press, Second edn.

[2] Singh, A. and Srinivasan, S. (2003) *Digital Signal Processing Implementations: Using DSP Microprocessors*, Cengage Learning.

[3] Kehtarnavaz, N. (2004) *Real-Time Digital Signal Processing Based on the TMS320C600*, Elsevier.

[4] Kuo, S.M., Lee, B.H., and Tian, W. (2013) *Real-Time Digital Signal Processing: Fundamentals, Implementations and Applications*, Wiley, Third edn.

10.11 Lab 10

10.11.1 Introduction

Fixed-point implementation is advantageous when the microcontroller at hand does not have a floating-point unit. Almost all CMSIS-DSP library functions have a fixed-point form. In this lab, we first consider these functions. Then, we redesign the three-band equalizer design from Chapter 8 using fixed-point functions.

10.11.2 Fixed-Point and Floating-Point Number Conversions

The CMSIS-DSP library has built-in functions for conversion between fixed-point and floating-point numbers. We consider them in this section. We explored conversion formulas for this purpose in Section 10.5.

Floating-Point to Fixed-Point Number Conversion

The CMSIS-DSP library functions for floating-point to fixed-point number conversion are given in Listing 10.1. Here, `pSrc` points to the floating-point input vector, `pDst` points to the fixed-point output vector (in Q1.7, Q1.15, or Q1.31 format), and `blockSize` is the length of the input vector.

```
void arm_float_to_q7(float32_t *pSrc, q7_t *pDst, uint32_t blockSize)
/*
pSrc: points to the floating-point input vector
pDst: points to the Q7 output vector
```

```
blockSize: length of the input vector
*/

void arm_float_to_q15(float32_t *pSrc, q15_t *pDst, uint32_t blockSize)
/*
pSrc: points to the floating-point input vector
pDst: points to the Q15 output vector
blockSize: length of the input vector
*/

void arm_float_to_q31(float32_t *pSrc, q31_t *pDst, uint32_t blockSize)
/*
pSrc: points to the floating-point input vector
pDst: points to the Q31 output vector
blockSize: length of the input vector
*/
```

Listing 10.1 CMSIS-DSP library functions for floating-point to fixed-point number conversion.

Fixed-Point to Floating-Point Number Conversion

The CMSIS-DSP library fixed-point to floating-point number conversion functions are given in Listing 10.2. Here, `pSrc` points to the fixed-point input vector (in Q1.7, Q1.15, or Q1.31 format), `pDst` points to the floating-point output vector, and `blockSize` is the length of the input vector.

```
void arm_q7_to_float(q7_t *pSrc, float32_t *pDst, uint32_t blockSize)
/*
pSrc: points to the Q7 input vector
pDst: points to the floating-point output vector
blockSize: length of the input vector
*/

void arm_q15_to_float(q15_t *pSrc, float32_t *pDst, uint32_t blockSize)
/*
pSrc: points to the Q15 input vector
pDst: points to the floating-point output vector
blockSize: length of the input vector
*/

void arm_q31_to_float (q31_t *pSrc, float32_t *pDst, uint32_t blockSize)
/*
pSrc: points to the Q31 input vector
pDst: points to the floating-point output vector
blockSize: length of the input vector
*/
```

Listing 10.2 CMSIS-DSP library fixed-point to floating-point number conversion functions.

Fixed-Point to Fixed-Point Number Conversion

The CMSIS-DSP library also has conversion functions between fixed-point representations. These are given in Listing 10.3. As in previous sections, `pSrc` points to the fixed-point input vector (in Q1.7, Q1.15, or Q1.31 format), `pDst` points to the fixed-point output vector (in Q1.7, Q1.15, or Q1.31 format), and `blockSize` is the length of the input vector.

```
void arm_q7_to_q15(q7_t *pSrc, q15_t *pDst, uint32_t blockSize)
/*
pSrc: points to the Q7 input vector
pDst: points to the Q15 output vector
blockSize: length of the input vector
*/

void arm_q7_to_q31(q7_t *pSrc, q31_t *pDst, uint32_t blockSize)
/*
pSrc: points to the Q7 input vector
pDst: points to the Q31 output vector
blockSize: length of the input vector
*/

void arm_q15_to_q7(q15_t *pSrc, q7_t *pDst, uint32_t blockSize)
/*
pSrc: points to the Q15 input vector
pDst: points to the Q7 output vector
blockSize: length of the input vector
*/

void arm_q15_to_q31(q15_t *pSrc, q31_t *pDst, uint32_t blockSize)
/*
pSrc: points to the Q15 input vector
pDst: points to the Q31 output vector
blockSize: length of the input vector
*/

void arm_q31_to_q7(q31_t *pSrc, q7_t *pDst, uint32_t blockSize)
/*
pSrc: points to the Q31 input vector
pDst: points to the Q7 output vector
blockSize: length of the input vector
*/

void arm_q31_to_q15(q31_t *pSrc, q15_t *pDst, uint32_t blockSize)
/*
pSrc: points to the Q31 input vector
pDst: points to the Q15 output vector
blockSize: length of the input vector
*/
```

Listing 10.3 CMSIS-DSP library fixed-point number conversion functions.

10.11.3 Fixed-Point Convolution, Correlation, and FIR Filtering

We can implement fixed-point convolution, correlation, and FIR filtering operations using predefined CMSIS-DSP library functions. We will now review each function set in detail.

Convolution Operation

The fixed-point convolution operation functions in the CMSIS-DSP library are given in Listing 10.4. Here, there are four function categories. The first category contains fast convolution operations that are applicable to Q1.15 and Q1.31 fixed-point numbers. These can be used to obtain the results in a fast but less precise way. The functions in this category are prone to intermediate overflow error. To avoid this, the CMSIS-DSP library suggests scaling inputs by $\log_2$ of the minimum signal length to be processed.

The second category contains a fast convolution operation function applicable to only Q1.15 fixed-point numbers. This function provides the result faster but uses more memory. The third category contains convolution operation functions that work in an optimal way. They are applicable to Q1.7 and Q1.15 fixed-point numbers. There is no intermediate overflow risk with these functions. Functions in this group are potentially faster compared to the regular fixed-point convolution operation. The fourth category contains regular convolution operation functions that are applicable to Q1.7, Q1.15, and Q1.31 fixed-point numbers. The function that processes Q1.31 fixed-point numbers is prone to intermediate overflow error. Therefore, inputs for this function should be scaled down by $\log_2$ of the minimum signal length to be processed.

```
void arm_conv_fast_q15(q15_t *pSrcA, uint32_t srcALen, q15_t *pSrcB,
uint32_t srcBLen, q15_t *pDst)
/*
pSrcA: points to the first input sequence.
srcALen: length of the first input sequence.
pSrcB: points to the second input sequence.
srcBLen: length of the second input sequence.
pDst: points to the location where the output result is written.
     Length srcALen+srcBLen-1.
*/

void arm_conv_fast_q31(q31_t *pSrcA, uint32_t srcALen, q31_t *pSrcB,
uint32_t srcBLen, q31_t *pDst)
/*
pSrcA: points to the first input sequence.
srcALen: length of the first input sequence.
pSrcB: points to the second input sequence.
srcBLen: length of the second input sequence.
pDst: points to the location where the output result is written.
     Length srcALen+srcBLen-1.
*/

void arm_conv_fast_opt_q15(q15_t *pSrcA, uint32_t srcALen, q15_t *pSrcB,
uint32_t srcBLen, q15_t *pDst, q15_t *pScratch1, q15_t *pScratch2)
/*
pSrcA: points to the first input sequence.
srcALen: length of the first input sequence.
pSrcB: points to the second input sequence.
srcBLen: length of the second input sequence.
pDst: points to the location where the output result is written.
     Length srcALen+srcBLen-1.
pScratch1: points to scratch buffer of size max(srcALen, srcBLen)
     + 2*min(srcALen, srcBLen) - 2.
pScratch2: points to scratch buffer of size min(srcALen, srcBLen).
*/


void arm_conv_opt_q7(q7_t *pSrcA, uint32_t srcALen, q7_t *pSrcB,
uint32_t srcBLen, q7_t *pDst, q15_t *pScratch1, q15_t *pScratch2)
/*
pSrcA: points to the first input sequence.
srcALen: length of the first input sequence.
pSrcB: points to the second input sequence.
srcBLen: length of the second input sequence.
pDst: points to the location where the output result is written.
     Length srcALen+srcBLen-1.
pScratch1: points to scratch buffer(of type q15_t) of size max(srcALen, srcBLen)
     + 2*min(srcALen, srcBLen) - 2.
pScratch2: points to scratch buffer (of type q15_t) of size min(srcALen, srcBLen).
*/

void arm_conv_opt_q15(q15_t *pSrcA, uint32_t srcALen, q15_t *pSrcB,
```

```
uint32_t srcBLen, q15_t *pDst, q15_t *pScratch1, q15_t *pScratch2)
/*
pSrcA: points to the first input sequence.
srcALen: length of the first input sequence.
pSrcB: points to the second input sequence.
srcBLen: length of the second input sequence.
pDst: points to the location where the output result is written.
     Length srcALen+srcBLen-1.
pScratch1: points to scratch buffer of size max(srcALen, srcBLen)
     + 2*min(srcALen, srcBLen) - 2.
pScratch2: points to scratch buffer of size min(srcALen, srcBLen).
*/

void arm_conv_q7(q7_t *pSrcA, uint32_t srcALen, q7_t *pSrcB, uint32_t srcBLen,
q7_t *pDst)
/*
pSrcA: points to the first input sequence.
srcALen: length of the first input sequence.
pSrcB: points to the second input sequence.
srcBLen: length of the second input sequence.
pDst: points to the location where the output result is written.
     Length srcALen+srcBLen-1.
*/

void arm_conv_q15(q15_t *pSrcA, uint32_t srcALen, q15_t *pSrcB, uint32_t srcBLen,
q15_t *pDst)
/*
pSrcA: points to the first input sequence.
srcALen: length of the first input sequence.
pSrcB: points to the second input sequence.
srcBLen: length of the second input sequence.
pDst: points to the location where the output result is written.
Length srcALen+srcBLen-1.
*/

void arm_conv_q31(q31_t *pSrcA, uint32_t srcALen, q31_t *pSrcB, uint32_t srcBLen,
q31_t *pDst)
/*
pSrcA: points to the first input sequence.
srcALen: length of the first input sequence.
pSrcB: points to the second input sequence.
srcBLen: length of the second input sequence.
pDst: points to the location where the output result is written.
Length srcALen+srcBLen-1.
*/
```

Listing 10.4 Fixed-point convolution operation functions.

We provide an example on the usage of fixed-point convolution operation functions in `Online_Student_Resources\Lab10\fixed_point_convolution`. Here, we repeat the convolution operation in Task 2.17. Note that the input signal is scaled between 0 and 3.3 in fixed-point form.

Task 10.1

To observe how the convolution operation can be done in fixed-point form, run the C code in the mentioned folder. Plot the obtained signal in different fixed-point formats. We should compare these with the result of the convolution operation done in floating-point form. Summarize your findings with a focus on the effect that fixed-point number representation has on operations.

Correlation Operation

The fixed-point correlation operation functions in the CMSIS-DSP library are given in Listing 10.5. Here, there are four function categories. The first category contains fast correlation operations that are applicable to Q1.15 and Q1.31 fixed-point numbers. These can be used to obtain fast correlation results in a less precise way. These functions are prone to intermediate overflow error. To avoid this, the CMSIS-DSP library suggests scaling one of the inputs by the minimum signal length to be processed. The second category contains a fast correlation operation function that works in an optimal way. This function (applicable to only fixed-point Q1.15 numbers) provides the result faster but uses more memory. The third category contains correlation operation functions that work in an optimal way. The functions in this category are applicable to Q1.7 and Q1.15 fixed-point numbers. There is no intermediate overflow risk here. These functions are potentially faster than regular fixed-point correlation operation functions. The fourth category contains regular correlation operation functions that are applicable to Q1.7, Q1.15, and Q1.31 fixed-point numbers. The function that processes Q1.31 fixed-point numbers has intermediate overflow risk. Therefore, one of the inputs for this function should be scaled down by the minimum signal length to be processed.

```
void arm_correlate_fast_q15(q15_t *pSrcA, uint32_t srcALen, q15_t *pSrcB,
uint32_t srcBLen, q15_t *pDst)
/*
pSrcA: points to the first input sequence.
srcALen: length of the first input sequence.
pSrcB: points to the second input sequence.
srcBLen: length of the second input sequence.
pDst: points to the location where the output result is written.
    Length 2 * max(srcALen, srcBLen) - 1.
*/

void arm_correlate_fast_q31(q31_t *pSrcA, uint32_t srcALen, q31_t *pSrcB,
uint32_t srcBLen, q31_t *pDst)
/*
pSrcA: points to the first input sequence.
srcALen: length of the first input sequence.
pSrcB: points to the second input sequence.
srcBLen: length of the second input sequence.
pDst: points to the location where the output result is written.
Length 2 * max(srcALen, srcBLen) - 1.
*/

void arm_correlate_fast_opt_q15(q15_t *pSrcA, uint32_t srcALen, q15_t *pSrcB,
uint32_t srcBLen, q15_t *pDst, q15_t *pScratch)
/*
pSrcA: points to the first input sequence.
srcALen: length of the first input sequence.
pSrcB: points to the second input sequence.
srcBLen: length of the second input sequence.
pDst: points to the location where the output result is written.
Length 2 * max(srcALen, srcBLen) - 1.
pScratch:points to scratch buffer of size max(srcALen, srcBLen)
     + 2*min(srcALen, srcBLen) - 2.
*/

void arm_correlate_opt_q7(q7_t *pSrcA, uint32_t srcALen, q7_t *pSrcB,
uint32_t srcBLen, q7_t *pDst, q15_t *pScratch1, q15_t *pScratch2)
/*
```

```
pSrcA: points to the first input sequence.
srcALen: length of the first input sequence.
pSrcB: points to the second input sequence.
srcBLen: length of the second input sequence.
pDst: points to the location where the output result is written.
Length 2 * max(srcALen, srcBLen) - 1.
pScratch1: points to scratch buffer(of type q15_t) of size max(srcALen, srcBLen)
      + 2*min(srcALen, srcBLen) - 2.
pScratch2: points to scratch buffer (of type q15_t) of size min(srcALen, srcBLen).
*/

void arm_correlate_opt_q15(q15_t *pSrcA, uint32_t srcALen, q15_t *pSrcB,
uint32_t srcBLen, q15_t *pDst, q15_t *pScratch)
/*
pSrcA: points to the first input sequence.
srcALen: length of the first input sequence.
pSrcB: points to the second input sequence.
srcBLen: length of the second input sequence.
pDst: points to the location where the output result is written.
      Length 2 * max(srcALen, srcBLen) - 1.
pScratch: points to scratch buffer of size max(srcALen, srcBLen)
      + 2*min(srcALen, srcBLen) - 2.
*/

void arm_correlate_q7(q7_t *pSrcA, uint32_t srcALen, q7_t *pSrcB,
uint32_t srcBLen, q7_t *pDst)
/*
pSrcA: points to the first input sequence.
srcALen: length of the first input sequence.
pSrcB: points to the second input sequence.
srcBLen: length of the second input sequence.
pDst: points to the location where the output result is written.
Length 2 * max(srcALen, srcBLen) - 1.
*/

void arm_correlate_q15(q15_t *pSrcA, uint32_t srcALen, q15_t *pSrcB,
uint32_t srcBLen, q15_t *pDst)
/*
pSrcA: points to the first input sequence.
srcALen: length of the first input sequence.
pSrcB: points to the second input sequence.
srcBLen: length of the second input sequence.
pDst: points to the location where the output result is written.
Length 2 * max(srcALen, srcBLen) - 1.
*/

void arm_correlate_q31(q31_t *pSrcA, uint32_t srcALen, q31_t *pSrcB,
uint32_t srcBLen, q31_t *pDst)
/*
pSrcA: points to the first input sequence.
srcALen: length of the first input sequence.
pSrcB: points to the second input sequence.
srcBLen: length of the second input sequence.
pDst: points to the location where the output result is written.
      Length 2 * max(srcALen, srcBLen) - 1.
*/
```

Listing 10.5 Fixed-point correlation operation functions.

We provide an example on the usage of fixed-point correlation operation functions in `Online_Student_Resources\Lab10\fixed_point_correlation`. Here, we use the correlation operation to calculate the delay value between the original and shifted signal.

Task 10.2

Run the correlation code in the mentioned folder to observe how the correlation operation can be done in fixed-point form. Plot the correlation result in different fixed-point formats. Compare your results with the results of correlation operations using floating-point form. Will there be a difference between the delay values calculated using different fixed-point formats?

FIR Filtering Operation

As with the convolution and correlation operations, the CMSIS-DSP library also has fixed-point FIR filtering functions. These are given in Listing 10.6. Here, there are two function categories. The first category contains fast FIR filtering operations that are applicable to Q1.15 and Q1.31 fixed-point numbers. These functions can be used to obtain fast FIR filtering results in a less precise way. However, they are prone to intermediate overflow error. To avoid this, the CMSIS-DSP library suggests scaling the inputs by $\log_2$ of the minimum signal length to be processed. The second category contains regular FIR filtering operations that are applicable to Q1.7, Q1.15, and Q1.31 fixed-point numbers. The function that processes Q1.31 fixed-point numbers has intermediate overflow risk. Therefore, one of the inputs for this function should be scaled down by $\log_2$ of the minimum signal length to be processed.

```
void arm_fir_fast_q15(const arm_fir_instance_q15 *S, q15_t *pSrc, q15_t *pDst,
uint32_tblockSize)
/*
S: points to an instance of the Q15 FIR filter structure.
pSrc: points to the block of input data.
pDst: points to the block of output data.
blockSize: number of samples to process per call.
*/

void arm_fir_fast_q31(const arm_fir_instance_q31 *S, q31_t *pSrc, q31_t *pDst,
uint32_t blockSize)
/*
S: points to an instance of the Q31 structure.
pSrc: points to the block of input data.
pDst: points to the block output data.
blockSize: number of samples to process per call.
*/

void arm_fir_init_q7(arm_fir_instance_q7 *S, uint16_t numTaps, q7_t *pCoeffs,
q7_t *pState, uint32_t blockSize)
/*
S: points to an instance of the Q7 FIR filter structure.
numTaps: Number of filter coefficients in the filter.
pCoeffs: points to the filter coefficients buffer.
pState: points to the state buffer.
blockSize: number of samples that are processed per call.
*/

void arm_fir_q7(const arm_fir_instance_q7 *S, q7_t *pSrc, q7_t *pDst,
uint32_t blockSize)
/*
S: points to an instance of the Q7 FIR filter structure.
pSrc: points to the block of input data.
pDst: points to the block of output data.
```

```
blockSize: number of samples to process per call.
*/

arm_status arm_fir_init_q15(arm_fir_instance_q15 *S, uint16_t numTaps,
q15_t *pCoeffs, q15_t *pState, uint32_t blockSize)
/*
S: points to an instance of the Q15 FIR filter structure.
numTaps: Number of filter coefficients in the filter. Must be even and
      greater than or equal to 4.
pCoeffs: points to the filter coefficients buffer.
pState: points to the state buffer.
blockSize: is number of samples processed per call.
*/

void arm_fir_q15(const arm_fir_instance_q15 *S, q15_t *pSrc, q15_t *pDst,
uint32_t blockSize)
/*
S: points to an instance of the Q15 FIR structure.
pSrc: points to the block of input data.
pDst: points to the block of output data.
blockSize: number of samples to process per call.
*/

void arm_fir_init_q31(arm_fir_instance_q31 *S, uint16_t numTaps, q31_t *pCoeffs,
q31_t *pState, uint32_t blockSize)
/*
S: points to an instance of the Q31 FIR filter structure.
numTaps: Number of filter coefficients in the filter.
pCoeffs: points to the filter coefficients buffer.
pState: points to the state buffer.
blockSize: number of samples that are processed per call.
*/

void arm_fir_q31(const arm_fir_instance_q31 *S, q31_t *pSrc, q31_t *pDst,
uint32_t blockSize)
/*
S: points to an instance of the Q31 FIR filter structure.
pSrc: points to the block of input data.
pDst: points to the block of output data.
blockSize: number of samples to process per call.
*/
```

Listing 10.6 Fixed-point FIR filtering functions.

We provide an example on the usage of fixed-point FIR filtering functions in `Online_Student_Resources\Lab10\fixed_point_FIR_filtering`. Again, we repeat Task 2.17 in the below exercise. As in the fixed-point convolution operation, we scaled the input signal between 0 and 3.3.

Task 10.3

The abovementioned project is a good starting point for understanding how to use the fixed-point FIR filtering functions. Run the FIR filtering code and plot the obtained signal in different fixed-point formats. Compare these results. Discuss your comparison results and give possible reasons for the differences you found.

10.11.4 Fixed-Point FFT Calculations

The CMSIS-DSP library has fixed-point complex FFT functions for Q1.15 and Q1.31 numbers, which are given in Listing 10.7. Here, the function structures for the fixed-point Q1.15 and Q1.31 numbers are the same. Therefore, we only discuss the FFT functions for fixed-point Q1.31 numbers.

```
void arm_cfft_q15(const arm_cfft_instance_q15 *S, q15_t *p1, uint8_t ifftFlag,
uint8_t bitReverseFlag)
/*
S: points to an instance of the Q15 CFFT structure.
p1: points to the complex data buffer of size 2*fftLen. Processing occurs in-place.
ifftFlag: flag that selects forward (ifftFlag=0) or inverse (ifftFlag=1) transform.
bitReverseFlag: flag that enables (bitReverseFlag=1) or disables
        (bitReverseFlag=0) bit reversal of output.
*/

void arm_cfft_q31(const arm_cfft_instance_q31 *S, q31_t *p1, uint8_t ifftFlag,
uint8_t bitReverseFlag)
/*
S: points to an instance of the fixed-point CFFT structure.
p1: points to the complex data buffer of size 2*fftLen. Processing occurs
    in-place.
ifftFlag: flag that selects forward (ifftFlag=0) or inverse (ifftFlag=1)
    transform.
bitReverseFlag: flag that enables (bitReverseFlag=1) or disables
    (bitReverseFlag=0) bit reversal of output.
*/
```

Listing 10.7 CMSIS-DSP library fixed-point complex FFT functions.

There is no initialization function for complex FFT calculations. Instead, there are predefined structures. Here, the instance `S` stores information about the FFT/IFFT structure. `p1` points to the complex ***data buffer*** with a size that is twice the length of the FFT signal to be calculated. The parameter `ifftFlagR` controls whether a forward or inverse transform is computed. It should be set when calculating the IFFT and reset when calculating the FFT. The parameter `bitReverseFlag` controls whether the output is in normal or bit reversed order. It should be set for the output to be in normal order and reset to obtain the output in bit reversed order.

> **data buffer**
> *A region of physical memory storage used to temporarily store data while they are being moved from one place to another.*

We can repeat the complex FFT calculations in Task 4.5 using fixed-point functions. The relevant example is provided in `Online_Student_Resources\Lab10\fixed_point_complex_FFT`. Here, the input signal is scaled between 0 and 3.3.

Task 10.4

Complex FFT calculations can also be done in fixed-point form. To observe how this can be done, run the C code in the mentioned folder. Now, plot the obtained FFT results in the different fixed-point formats. Compare your results with the results of the complex FFT operations using floating-point form. What do you observe?

10.11.5 Fixed-Point Downsampling and Interpolation

Downsampling and interpolation on fixed-point numbers can be done using CMSIS-DSP library functions. We review them now.

Downsampling

The CMSIS-DSP library has two fixed-point downsampling function types, which are fast and regular. Both require an initialization function to construct the structure needed for operation. These functions are defined for Q1.15 and Q1.31 fixed-point numbers. We provide the CMSIS-DSP library fixed-point downsampling functions in Listing 10.8.

```
arm_status arm_fir_decimate_init_q15(arm_fir_decimate_instance_q15 *S,
uint16_t numTaps, uint8_t M, q15_t *pCoeffs, q15_t *pState, uint32_t blockSize)
/*
S: points to an instance of the Q15 FIR decimator structure.
numTaps: number of coefficients in the filter.
M: decimation factor.
pCoeffs: points to the filter coefficients.
pState: points to the state buffer.
blockSize:number of input samples to process per call.
*/

void arm_fir_decimate_fast_q15(const arm_fir_decimate_instance_q15 *S,
q15_t *pSrc, q15_t *pDst, uint32_t blockSize)
/*
S: points to an instance of the Q15 FIR decimator structure.
pSrc: points to the block of input data.
pDst: points to the block of output data
blockSize: number of input samples to process per call.
*/

void arm_fir_decimate_q15(const arm_fir_decimate_instance_q15 *S, q15_t *pSrc,
q15_t *pDst, uint32_t blockSize)
/*
S: points to an instance of the Q15 FIR decimator structure.
pSrc: points to the block of input data.
pDst: points to the location where the output result is written.
blockSize: number of input samples to process per call.
*/

arm_status arm_fir_decimate_init_q31(arm_fir_decimate_instance_q31 *S,
uint16_t numTaps, uint8_t M, q31_t *pCoeffs, q31_t *pState, uint32_t blockSize)
/*
S: points to an instance of the Q31 FIR decimator structure.
numTaps: number of coefficients in the filter.
M: decimation factor.
pCoeffs: points to the filter coefficients.
pState: points to the state buffer.
blockSize: number of input samples to process per call.
*/

```

```
void arm_fir_decimate_fast_q31(arm_fir_decimate_instance_q31 *S,
q31_t *pSrc, q31_t *pDst, uint32_t blockSize)
/*
S: points to an instance of the Q31 FIR decimator structure.
pSrc: points to the block of input data.
pDst: points to the block of output data
blockSize: number of input samples to process per call.
*/

void arm_fir_decimate_q31(const arm_fir_decimate_instance_q31 *S, q31_t *pSrc,
q31_t *pDst, uint32_t blockSize)
/*
S: points to an instance of the Q31 FIR decimator structure.
pSrc: points to the block of input data.
pDst: points to the block of output data
blockSize: number of input samples to process per call.
*/
```

Listing 10.8 CMSIS-DSP library fixed-point downsampling functions.

In order to demonstrate the usage of fixed-point downsampling functions, we provide an example in `Online_Student_Resources\Lab10\fixed_point_downsampling`. Here, we repeat Task 5.7 by scaling input between 0 and 3.3.

Task 10.5

This exercise aims to demonstrate how downsampling can be done in fixed-point form and how it differs from the floating-point version. The C code in the mentioned folder is the starting point for this operation. Run the downsampling code and plot the obtained results in different fixed-point formats. What do you observe from the obtained results? Comment on your observations.

Interpolation

The CMSIS-DSP library has only a regular fixed-point interpolation function. It requires an initialization function to construct the structure required for operation. We provide these functions in Listing 10.9.

```
arm_status arm_fir_interpolate_init_q15(arm_fir_interpolate_instance_q15 *S,
uint8_t L, uint16_t numTaps, q15_t *pCoeffs, q15_t *pState, uint32_t blockSize)
/*
S: points to an instance of the Q15 FIR interpolator structure.
L: upsample factor.
numTaps: number of filter coefficients in the filter.
pCoeffs: points to the filter coefficient buffer.
pState: points to the state buffer.
blockSize: number of input samples to process per call.
*/

void arm_fir_interpolate_q15(const arm_fir_interpolate_instance_q15 *S,
q15_t *pSrc, q15_t *pDst, uint32_t blockSize)
/*
S: points to an instance of the Q15 FIR interpolator structure.
pSrc: points to the block of input data.
pDst: points to the block of output data.
blockSize: number of input samples to process per call.
*/

arm_status arm_fir_interpolate_init_q31(arm_fir_interpolate_instance_q31 *S,
```

```
uint8_t L, uint16_t numTaps, q31_t *pCoeffs, q31_t *pState, uint32_t blockSize)
/*
S: points to an instance of the Q31 FIR interpolator structure.
L: upsample factor.
numTaps: number of filter coefficients in the filter.
pCoeffs: points to the filter coefficient buffer.
pState: points to the state buffer.
blockSize: number of input samples to process per call.
*/

void arm_fir_interpolate_q31(const arm_fir_interpolate_instance_q31 *S,
q31_t *pSrc, q31_t *pDst, uint32_t blockSize)
/*
S: points to an instance of the Q31 FIR interpolator structure.
pSrc: points to the block of input data.
pDst: points to the block of output data.
blockSize: number of input samples to process per call.
*/
```

Listing 10.9 CMSIS-DSP library fixed-point interpolation functions.

As with downsampling, we provide an example in the folder `Online_Student_Resources\Lab10\fixed_point_interpolation` to demonstrate the usage of fixed-point interpolation functions. Here, we repeat Task 5.8 by scaling input between 0 and 3.3.

Task 10.6

As with downsampling, this exercise aims to show how interpolation can be done in fixed-point form and how it differs from the floating-point version. The C code in the mentioned folder is the starting point for this operation. Run the interpolation code and plot the obtained results in different fixed-point formats. What do you observe from the obtained results? Comment on your observations.

10.11.6 Fixed-Point Implementation of Structures for LTI Filters

As mentioned in Chapter 8, a digital filter can be implemented in more than one way. In this section, we reconsider digital filter structures from a fixed-point perspective. Here, we benefit from CMSIS-DSP library functions.

Biquad Cascade Direct Form I

The first structure to be considered in fixed-point representation is the biquad cascade direct form I. Predefined CMSIS-DSP library functions for this representation are given in Listing 10.10.

```
void arm_biquad_cascade_df1_init_q15(arm_biquad_casd_df1_inst_q15 *S,
uint8_t numStages, q15_t *pCoeffs, q15_t *pState, int8_t postShift)
/*
S: points to an instance of the Q15 Biquad cascade structure.
numStages: number of 2nd order stages in the filter.
pCoeffs: points to the filter coefficients.
pState: points to the state buffer.
```

```
postShift: Shift to be applied to the accumulator result. Varies according
      to the coefficients format
*/

void arm_biquad_cascade_df1_q15(const arm_biquad_casd_df1_inst_q15 *S,
q15_t *pSrc, q15_t *pDst, uint32_t blockSize)
/*
S: points to an instance of the Q15 Biquad cascade structure.
pSrc: points to the block of input data.
pDst: points to the location where the output result is written.
blockSize: number of samples to process per call.
*/

void arm_biquad_cascade_df1_init_q31(arm_biquad_casd_df1_inst_q31 *S,
uint8_t numStages, q31_t *pCoeffs, q31_t *pState, int8_t postShift)
/*
S: points to an instance of the Q31 Biquad cascade structure.
numStages: number of 2nd order stages in the filter.
pCoeffs: points to the filter coefficients buffer.
pState: points to the state buffer.
postShift: Shift to be applied after the accumulator. Varies according to the
     coefficients format
*/

void arm_biquad_cascade_df1_q31(const arm_biquad_casd_df1_inst_q31 *S,
q31_t *pSrc, q31_t *pDst, uint32_t blockSize)
/*
S: points to an instance of the Q31 Biquad cascade structure.
pSrc: points to the block of input data.
pDst: points to the block of output data.
blockSize: number of samples to process per call.
*/

void arm_biquad_cas_df1_32x64_init_q31(arm_biquad_cas_df1_32x64_ins_q31 *S,
uint8_t numStages, q31_t *pCoeffs, q63_t *pState, uint8_t postShift)
/*
S: points to an instance of the high precision Q31 Biquad cascade filter
  structure.
numStages: number of 2nd order stages in the filter.
pCoeffs: points to the filter coefficients.
pState: points to the state buffer.
postShift: Shift to be applied after the accumulator. Varies according to the
      coefficients format.
*/


void arm_biquad_cas_df1_32x64_q31(const arm_biquad_cas_df1_32x64_ins_q31 *S,
q31_t *pSrc, q31_t *pDst, uint32_t blockSize)
/*
S: points to an instance of the high precision Q31 Biquad cascade filter.
pSrc: points to the block of input data.
pDst: points to the block of output data.
blockSize: number of samples to process.
*/
```

Listing 10.10 CMSIS-DSP library fixed-point biquad cascade direct form I filter functions.

Coefficients are stored in the array `pCoeffs` for the `arm_biquad_cascade_df1_init_q15` filter initialization function in Listing 10.10. The order for these is {`b10, 0, b11, b12, a11, a12, b20, 0, b21, b22, a21, a22,`⋯}. Here, `b1x` and `a1x` are the coefficients for the first stage, `b2x` and `a2x` are the coefficients for the second stage, and so on. The array `pCoeffs` contains a total of 6×`numStages` values. State variables are stored in the array `pState`. Each biquad stage has four state variables, which are `x[n-1], x[n-2], y[n-1]`, and y[n-2]. They are arranged in the array pState as {x[n-1], x[n-2], y[n-1], y[n-2]}. The state variable array has a total length of 4×numStages. State variables are updated after each data block is processed, while coefficients are untouched.

There is no internal overflow risk for the fixed-point filter implementation function, arm_biquad_cascade_df1_q15, given in Listing 10.10, but there is an internal overflow risk for the fixed-point function arm_biquad_cascade_df1_q31. The CMSIS-DSP library suggests scaling the input signal down by two bits. Therefore, it will lie in the range [-0.25, +0.25). The CMSIS-DSP library also has a high precision version of the Q1.31 fixed-point biquad cascade direct form I filtering function given in Listing 10.10. Care must be taken while using it. We suggest that you refer to the CMSIS-DSP library website, where its usage is described.

We provide an example on the usage of fixed-point biquad cascade direct form I filtering functions in Online_Student_Resources\Lab10\fixed_point_BCDFI. Here, we repeat Task 8.1 by scaling the input signal between 0 and 3.3.

Task 10.7

To observe how biquad cascade direct form I filtering can be done in fixed-point form, run the C code in the mentioned folder. Plot the obtained signal in different fixed-point formats. Compare these with the results of biquad cascade direct form I filtering done in floating-point form. Summarize your findings with a focus on the effect that fixed-point number representation has on operations.

FIR Lattice Filter

The second structure to be considered in fixed-point representation is the FIR lattice filter. Predefined CMSIS-DSP library functions for this representation are given in Listing 10.11. As can be seen here, FIR lattice filters are applicable to fixed-point Q1.15 and Q1.31 numbers. There is no internal overflow risk for the function with Q1.15 number format, but there is an internal overflow risk for the function with Q1.31 number format. The CMSIS-DSP library suggests scaling the input signal down by $2 \times \log_2$(numStages) bits to overcome the internal overflow risk for this function.

```
void arm_fir_lattice_init_q15(arm_fir_lattice_instance_q15 *S,
uint16_t numStages, q15_t *pCoeffs, q15_t *pState)
/*
S: points to an instance of the Q15 FIR lattice structure.
numStages: number of filter stages.
pCoeffs: points to the coefficient buffer. The array is of length numStages.
pState: points to the state buffer. The array is of length numStages.
*/

void arm_fir_lattice_init_q31(arm_fir_lattice_instance_q31 *S,
uint16_t numStages, q31_t *pCoeffs, q31_t *pState)
/*
S: points to an instance of the Q31 FIR lattice structure.
numStages: number of filter stages.
pCoeffs: points to the coefficient buffer. The array is of length numStages.
```

```
pState: points to the state buffer. The array is of length numStages.
*/

void arm_fir_lattice_q15(const arm_fir_lattice_instance_q15 *S, q15_t *pSrc,
q15_t *pDst, uint32_t blockSize)
/*
S: points to an instance of the Q15 FIR lattice structure.
pSrc: points to the block of input data.
pDst: points to the block of output data
blockSize: number of samples to process.
*/

void arm_fir_lattice_q31(const arm_fir_lattice_instance_q31 *S, q31_t *pSrc,
q31_t *pDst, uint32_t blockSize)
/*
S: points to an instance of the Q31 FIR lattice structure.
pSrc: points to the block of input data.
pDst: points to the block of output data
blockSize: number of samples to process.
*/
```

Listing 10.11 CMSIS-DSP library fixed-point FIR lattice filters.

We provide an example on the usage of fixed-point FIR lattice filtering functions in `Online_Student_Resources\Lab10\fixed_point_FIR_lattice_filter`. Here, we use the obtained structure for lowpass filtering.

Task 10.8

Run the FIR lattice filtering code in the mentioned folder to observe how this operation can be done in fixed-point form. Plot the obtained result in different fixed-point formats. Compare your results with the results of FIR lattice filtering operations using floating-point form. Does the fixed-point FIR lattice structure perform the required filtering operation?

IIR Lattice Filter

The third and final structure to be considered in fixed-point representation is the IIR lattice filter. Predefined CMSIS-DSP library functions for this representation are given in Listing 10.12. As with FIR filtering, fixed-point IIR lattice filters are applicable to fixed-point Q1.15 and Q1.31 numbers. The overflow characteristics of fixed-point IIR lattice filters are the same as for the fixed-point FIR lattice filters mentioned in the previous section

```
void arm_iir_lattice_init_q15(arm_iir_lattice_instance_q15 *S,
uint16_t numStages, q15_t *pkCoeffs, q15_t *pvCoeffs, q15_t *pState,
uint32_t blockSize)
/*
S: points to an instance of the Q15 IIR lattice structure.
numStages: number of stages in the filter.
pkCoeffs: points to reflection coefficient buffer. The array is of length
      numStages.
pvCoeffs: points to ladder coefficient buffer. The array is of length
      numStages+1.
pState: points to state buffer. The array is of length numStages+blockSize.
blockSize: number of samples to process per call.
*/
```

```

void arm_iir_lattice_q15(const arm_iir_lattice_instance_q15 *S, q15_t *pSrc,
q15_t *pDst, uint32_t blockSize)
/*
S: points to an instance of the Q15 IIR lattice structure.
pSrc: points to the block of input data.
pDst: points to the block of output data.
blockSize: number of samples to process.
*/

void arm_iir_lattice_init_q31(arm_iir_lattice_instance_q31 *S,
uint16_t numStages, q31_t *pkCoeffs, q31_t *pvCoeffs, q31_t *pState,
uint32_t blockSize)
/*
S: points to an instance of the Q31 IIR lattice structure.
numStages: number of stages in the filter.
pkCoeffs: points to the reflection coefficient buffer. The array is of length
      numStages.
pvCoeffs: points to the ladder coefficient buffer. The array is of length
      numStages+1.
pState: points to the state buffer. The array is of length numStages+blockSize.
blockSize: number of samples to process.
*/

void arm_iir_lattice_q31(const arm_iir_lattice_instance_q31 *S, q31_t *pSrc,
q31_t *pDst, uint32_t blockSize)
/*
S: points to an instance of the Q31 IIR lattice structure.
pSrc: points to the block of input data.
pDst: points to the block of output data.
blockSize: number of samples to process.
*/
```

Listing 10.12 CMSIS-DSP library fixed-point IIR lattice filters.

We provide an example on the usage of fixed-point IIR lattice filtering functions in `Online_Student_Resources\Lab10\fixed_point_IIR_lattice_filter`. Here, we repeat Task 8.6 by scaling the input signal between 0 and 3.3.

Task 10.9

As in FIR lattice filtering, this exercise aims to show how IIR lattice filtering can be done in fixed-point form and how it differs from the floating-point version. The C code in the mentioned folder is the starting point for this operation. Run the code and plot the obtained results in different fixed-point formats. What do you observe from the plots? Comment on your observations.

10.11.7 Fixed-Point Adaptive Filtering

The CMSIS-DSP library offers fixed-point LMS and normalized LMS functions for adaptive filtering. We explore them in this section.

LMS

The fixed-point LMS filtering functions in the CMSIS-DSP library are given in Listing 10.13. `pCoeffs` points to the array of filter coefficients stored in time reversed order. Initial filter coefficients serve as a starting point for iterations. `pState` points to the array of state variables. The size of this array is `numTaps+blockSize-1`. `blockSize` is the number of input samples processed by each call of the LMS function.

```
void arm_lms_init_q15(arm_lms_instance_q15 *S, uint16_t numTaps,
q15_t *pCoeffs, q15_t *pState, q15_t  mu, uint32_t blockSize,
uint32_t postShift)
/*
S: points to an instance of the Q15 LMS filter structure.
numTaps: number of filter coefficients.
pCoeffs: points to the coefficient buffer.
pState: points to the state buffer.
mu: step size that controls filter coefficient updates.
blockSize: number of samples to process.
postShift: bit shift applied to coefficients.
*/

void arm_lms_q15(const arm_lms_instance_q15 *S, q15_t *pSrc, q15_t *pRef,
q15_t *pOut, q15_t *pErr, uint32_t  blockSize)
/*
S: points to an instance of the Q15 LMS filter structure.
pSrc: points to the block of input data.
pRef: points to the block of reference data.
pOut: points to the block of output data.
pErr: points to the block of error data.
blockSize: number of samples to process.
*/

void arm_lms_init_q31(arm_lms_instance_q31 *S, uint16_t numTaps, q31_t *pCoeffs,
q31_t *pState, q31_t  mu, uint32_t blockSize, uint32_t postShift)
/*
S: points to an instance of the Q31 LMS filter structure.
numTaps: number of filter coefficients.
pCoeffs: points to coefficient buffer.
pState: points to state buffer.
mu: step size that controls filter coefficient updates.
blockSize: number of samples to process.
postShift: bit shift applied to coefficients.
*/

void arm_lms_q31(const arm_lms_instance_q31 *S, q31_t *pSrc, q31_t *pRef,
q31_t *pOut, q31_t *pErr, uint32_t blockSize)
/*
S: points to an instance of the Q15 LMS filter structure.
pSrc: points to the block of input data.
pRef: points to the block of reference data.
pOut: points to the block of output data.
pErr: points to the block of error data.
blockSize: number of samples to process.
*/
```

Listing 10.13 CMSIS-DSP library fixed-point LMS functions.

There is no internal overflow risk for the LMS function `arm_lms_q15` in Listing 10.13, but there is an internal overflow risk for the fixed-point function `arm_lms_q31`. The CMSIS-DSP library suggests scaling the input signal down by $2 \times \log_2$(`numStages`) bits to overcome this risk.

We now reconsider Task 9.7 (prediction using adaptive filtering) to demonstrate the usage of fixed-point LMS functions. We provide the modified project in `Online_Student_Resources\Lab10\fixed_point_LMS`. Here, we scaled the input signal between 0 and 3.3.

Task 10.10

We can apply the prediction operation using fixed-point LMS functions. Run the C code in the mentioned folder. Plot the obtained signal in different fixed-point formats. Compare these with the results of the prediction operation done in floating-point form. Summarize your findings with a focus on the effect that fixed-point number representation has on operations.

Normalized LMS

The fixed-point normalized LMS filtering functions in the CMSIS-DSP library are given in Listing 10.14. Here, the same variables are used for the initialization functions as in the previous section.

There is no internal overflow risk for the fixed-point normalized LMS function `arm\_lms\_norm\_q15` in Listing 10.14, but there is an internal overflow risk for the fixed-point function `arm_lms_norm_q31`. The CMSIS-DSP library suggests scaling the input signal down by $2 \times \log_2($`numStages`$)$ bits to overcome this risk.

As with the LMS case, we provide an example on the usage of fixed-point normalized LMS functions in `Online_Student_Resources\Lab10\fixed_point_normalized_LMS`. Here, we repeat Task 9.7 but with normalized LMS and with the input signal scaled between 0 and 3.3.

Task 10.11

We can also apply the prediction operation using fixed-point normalized LMS functions. Run the C code in the mentioned folder. Plot the obtained signal in different fixed-point formats. Compare these with the results of the prediction operation done in floating-point form. Summarize your findings with a focus on the effect that fixed-point number representation has on operations.

10.11.8 Three-band Audio Equalizer Design using Fixed-Point Arithmetic

We designed a three-band audio equalizer in Section 8.11.4. Here, we will redesign this equalizer using fixed-point functions. To do so, we start with the chirp signal in fixed-point form. We provide the C code to generate this signal in `Online_Student_Resources\Lab10\fixed_point_chirp_signal`.

Task 10.12

Generate the fixed-point chirp signal using the C code in `Online_Student_Resources\Lab10\fixed_point_chirp_signal`. Plot the generated signal. Comment on the differences between this signal and its floating-point counterpart.

```
void arm_lms_norm_init_q15(arm_lms_norm_instance_q15 *S, uint16_t numTaps,
q15_t *pCoeffs, q15_t *pState, q15_t mu, uint32_t blockSize, uint8_t postShift)
/*
S: points to an instance of the Q15 normalized LMS filter structure.
numTaps: number of filter coefficients.
pCoeffs: points to coefficient buffer.
pState: points to state buffer.
mu: step size that controls filter coefficient updates.
blockSize: number of samples to process.
postShift: bit shift applied to coefficients.
*/

void arm_lms_norm_q15(arm_lms_norm_instance_q15 *S, q15_t *pSrc, q15_t *pRef,
q15_t *pOut, q15_t *pErr, uint32_t blockSize)
/*
S: points to an instance of the Q15 normalized LMS filter structure.
pSrc: points to the block of input data.
pRef: points to the block of reference data.
pOut: points to the block of output data.
pErr: points to the block of error data.
blockSize: number of samples to process.
*/

void arm_lms_norm_init_q31(arm_lms_norm_instance_q31 *S, uint16_t numTaps,
q31_t *pCoeffs, q31_t *pState, q31_t mu, uint32_t blockSize, uint8_t postShift)
/*
S: points to an instance of the Q31 normalized LMS filter structure.
numTaps: number of filter coefficients.
pCoeffs: points to coefficient buffer.
pState: points to state buffer.
mu: step size that controls filter coefficient updates.
blockSize: number of samples to process.
postShift: bit shift applied to coefficients.
*/

void arm_lms_norm_q31(arm_lms_norm_instance_q31 *S, q31_t *pSrc, q31_t *pRef,
q31_t *pOut, q31_t *pErr, uint32_t blockSize)
/*
S: points to an instance of the Q31 normalized LMS filter structure.
pSrc: points to the block of input data.
pRef: points to the block of reference data.
pOut: points to the block of output data.
pErr: points to the block of error data.
blockSize: number of samples to process.
*/
```

Listing 10.14 CMSIS-DSP library fixed-point normalized LMS functions.

Task 10.13

Redesign the three-band audio equalizer system using fixed-point direct form FIR filters. Use the Hamming window in its design. For the first band, design a 127th order lowpass FIR filter with a normalized angular cut-off frequency of $\pi/24$ rad/sample. For the second band, design a 127th order bandpass FIR filter with normalized angular cut-off frequency values of $\pi/24$ rad/sample and $\pi/6$ rad/sample. For the third band, design a 127th order highpass FIR filter with a normalized angular cut-off frequency value of $\pi/6$ rad/sample. Implement the designed equalizer using Q1.15 fast fixed-point functions. Test your system with the fixed-point chirp signal generated in Task 10.12.

The audio equalizer exercise should emphasize the similarities and differences between fixed-point and floating-point implementations. We can go one step further to observe the differences between these two representations by considering their timing, memory usage, and accuracy. Reconsidering the FIR filtering operation in Task 10.3, we propose the following experiment. We can implement the same filter with different fixed-point representations, and compare each implementation in terms of timing, memory usage, and accuracy. For comparison purposes, we should also consider the floating-point implementation of the system.

Given an analog system, its input and output should also be analog. However, we are using a digital system for implementation. Therefore, the closest floating-point representation is the `float` data type. We will use the floating-point implementation with floating-point input and output signals as a benchmark. We will compare all our results with this benchmark signal. Actual signal filtering implementation requires signal samples to be obtained from a 12-bit ADC module. Hence, we generate a new input with fixed-point values in the range 0 to $2^{12} - 1$. We process this input signal with all filtering functions. Then, we obtain the final output by running the obtained output through a 15-bit DAC module. For comparison with the benchmark, we convert the obtained result to float form. We provide part of this implementation in `Online_Student_Resources\Lab10\fixed_point_comparison`.

Task 10.14

Extend the project in `Online_Student_Resources\Lab10\fixed_point_comparison` so that we can compare all fixed-point and floating-point implementations. Specifically, form comparison tables for memory usage, execution time, and accuracy. Then, comment on the results.

Fixed-point implementation becomes more important when the microcontroller at hand does not have a floating-point unit. As the microcontroller considered in this lab has a floating-point unit, we must disable it to observe the effect that fixed-point implementation has on operations. To do so, we must adjust project properties with Keil before compiling it. In Keil, open Options for Target – > Target window, and select "Not Used" from the "Floating Point Hardware" dropdown menu. Now, the project will be compiled without the use of the floating-point unit.

Task 10.15

Repeat the floating-point part of Task 10.14 to observe the effect of the floating-point unit in calculations. In relation to this, observe how fixed-point calculations can be beneficial when a microcontroller without a floating-point unit is used.

Real-Time Digital Signal Processing

Contents

11.1 Introduction

Up to now, we have not placed a time constraint on signal processing. We deliberately followed this approach to explain DSP concepts in detail. However, real-time DSP applications will always have a time constraint. In other words, there will be limited time to process data. Moreover, data to be processed will be continuously streaming and thus will not have an end (at least from our perspective). Moreover, we know that the digital system will always have limited resources. This chapter is about the methodologies proposed for these problems. We start with the definition of "real-time" digital signal processing. Then, we explore buffer structures. Next, we introduce overlap methods, which can be used in frame-based signal processing. The accompanying lab will clarify these issues further.

11.2 What is Real-Time Signal Processing?

For a clear understanding of operations to be performed, we first define real-time signal processing. Two fundamental algorithm implementation methods that may also be used in real-time signal processing are sample and frame-based processing, which we briefly introduced in Chapter 2. Input signals are acquired one by one in sample-based signal processing, whereas in frame-based processing, input signals are stored in a buffer (limited memory space). Then, they are processed and fed to the output in block form. Some signal processing algorithms, such as the DFT or FFT, by their nature, have frame-based implementation.

11.2.1 Sample-based Processing

The first implementation method that can be used in real-time digital signal processing is based on the sample-based approach. Let us define what we understand from the real-time phrase for sample-based signal processing. We describe a sample-based system as real-time if the acquired sample is processed and the corresponding output is generated before the next sample arrives. In other words, data do not accumulate in a real-time sample-based system.

11.2.2 Frame-based Processing

The second implementation method that can be used in real-time digital signal processing is based on the frame-based approach. Different from the sample-based approach, input samples are stored in a buffer before frame-based processing. Then, the buffer (frame) is processed and fed to the output in block form. We can update the definition of real-time signal processing based on buffers as follows. We describe a frame-based system as real-time if the input buffer is processed and fed to the output before the next input buffer is filled.

Latency within frame-based systems is always greater than the time needed to fill the input buffer. Hence, for a system with buffer size L and sampling period $T_{sampling}$, latency will be at least $L \times T_{sampling}$. The user should take this into account in operation.

Frame-based signal processing is essential for applications such as the FFT. Furthermore, CMSIS-DSP library transform and filtering functions are built for frame-based processing. Therefore, this method is dominant in real-time signal processing applications. Now, we will compare sample and frame-based processing to emphasize why this is the case.

11.2.3 Differences between Sample and Frame-based Processing

Sample and frame-based processing can be compared based on three main properties: CPU usage, memory usage, and latency. In sample-based processing, the CPU must be invoked for each sample acquired by the ADC module. If this operation is done by an interrupt, then its overhead will be in terms of clock cycles used in operation. However, in frame-based processing, the ADC module can work without interrupting the CPU. Hence, the net overhead will be minimal. This also allows parallel processing of sampling and CPU operations. Moreover, if direct memory access is used, the advantage of parallel processing will become more dominant. Frame-based processing requires extra memory space for operation. However, recent microcontrollers can handle such memory requirements. If memory space is limited, then the only option left is to use sample-based processing. Finally, latency for frame-based processing is directly proportional to the buffer size. This may negatively affect operation. If latency is of utmost importance in operation, then sample-based processing should be used. Otherwise, a balance between an acceptable latency and buffer size should be established and frame-based processing should be used.

11.3 Basic Buffer Structures

Filtering operations require more than one sample to operate, which is clear from the difference equation in Eqn. 2.24. This is the case for sample-based processing also. Therefore, data must be stored in a buffer. We will now introduce two fundamental buffer structures, which are linear and circular.

11.3.1 Linear Buffer

The linear buffer has a simple form, where the newest entry is always appended to the start of the buffer. Therefore, when a new sample is added, all samples are shifted by one within the buffer, and the last entry is discarded. Based on this structure, the filtering operation in Eqn. 2.24 can be applied on a fixed array structure.

$$y[n] = \sum_{k=0}^{K-1} h[k]x[n-k] \tag{2.24}$$

Only the array entries change, so filter implementation is relatively easy when the linear buffer structure is used. However, the main drawback of this structure is the shifting operation for each sample entry, and this operation may be time consuming when the buffer size is large.

11.3.2 Circular Buffer

The circular buffer offers an alternative to the time consuming shifting operation in the linear buffer structure. The circular buffer is formed by connecting the beginning of the linear buffer to its endpoint. This is called wrapping. Within the circular buffer, a pointer (or index) is used to point to the newest entry. This index value is incremented with each data entry. Therefore, instead of shifting each buffer entry, only the pointer is changed. At the same time, if the pointer exceeds the range of the buffer, it is restarted by the wrapping operation.

11.4 Usage of Buffers in Frame-based Processing

In frame-based processing, more than one buffer is required for the ADC, processing, and DAC operations. The idea here is setting up a framework to apply all these operations in a parallel manner. Although there may be several options to perform this, only the triple and ***ping-pong buffer*** structures are considered herein.

> **ping-pong buffer**
> *A double buffering technique in which one buffer is being read while the other continues to receive data.*

11.4.1 Triple Buffer

This structure requires three equal-sized buffers in operation. We will refer to these as A, B, and C. The triple buffer works as follows. While incoming data fill buffer A, the already filled buffer B is processed. At the same time, the output is obtained from the processed buffer C. Once all the buffers are filled, processed, and emptied, they are rotated, where A becomes the processing buffer, B becomes the output buffer, and C becomes the input buffer. A timing diagram for the triple buffer is shown in Figure 11.1. This diagram demonstrates how each buffer behaves in time.

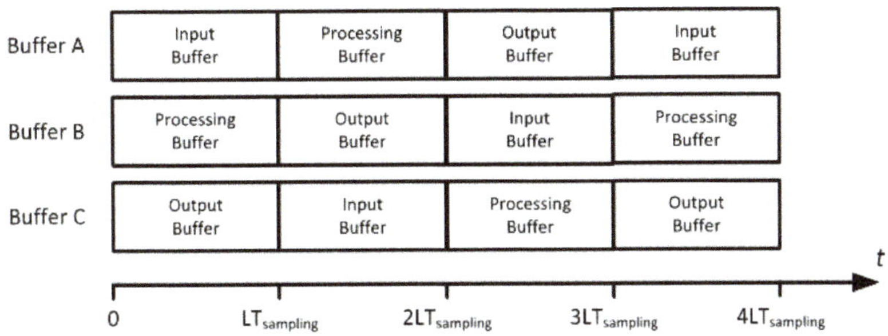

Figure 11.1 Triple buffer timing diagram.

11.4.2 Ping-Pong Buffer

The ping-pong buffer offers an alternative for frame-based signal processing. This structure has either four buffers (one for input, one for output, and two for processing) or two buffers (one for input/output and one for processing). Let us consider the four buffer case. Assume that we have the four buffers A_1, B_1, A_2, and B_2. All of these buffers have size L.

The idea behind the ping-pong buffer is to switch specific buffers in operation. Hence, the name ping-pong. Let us assume that buffers A_1 and B_1 are in the ping state, and the buffers A_2 and B_2 are in the pong state. While the ping buffer A_1 is filled from the ADC module, data in B_1 are sent to the DAC module. Meanwhile, the pong buffer A_2 is processed by the CPU and the obtained output is stored in B_2. When these operations end, the ping and pong states swap. Hence, A_2 and B_2 switch to the ping state, and A_1 and B_1 switch to the pong state. This switching operation will continue until the overall operation ends. We provide the timing diagram of the ping-pong buffer in Figure 11.2. This diagram demonstrates how each buffer behaves in time.

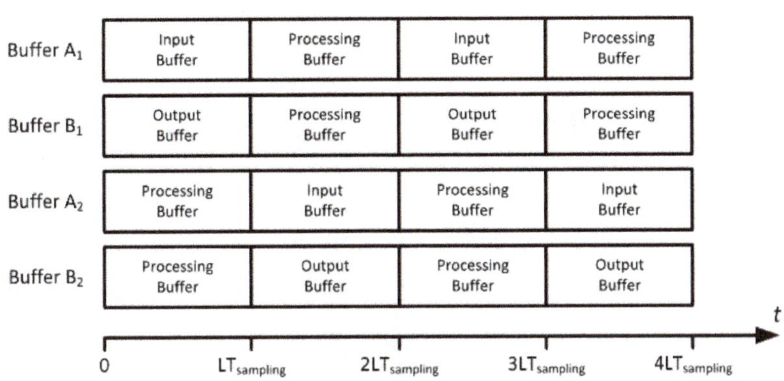

Figure 11.2 Ping-pong buffer timing diagram.

Different from the four-buffer ping-pong structure, the two-buffer ping-pong structure simultaneously uses one buffer for the input and output operation. The other buffer is used for the processing operation by the CPU and the obtained output is stored in the buffer itself. Hence, memory usage is decreased compared to the four-buffer structure. The drawback of this structure becomes evident when accessing both buffers simultaneously. Accessing the same memory region for reading and writing operations simultaneously may require some additional measures.

11.4.3 Differences between Triple and Ping–Pong Buffers

Triple and ping–pong buffers can be compared based on their memory and CPU usage. Three buffers, each with size L, are required for the triple buffer structure. Four buffers, each with size L, (or two buffers with size L) are required to process the same data in the ping–pong buffer structure. Therefore, the triple buffer structure offers an advantage in terms of lower memory usage.

11.5 Overlap Methods for Frame-based Processing

Data to be processed may be extremely large. Hence, single frame-based processing may not be possible due to hardware or time limitations. One solution for such a case is to divide the data into adjacent blocks (frames), each with length L. Then, each block can be processed separately. The final result may be obtained by appending the partial outputs. However, some portion of data in a block may be required in adjacent blocks for operations such as filtering or convolution. There are two methods to achieve this. These are called overlap-add and overlap-save. In this section, we will introduce them. To note here, the overlap-add and overlap-save methods are not specific to real-time signal processing. They can also be used in offline signal processing when data size is large. However, their usage is necessary for real-time operations, so they are included in this chapter.

11.5.1 Overlap-Add Method

The overlap-add method works as follows. Assume that there is an input signal $x[n]$ with length $K \times L$. This signal can be partitioned into K non-overlapping adjacent blocks, each with length L, as in Eqn. 11.1.

$$x[n] = \sum_{k=0}^{K-1} x_k[n] \tag{11.1}$$

where $x_k[n] = x[n](u[n-kL] - u[n-(k+1)L-1])$ is the kth block. Let us assume that we convolve $x[n]$ with an FIR filter, $h[n]$, having length N. Then, the convolution operation can be expressed as

$$y[n] = x[n] * h[n] \tag{11.2}$$

$$= \sum_{k=0}^{K-1} x_k[n] * h[n] \tag{11.3}$$

$$= \sum_{k=0}^{K-1} y_k[n] \tag{11.4}$$

where each partial output, $y_k[n]$, has length $L+N-1$. Each successive output, $y_k[n]$ and $y_{k+1}[n]$, will overlap by $N-1$ samples. Let us start with the first output block, $y_0[n]$, with length $L+N-1$. The overlapping part of $y_0[n]$ for $n = L-1, \cdots, L+N-2$ should be added to $y_1[n]$. This addition operation should continue for all subsequent blocks. Hence, this method is called overlap-add. We graphically illustrate the working principle of the overlap-add method in Figure 11.3.

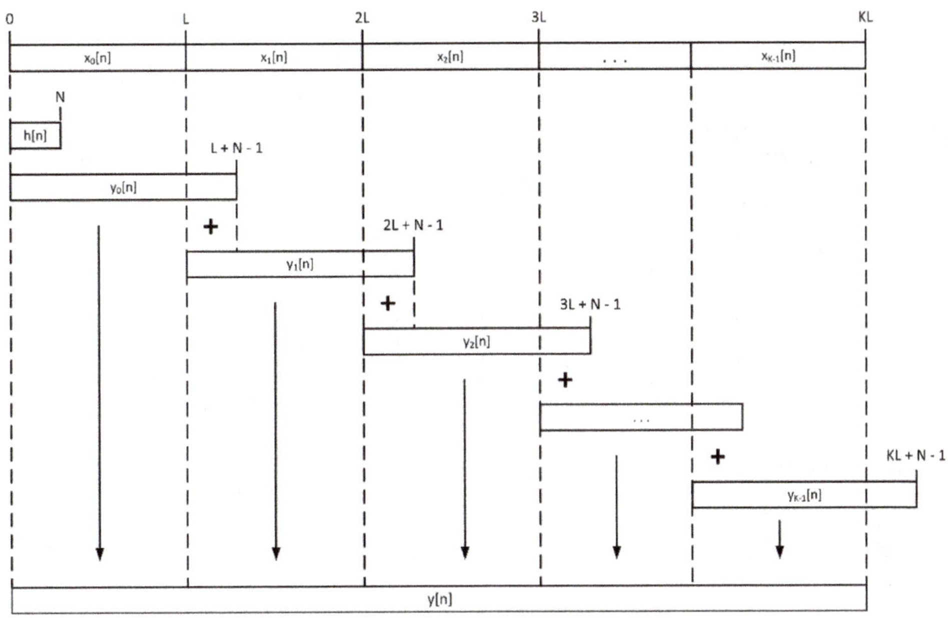

Figure 11.3 Working principle of the overlap-add method.

The convolution sum in Eqn. 11.2 can also be performed in the frequency domain. Please check Section 4.6 for this option.

11.5.2 Overlap-Save Method

The overlap-save method takes a different approach in processing large data. For this method, the input signal is divided into overlapping blocks, each with length $L+N-1$.

Hence, each block overlaps with its adjacent block by the last $N-1$ samples. As each block is filtered with an FIR filter with length N, the output is of length $L+2N-2$. The overlapping output within this block is the first and last $N-1$ samples, which are discarded. This operation is performed for all input blocks. The output is formed by the non-overlapping parts. We graphically illustrate the working principle of the overlap-save method in Figure 11.4.

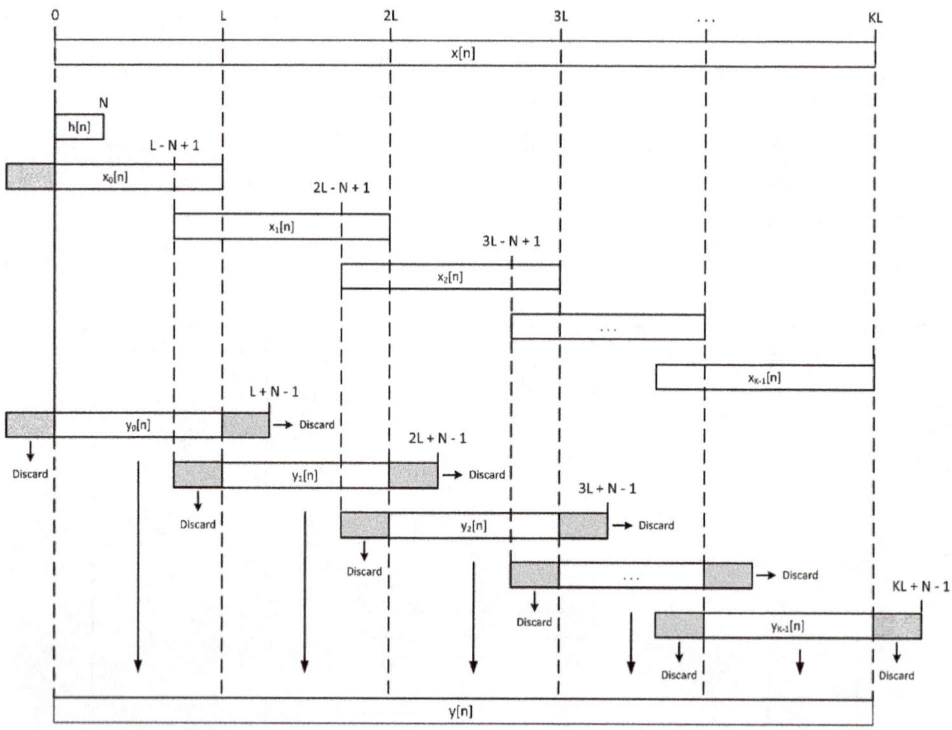

Figure 11.4 Working principle of the overlap-save method.

11.6 Chapter Summary

This chapter was about real-time signal processing techniques. Therefore, we started with the definition of real-time signal processing. Then, we introduced buffer structures for sample and frame-based signal processing. Related to this, we explored buffer usage in frame-based processing. Finally, we considered overlap methods for frame-based signal processing. All of these techniques set a framework for real-time signal processing. Therefore, they can be applied to all topics introduced in previous chapters.

11.7 Further Reading

Real-time digital signal processing heavily depends on the hardware and implementation method selected. Therefore, referring to the related literature is fundamental. A real-time digital signal processing application may also include asynchronous concerns such as a user interface. Then, scheduling the operations within the system becomes important. A real-time operating system (RTOS) is often used for such applications. You may want to consult the wider literature to delve more into this area, e.g., [1]. Although this is an elegant solution, it is beyond the concepts addressed in this book. Therefore, we strongly encourage you to consult the given reference on the usage of RTOS for a real-time signal processing application with a user interface.

11.8 Exercises

11.8.1 What happens when real-time constraints are not satisfied in a digital system?

11.8.2 Construct a linear buffer to revisit the samples of the signal $x[n] = \sin(\pi n/40)$. The buffer will hold 240 samples. Add a counter to count the shifting operation until the first data are fed to output. Observe the operation.

11.8.3 Repeat Exercise 11.8.2 using a circular buffer. Compare the shifting operation count values of the two buffer structures.

11.8.4 Construct a triple buffer structure to calculate the square of a sinusoidal signal. Use the output of the buffer in Exercise 11.8.3 as the input signal. Each buffer will hold 80 samples. Determine the latency in terms of the shifting operation.

11.8.5 Repeat Exercise 11.8.4 using a four-buffer ping–pong buffer.

- **(a)** Compare the latencies of this implementation and the triple buffer.
- **(b)** Increase the buffer size to 120 samples. Repeat the exercise and compare latencies again.

11.8.6 Design a digital system to sample and filter an analog signal using the overlap-add method in real time. The maximum frequency component of the analog signal to be processed is 24 kHz.

11.8.7 Design a digital system to sample and filter an analog signal using the overlap-add method in real time. The maximum frequency component of the analog signal to be processed is 24 kHz.

- **(a)** Adjust the ADC sampling rate to sample the analog signal without aliasing.
- **(b)** Store the samples in a triple buffer. Maximum latency is expected to be 128 ms. Adjust the maximum buffer size accordingly.
- **(c)** Design a 64th-order Hamming window FIR stop-band filter with cut-off frequencies 1 kHz and 5 kHz.
- **(d)** Use CMSIS-DSP library functions to filter a given analog signal in real time.

11.8.8 Repeat Exercise 11.8.7 using the overlap-save method.

11.8.9 Use the four-buffer ping–pong buffer structure to complete Exercises 11.8.7 and 11.8.8.

11.9 References

[1] Arm, *Arm University Program RTOS Education Kit*, available via www.arm.com/education.

11.10 Lab 11

11.10.1 Introduction

This lab discusses the practical implementation of real-time digital signal processing. The operations we perform in this chapter are explained in detail in Chapter 11. Because the hardware concepts considered in this chapter depend on the STM32F4 Discovery kit and the AUP audio card, we strongly suggest that you review Chapter 1 before continuing. In this chapter, we first explore buffer structures. Then, we consider overlap methods for frame-based signal processing. Finally, we focus on digital signal processing concepts in real time. To do so, we start with audio effects, and then we focus on real-time digital signal filtering via the equalizer system introduced in Chapter 8.

11.10.2 Setup for Digital Signal Processing in Real Time

We start with the basic setup for processing audio signals. We will use the AUP audio card for this purpose. Please consult Section 1.4.7 for the details of this card. There, we provide a sample C code for acquiring audio samples and regenerating them using the setup in Section 1.4.7.

Task 11.1

Select the onboard digital MEMS microphone as the input channel. Set the sampling rate to 32 kHz. The code in the project simply forms a loop between ADC input and DAC output on the AUP audio card. Data coming from the microphone input are directly fed to the headphone output. Run the code provided in `Online_Student_Resources\Lab1\LTEK_Example` and listen to your voice. You should become familiar with acquiring and generating a real-time audio signal by applying these steps.

11.10.3 Audio Effects

The sounds that we perceive depends on the listening conditions and the acoustic properties of the environment. For example, the same sound signal played in a concert hall, a small room, and a large room will be perceived differently in each space. Because these effects are important for musicians, digital signal processing can be used to simulate different environments. In this section, we will implement delay, echo, and reverberation effects using real-time signal processing. These effects can be taken as realizations of sample-based signal processing.

Delay

A delay is the simplest audio effect to apply. It holds the input signal and then plays it back after a period of time. Delays are used very often by musicians. It is also the main block for other audio effects such as echo, reverberation, chorus, and flanging. The difference equation for the delay operation is $y[n] = x[n - n_0]$, where n_0 is the delay time. The basic delay block is shown in Figure 11.5. The linear and circular buffer structures given in the previous section can be used to generate the delay effect.

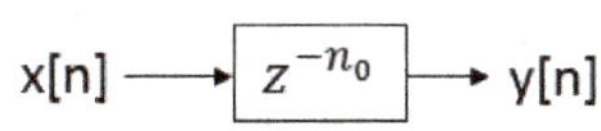

Figure 11.5 The basic delay block.

Echo

An echo block simply takes an audio signal and plays the sum of the signal and a delayed version of the signal. The delay time can range from several milliseconds to several seconds. The basic echo block is shown in Figure 11.6.

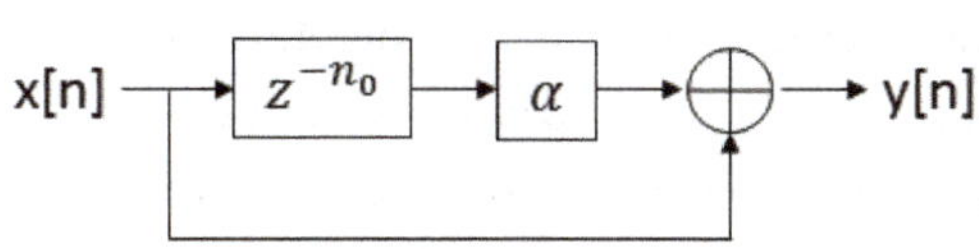

Figure 11.6 The basic echo block.

The difference equation for this system is $y[n] = x[n] + \alpha x[n - n_0]$, where α is the delay mix parameter, and n_0 is the delay. As can be seen in this equation, to generate the echo effect we need both the present and the delayed version of the signal. We provide the C code for the echo effect in `Online_Student_Resources\Lab11\Echo_effect`. Here, we set the sampling rate to 32 kHz and circular buffer size to 16000.

Task 11.2

We can observe the echo effect in real time by running the code in `Online_Student_Resources\Lab11\Echo_effect`. To do so, talk through the digital microphone of the AUP audio card and listen your voice through the headphones. The headphones should be plugged into the *Headphone Out* jack of the AUP audio card. As the code is run, you should hear your voice fed to the microphone at the same time from the right headphone output. Each sound will be heard once immediately after it is fed to the microphone, and again after a 0.5 s delay.

Reverberation

Reverberation is one of the most heavily used effects in music. In this exercise, we introduce the simplest form of reverberation using a simple comb filter. Reverberation combines the original signal with a very short time-delayed version (less than 100 ms) of itself. In fact, when a signal is delayed for a very short time and then added to the original, the ear perceives a fused sound rather than two separate sounds. If the number of very short delays (that are fed back) are mixed together, a crude form of reverberation can be achieved. The block diagram of our reverberation module (formed by a simple comb filter) is shown in Figure 11.7.

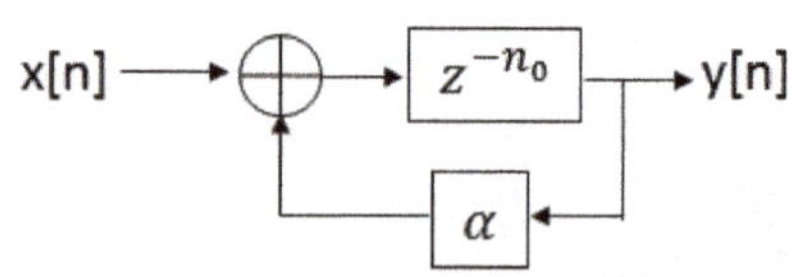

Figure 11.7 Block diagram of a reverberation module (formed by a simple comb filter).

The difference equation for this system is $y[n]=\alpha y[n-n_0]+x[n-n_0]$. As can be seen in this difference equation and Figure 11.7, we need both the delayed input and output signals to implement a reverberation system. Therefore, we must store the sum of two signals for reverberation. We can reach to $x[n-n_0]$ and $y[n-n_0]$ signals by storing input and output signals in an array. Then, the reverberation effect is obtained by summing the weighted delayed output, $\alpha y[n-n_0]$, with the delayed input, $x[n-n_0]$. By increasing α, we can increase the effect that previous outputs have on the most recent output signal. To prevent overflow of the output signal, the α value should be in the range [0, 1].

Task 11.3

The reverberation effect can be obtained by modifying any available project. Here, change the echo code in `Online_Student_Resources\Lab11\Echo_effect` to generate a reverberation effect. Set a buffer size less than 3200 to create a very short time-delayed version of the input. Talk through the digital microphone of the AUP audio card and listen to your voice through the headphones. Did you observe a reverberation effect?

These applications emphasize how sample-based operations can be achieved in real time. Moreover, they provide a good example of real-time signal processing on a system represented by a difference equation.

11.10.4 Usage of Buffers in Frame-based Processing

We considered the usage of buffer structures in frame-based signal processing in Section 11.4. We will handle the practical aspects of triple and ping-pong buffers in this section.

Let us start with the triple buffer. We provide a sample project on the usage of the triple buffer in `Online_Student_Resources\Lab11\FIR_filtering_triple_buffer`. Here, an FIR filter with the frame-based structure is implemented using the triple buffer and CMSIS-DSP library functions.

Task 11.4

The usage of the triple buffer can best be understood in an application. Therefore, run the code in `Online_Student_Resources\Lab11\FIR_filtering_triple_buffer`. To observe the memory usage of the triple buffer structure, apply the procedure given in Section A.6. Comment on the result. Now, talk through the digital microphone of the AUP Audio Board and listen to your voice through the headphones. What do you observe from the two audio channels?

We provide a project on the usage of the four-buffer ping-pong structure in `Online_Student_Resources\Lab11\FIR_filtering_ping_pong_buffer`. Here, a frame-based FIR filter is implemented using the ping-pong buffer structure.

Task 11.5

As in the triple buffer case, the usage of the four-buffer ping-pong structure can best be understood in an application. Therefore, run the code in `Online_Student_Resources\Lab11\FIR_filtering_ping_pong_buffer`. To observe the memory usage of the triple buffer structure, apply the procedure given in Section A.6. Compare the memory usage of this structure with that of the triple buffer structure. Which one uses the least memory?

Task 11.6

The four-buffer ping–pong structure considered in Task 11.5 can be modified to form a two-buffer ping–pong structure. In order to observe its advantages and disadvantages, please modify the code in `Online_Student_Resources\Lab11\FIR_filtering_ping_pong_buffer` to create a two-buffer ping–pong structure. Rerun the code, measure the memory usage as given in Section A.6, and compare it with the triple buffer structure. Which one uses the least memory this time?

11.10.5 Overlap Methods for Frame-based Processing

We considered overlap methods in Section 11.5. The CMSIS-DSP library filtering functions are suitable for the overlap-add method. The only constraint here is that the filtering function must be used without initialization. This way, the function stores the last $N - 1$ output samples internally to be used for the next frame. Addition of the overlapped frames is done internally.

Let us explain this operation in detail. The CMSIS-DSP library explains the FIR filtering operation as follows: "The FIR filter algorithm is based upon a sequence of multiply-accumulate (MAC) operations. Each filter coefficient b[n] is multiplied by a state variable which equals a previous input sample x[n]." This indicates that the previous input samples are taken into account if the filter is not reinitialized. The explanation in the `arm_fir_f32` function also gives us information about the internal working principles of the filtering operation.

We have already described the overlap-add method using CMSIS-DSP library filtering functions in Section 11.10.4. These three examples intrinsically used the overlap-add method. Therefore, there is no need to provide further examples of this method. However, we can modify these previous examples for the overlap-save method. This modification can be found in `Online_Student_Resources\Lab11\FIR_filtering_overlap_save`. This project uses the triple buffer structure.

Task 11.7

The overlap-save-based FIR filtering operation can be observed by running the code in `Online_Student_Resources\Lab11\FIR_filtering_overlap_save`. Talk through the digital microphone of the AUP audio card and listen to your voice through the headphones. What is the difference between the two audio channels?

11.10.6 Implementing the Three-Band Audio Equalizer in Real Time

In this section, we will implement the three-band equalizer designed in Chapter 8 in real time. We will use the FIR filter-based implementation. We provide the project on the three-band audio equalizer using triple buffer structure and the overlap-add method in `Online_Student_Resources\Lab11\Equalizer_overlap_add_triple_buffer`.

Task 11.8

In order to observe the working principles of the triple buffer structure and overlap-add method, run the code in `Online_Student_Resources\Lab11\Equalizer_overlap_add_triple_buffer`. Talk through the digital microphone of the AUP Audio Board and listen to your voice through the headphones. Comment on the difference between the two audio channels.

We provide the project on the three-band audio equalizer using the triple buffer structure and overlap-save method in `Online_Student_Resources\Lab11\Equalizer_overlap_save_triple_buffer`.

Task 11.9

Task 11.8 can be repeated using the triple buffer structure and overlap-save method. To do so, run the code in `Online_Student_Resources\Lab11\Equalizer_overlap_save_triple_buffer`. Talk through the digital microphone of the AUP Audio Board and listen to your voice through the headphones. Comment on the difference between the two audio channels.

A

Appendices

Appendix A Getting Started with KEIL and CMSIS

Contents

Lab 0

A.1 Introduction

The Keil MDK (Microcontroller Development Kit) is a complete ***integrated development environment*** (IDE) from Arm. It combines Keil μVision (project manager, editor, ***debugger***, and simulator) with the Arm-specific C/C++ ***compiler*** and other components (libraries, CMSIS, and examples). We will use the Keil μVision MDK throughout this lab, so we will refer to it simply as Keil μVision from this point on. We use its most recent version (MDK-Arm v5.18) in this lab. We believe that any new version of Keil

μVision will be similar to the one used herein. Therefore, it is important for you to become familiar with the working principles of Keil μVision. We will begin with the installation process. Then, we will explain how to create and manage a project in Keil μVision. Next, we will demonstrate how to measure the execution time and memory usage of a given code block. Finally, we will introduce the CMSIS-DSP library, which will be used extensively throughout this lab.

> **integrated development environment**
> *A software application that combines the tools for software development.*

> **debugger**
> *A software program that allows users to detect errors in other software programs.*

> **compiler**
> *A software program that translates codes written in a particular programming language into machine language.*

A.2 Downloading and Installing Keil μVision

The latest version of Keil μVision can be downloaded from `https://www.keil.com/download/product/` after submitting the contact information form. When the download is complete, you should follow the installation steps given below.

1. Click on the executable file and begin the installation.
2. Accept the license agreement, and click **Next** in the pop-up window.
3. A window asking for the installation directory will appear. Install the program to the default location if possible.
4. When the customer information window appears, fill in the form and click **Next**. The installation should then start.
5. It is important to install all device *drivers* when prompted.
6. When the installation is complete, the pack installer window will appear as in Figure A.1.
7. At the left-hand side of the pack installer, select STMicroelectronics - STM32F4 Series - STM32F407 - STM32F407VG from the device window. Install or update all related software packs from the right-hand side of the pack installer.
8. Finally, install the ST-Link drivers for the STM32F4 Discovery kit from Keil's installation directory. Specifically, search for the subfolder `\Arm\STLink\USBDriver`.

driver
A program that controls the interface between a computer and a peripheral device.

Here, we will assume the Keil installation directory to be `C:\` and provide example project settings according to this location. If your installation directory is different, all related settings should be adjusted.

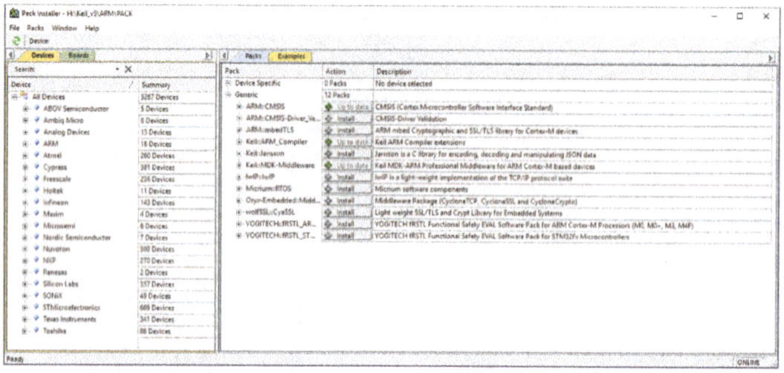

Figure A.1 Keil pack installer window.

A.3 Creating a New Project

A project typically contains source, header, and include files. Keil μVision generates an executable output file (with extension `.axf`) from these. This file is used by the STM32F4 microcontroller. This section concerns the creation of a project.

A.3.1 Creating a New Project

To create a new project, click **Project** and **New μVision Project**. In the opened window, select the location of the project folder. Give the project an appropriate name and click **Save**. The window shown in Figure A.2 will appear. It asks for the device you wish to use.

From the opened window, select **STMicroelectronics, STM32F4 Series, STM32F407, STM32F407VG**, and **STM32F407VGTx**. Then, click **OK**. A new window will appear, which is called **Manage Run-Time Environment**. In this window, select the necessary software components to be used, such as standard peripheral drivers, CMSIS libraries, core configuration, and initialization functions of the STM32F4 Discovery kit. The basic configuration should be as in Figure A.3.

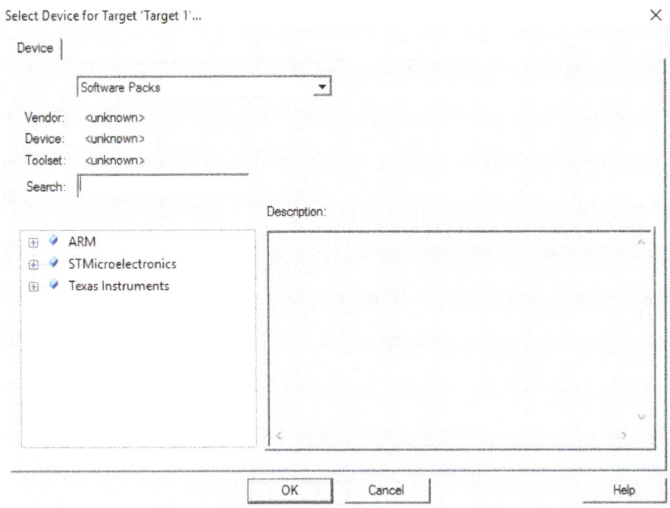

Figure A.2 Selecting the device for your project.

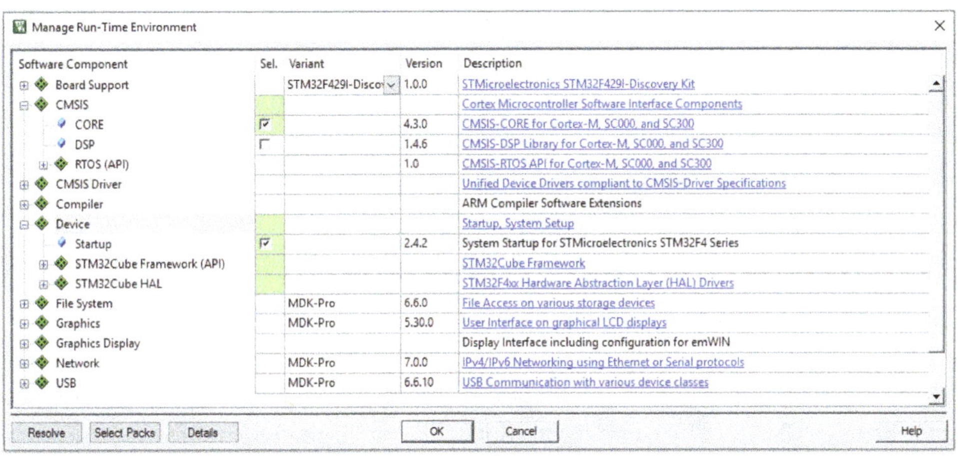

Figure A.3 Managing the run-time environment. The basic configuration.

In the Project window, the **Target 1** folder should have been created, including a subfolder named **Source Group 1**. We should add a main source file to the project. To do so, right click on the **Source Group 1** folder and select **"Add New Item to Group 'Source Group 1'...."** From the pop-up window, select **C File (.c)**. Give the file a name (such as `main.c`) and click **Add**. In this section, we will use the code given in Listing A.1 as an example. Open the `main.c` file by double clicking on it, and add your code to this file. Then, save it by clicking the **Save** button, which is in the upper-left corner of the menu.

```
int a[5]={1,2,3,4,5};
int b[5]={1,2,3,4,5};

int main(void) {
        int c[5];
        int cnt;

        for(cnt=0;cnt<5;cnt++)
        c[cnt]=a[cnt]+b[cnt];

        while(1);
        return(0);
}
```

Listing A.1 The first C code for the STM32F4 microcontroller.

A.3.2 Creating a Header File

A header file may be required for some projects. To add one to the project, right click on the **Source Group 1** folder and select **"Add New Item to Group 'Source Group 1'...."** From the pop-up window, choose **Header File (.h)**, give the file a name (such as `header.h`), and click **Add**. An empty window will appear for the ***header file***. Do not forget to add the line `#include` "`header.h`" to your main C code after including this file in your project.

> **header file**
> *A file that contains function and macro declarations shared among source files.*

A.3.3 Building and Loading the Project

In order to execute the code on a target device, we must first build it. Then, we must load it to the target device. The target device options must be configured before these steps. Click on the **Options for Target** button, which is located at the toolbar, or select it from the **Project** tab. From the **Target** tab, change the **Xtal (MHz)** value to `8 MHz` because the external ***crystal resonator*** in the STM32F4 Discovery kit has this value. Ensure that the operating system is set to **None**. Next, define the *Preprocessor Symbols* as `HSE_-VALUE=8000000` from the **C/C++** tab as in Figure A.4.

> **crystal resonator**
> *An oscillator circuit that uses the mechanical resonance of a vibrating crystal of piezoelectric material.*

Select either **Use Simulator** or a debugger tool from the **Debug** tab. Choose **ST-Link Debugger** from the list, as in Figure A.5, to run the code on the STM32F4 Discovery

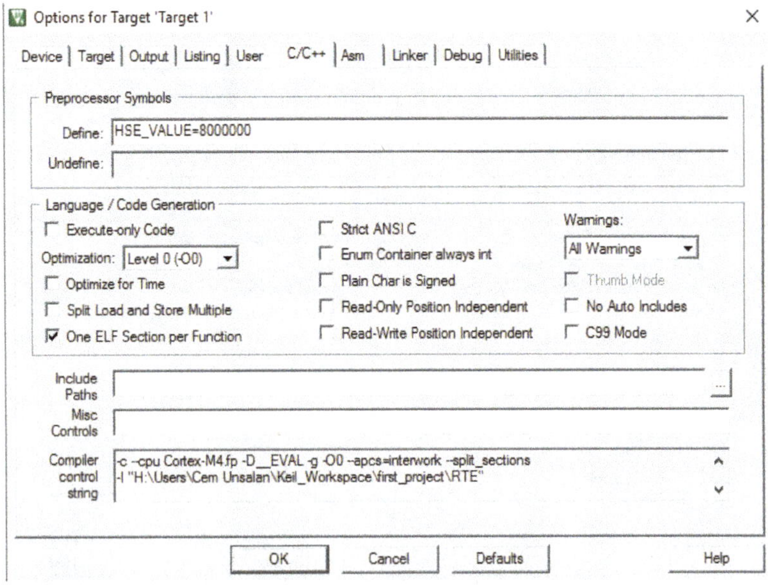

Figure A.4 Defining the HSE_VALUE preprocessor symbol.

kit. Then, click on the **Settings** button. Choose the *port* **SW** from the **Cortex-M Target Driver Setup** window and click **OK**. If you plan to use the simulator to *debug* your application, there are no additional settings. You can close the **Target Options Window** by clicking **OK**.

> **port**
> *An interface through which data can be passed into or out of a CPU or peripheral.*

> **debug**
> *The process of detecting and solving errors in software programs.*

Now, let us explain the build and load steps in detail. The first step in executing the code is building it. There is a **Build** button on Keil μVision's horizontal toolbar. It can also be reached from the **Project** tab. The main source code is linked to all other source and header files when the build button is clicked. Then, they are compiled. The running steps of the code (such as warnings and errors) can be observed in the **Build Output** window. Code sections with warnings and errors can be reached by double clicking on them as the code is built. Sometimes, one mistake can generate multiple errors. Double clicking on the error will direct you to an error free code line. In such cases, examine the code carefully to find the main source of the mistake.

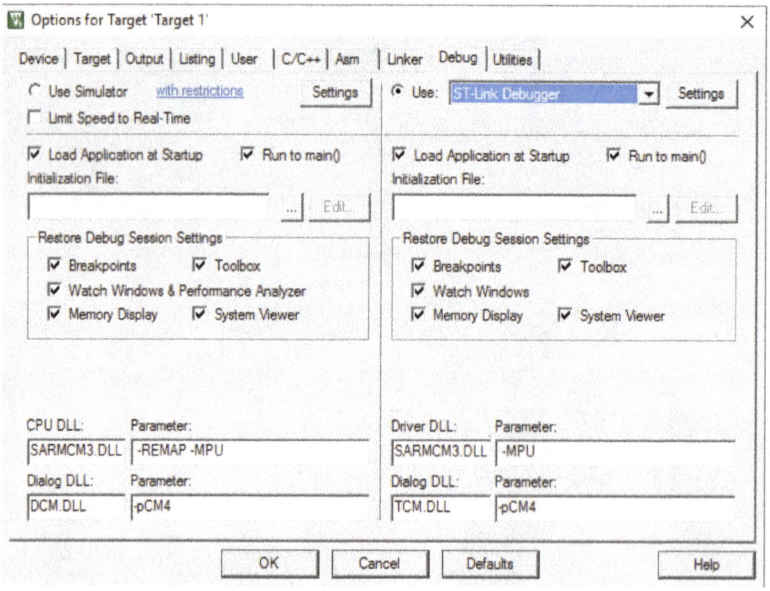

Figure A.5 Choosing the ST-Link/V2 debugger.

The second step in executing the code is loading it to the target device. There is a **Download** button on Keil μVision's horizontal toolbar. This can also be reached from the **Flash** tab. Load your application to the target device (the STM32F4 microcontroller in this case) by clicking this button. The STM32F4 Discovery kit must be connected to the host computer to perform this operation.

A.4 Program Execution

Once the main code is downloaded to the microcontroller, the next step is to execute it. This can be done by either pressing the reset button on the STM32F4 Discovery kit or by using the debug tool to follow the execution steps from Keil μVision. The debug tool can be accessed from the **Debug** tab. Click on the **Start/Stop Debug Session** button or use **Ctrl+F5** on the keyboard to start a debug session. There will be a warning that mentions the `32 KB` code size limitation. Click **OK** to skip this warning.

The code is executed up to the beginning of the main function. Then, the microcontroller waits for further commands. Buttons for program execution are located on the toolbar shown in Figure A.6.

The name of each button can be observed by hovering the cursor over it. These buttons and their functions are explained briefly below.

- **Reset**: Resets the target microcontroller. It works in a similar way to the reset pin. Device *registers* return to their default state when this button is clicked.

Figure A.6 The debug session menu.

- **Run**: Resumes execution of the code from the last location of the *Program Counter*. When this button is clicked, execution of the code continues until a breakpoint or a stop button press.
- **Stop**: Halts execution of the code. All windows used to observe software and hardware parts are updated with recent data.
- **Step**: Executes the next line of the code. If this line calls a subroutine, the compiler executes the next line in the subroutine. Then, it stops.
- **Step Over**: Executes the next line of the code. If this line calls a subroutine, the compiler executes the whole subroutine. Then, it stops.
- **Step Out**: Completes the execution of a subroutine.
- **Run to Cursor Line**: Executes the code until the line the cursor is positioned at.

> register
> *A memory location tightly coupled to the CPU. Typically a CPU will have a small number of registers and these will be accessible using specific instructions.*

> Program Counter
> *A CPU register that contains the memory location of the instruction being executed at the present time.*

A.4.1 Inserting a Break Point

Observing variables, registers, and memory is important while debugging the project. Code execution must halt for this operation. A breakpoint can be added to stop the execution on a specific line of code. To do so, right click on the desired code line and select **Insert/Remove Breakpoint**. The breakpoint buttons on the toolbar and double-clicking next to the code-line number are alternative ways to add breakpoints. When a breakpoint is added, a red circle will appear by the code line to indicate this. A breakpoint can be deleted by clicking on it.

A.4.2 Adding a Watch Expression

The **Watch** window (shown in Figure A.7) can be used to observe selected variables. In order to add a variable to this window, right click on the variable you wish to observe. Then, click **Add "variable" to**. Furthermore, the **Enter expression** button can be double-clicked in the **Watch** window and the name of the variable can be entered in the box that appears. Both global and *local variables* can be added to the **Watch** window.

> **local variable**
> *A variable that is accessible only from the function or block in which it is declared.*

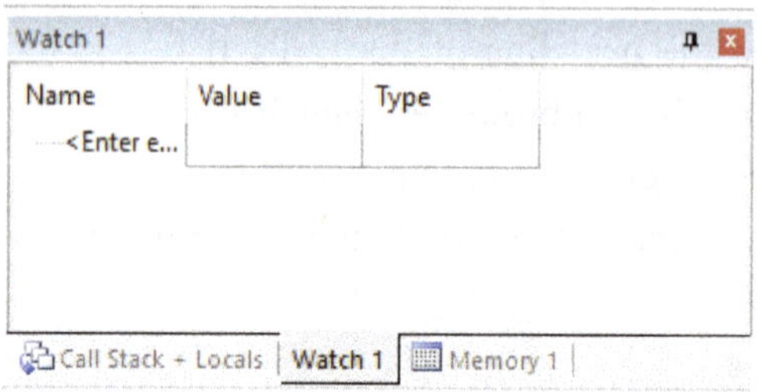

Figure A.7 The Watch window.

Local variables can also be observed in the **Call Stack + Locals** window as shown in Figure A.8. The **Watch** and **Call Stack + Locals** windows are only available in debug mode. If they are not shown by default, the user can enable them from the **View** tab.

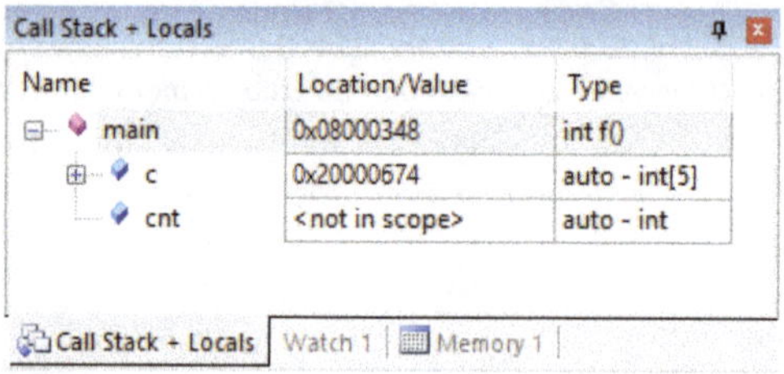

Figure A.8 Observing local variables in the Call stacks + Locals window.

A.4.3 Exporting Variables to MATLAB

You may wish to export variables from Keil μVision to some other applications. Because we are dealing with *digital signals*, exported variables will often be in the form of an *array*. This exported array can be analyzed in MATLAB. The method uses the debug function editor. To do so, you should first create a **.ini** file containing debug functions to save the target data. We provide such a file in Listing A.2.

> **digital signal**
> *A discrete-time signal comprising a sequence of numerical values from a finite, discrete set in a digital computer or digital hardware.*

> **array**
> *A data structure comprising a number of variables of the same type.*

```
FUNC void save_array() {
int array_length;
int idx;
array_length = 100;

  exec("log > Values.dat");
  for (idx = 0; idx < array_length; idx++) {
    printf ("%6.4f\n",array_name[idx]);
 // Users will need to change 'array_name' according to circumstances.
 // In this case it would be 'c'.
  }
  exec("log off");
 }
```

Listing A.2 The Export.ini file for exporting variables.

The main program must be run in debug mode in order to use this function. Then, it can be halted by clicking the **Stop** button. Next, you should click the **Function Editor (Open Ini File)** button in the **Debug** menu and select the **.ini** file. The setup should be as in Figure A.9.

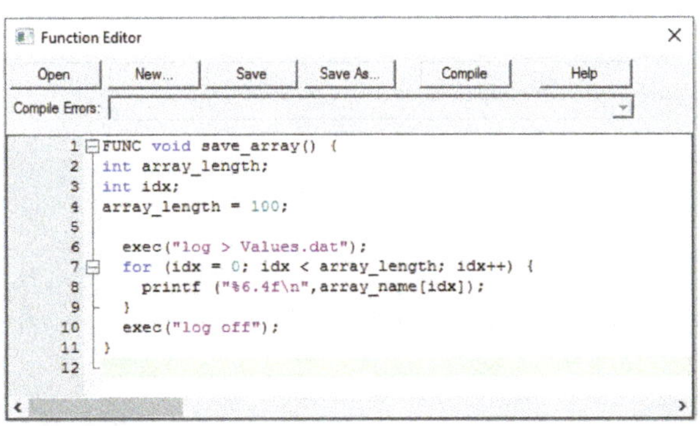

Figure A.9 Debug function editor window.

Please note that the debug function editor does not support functions that take arrays as input. Hence, the array name and its length should be set as the signal array name and length to be saved. Then, click the **Compile** button. Once the debug function is compiled without error, you can call this function (`save_array()` in this case) from the command window in Keil μVision. For the function given in Listing A.2, data are stored to the `Values.dat` file in the project directory. This file can be processed using software such as MATLAB.

A.4.4 Closing the Project

The project can be closed by clicking **Close Project** from the **Project** tab. This project can be re-opened by selecting **Open Project**, navigating to the project folder, and opening the project file (`.uvprojx` file).

A.5 Measuring the Execution Time

Measuring the execution time of a code block is crucial, especially in real-time applications. This information can be used to compare the performance of different problem-solving approaches. In general, execution time is measured in terms of the total number of clock cycles spent on executing a code block. The ***Data Watchpoint and Trace*** Unit (DWT) in Keil μVision is used for this purpose. This unit has a specific DWT cycle count register (`DWT_CYCCNT`), which holds the total number of clock cycles spent executing a code block. Normally, the core clock should be set for this purpose. However, this operation requires interaction with hardware, which we will discuss in Chapter 1. In this chapter, we will use the default 16-MHz internal ***RC oscillator*** as the core clock source. We will now introduce two methods for measuring the total number of clock cycles spent when executing the target code block.

> Data Watchpoint and Trace
> *Trace history of executed instructions and data for Arm Cortex-M devices.*

> oscillator
> *A circuit that produces a periodic, oscillating electronic signal, often a sine wave or a square wave.*

> **RC oscillator**
> *An oscillator circuit that produces a periodic signal with its frequency determined by resistance and capacitance.*

A.5.1 Using the DWT_CYCCNT Register in the Code

The first method for measuring the execution time involves using the `DWT_CYCCNT` register in the code. To do this, we should first enable the **DWT** unit by setting the `TRCENA` bit of the debug exception and monitor control register, (`DEMCR`). This is a 32-bit register (`TRCENA` being its 24th bit). Then, we should enable the cycle ***counter*** by setting the `CYCCNTENA` bit of the DWT control register, `DWT_CTRL`. This is a 32-bit register (`CYCCNTENA` being its zeroth bit). After making these adjustments, we can then use the `DWT_CYCCNT` register. To do so, we should reset it (by setting it to 0) at the beginning of the code block. Then, the value in the `DWT_CYCCNT` register can be copied to a variable at the end of the code block. This value will be the total number of clock cycles spent when executing the code block.

> **counter**
> *A device, module or variable that stores the number of times a particular event or process has occurred.*

To simplify these operations, we provide the header file `time.h` in `Online_Student_Resources\Lab0`. This header file should be included in the project. There are three functions in the `time.h` header file. These are `StartTiming`, `StopTiming`, and `CalculateTime`. The `StartTiming` function enables the DWT unit and then enables and resets the cycle counter. This function must be called before the code block that is to be measured. The `StopTiming` function reads the value of the cycle counter and gives it as an output. This function must be called after the code that is to be measured. The output of this function should be written to a variable as well. The `CalculateTime` function can be used to calculate the execution time in seconds using clock cycles and core clock frequency. This function must be called after the `StopTiming` function. Its output should also be written to a variable.

We provide a sample project on the usage of the `time.h` header file in `Online_Student_Resources\Lab0\Measure_Execution_Time`. You can add the code block to be measured in the indicated point in the main file of the project. Next, add a `for` loop with 2000000 iterations in the indicated place in the main file. Observe the number of clock cycles and execution time.

A.5.2 Observing the DWT_CYCCNT Register

The second method for measuring execution time involves observing the `DWT_CYCCNT` register from the **Registers** window shown in Figure A.10. This window can be accessed in the Debug session by clicking the **View and Registers** window.

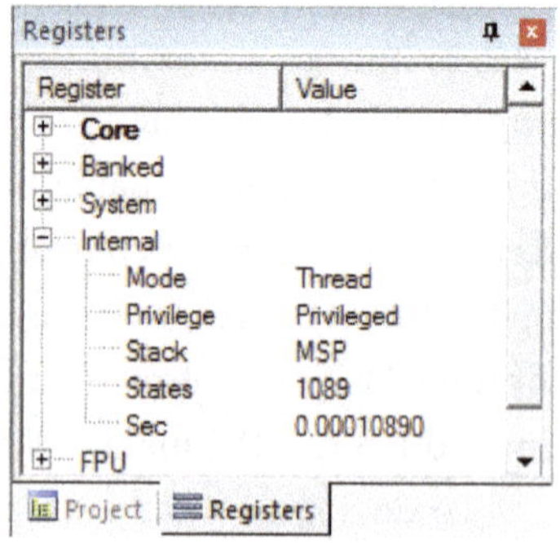

Figure A.10 The Registers window.

The **States** register represents the target cycle counter in the Registers window. This counter holds the total number of clock cycles spent from the beginning of the debug operation. The **Sec** register represents the elapsed time in seconds from the beginning of the debug operation. This value is updated based on the States register. Before using the Sec register, you should go to the **Trace** window (1. Stop the current debug session 2. Go to "Project"->"Options for Target"->"Target 1"->"Debug"->"Settings" on the right-hand side->Trace) and enter the core clock frequency of the microcontroller in the **Core Clock** box as in Figure A.11.

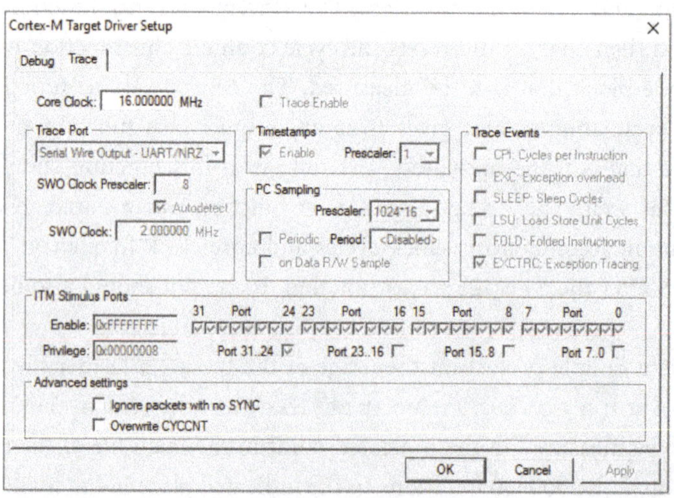

Figure A.11 The Trace window.

We will employ breakpoints in order to use the States and Sec registers to measure the execution time. We place a breakpoint at both the start and end of the code block to be measured. As we run the code, we note the values of the States and Sec registers at each breakpoint. The difference between these values will give the total number of clock cycles spent and the execution time in seconds for the target code block.

We can repeat the test given in the previous section to measure the execution time of the `for` loop with 2000000 iterations. You should obtain values of 36000014 clock cycles and 2.25000087 s between breakpoints. As can be seen here, the value observed with the second method is nine clock cycles less compared to the first method. This difference occurs while writing the number of total clock cycles to the variable in the first method.

A.6 Measuring Memory Usage

Microcontrollers have hard constraints in memory space. Therefore, the memory usage of a given code block may be crucial in some applications. We can use the **Build Output** window (under **View**) for this purpose. The program size is given in this window, as in Figure A.12, when the program is built.

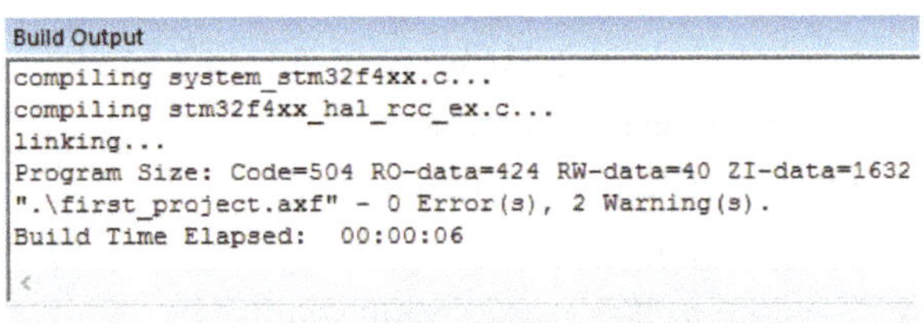

Figure A.12 The Build Output window for measuring memory usage.

As can be seen in Figure A.12, there are four parts to the program size line, namely `Code`, `RO-data` (Read Only), `RW-data` (Read Write), and `ZI-data` (Zero Initials). To find the total ROM usage, we sum the `Code`, `RO-data`, and `RW-data` values. We should sum `ZI-data` and `RW-data` to find the RAM usage in the code. These values can also be obtained from the `.map` file that is automatically generated in the project folder when it is built. This file contains information about cross references, input modules, memory map, program size, etc. Here, we will only use the part where the program size information is provided. In the `.map` file, the `Total RW Size` line indicates the RAM usage, and the `Total ROM Size` line indicates the ROM usage.

Let us consider the sample code given in Listing A.1. As the code is built, we obtain `Code`=504, `RO-data`=424, `RW-data`=40, and `ZI-data`=1632 from the build window. Therefore, the total ROM and RAM usages are 968 and 1672 bytes, respectively. We can obtain the same data from the `.map` file given in Listing A.3.

```
Total RO Size (Code + RO Data) 928 (0.91kB)
Total RW Size (RW Data + ZI Data) 1672 (1.63kB)
Total ROM Size (Code + RO Data + RW Data) 968 (0.95kB)
```

Listing A.3 The .map file.

A.7 CMSIS

The Cortex Microcontroller Software Interface Standard (CMSIS) is a vendor independent *hardware abstraction layer* for Cortex-M processors. As of the writing of this book, the version of CMSIS is 4.5.

> **hardware abstraction layer**
> *A low level software program that allows a high level program to interact with a hardware device at a general or abstract level, rather than at a detailed hardware level.*

A.7.1 CMSIS Components

CMSIS has seven components, which are as follows.

CMSIS-CORE: Interface to the processor cores and peripherals.
CMSIS-Driver: Interface to generic peripherals.
CMSIS-DSP: DSP library with over 60 functions.
CMSIS-RTOS API: Common *API* for *real-time operating systems*.
CMSIS-Pack: *XML*-based package description standard.
CMSIS-SVD: System view description for peripherals.
CMSIS-DAP: Interface to debug unit.

> **XML (Extensible Markup Language)**
> *A software and hardware independent file format for storing and transporting data.*

> **API (application programming interface)**
> *A set of subroutine definitions, protocols, and tools for building software and applications.*

> **real-time operating system**
> *An operating system designed for real-time systems.*

We use only the CMSIS-DSP library in this lab, so we will not focus on the remaining components.

A.7.2 CMSIS-DSP Library

We extensively use the CMSIS-DSP library throughout this lab. This library includes functions for vector operations, matrix computation, complex arithmetic, ***filter*** functions, control functions, ***proportional-integral-derivative (PID) controller***, ***Fourier transform***, and other frequently used DSP algorithms. Most of these functions can be used on both ***floating-point*** and fixed-point number representations. Moreover, the functions are optimized for Cortex-M series microcontrollers. Specifically, the CMSIS-DSP library contains the following functions that work with 8-, 16-, and 32-bit integers and 32-bit floating-point numbers.

> **filter**
> *A system. In the context of signal processing, it is conventional to refer to systems as filters and the frequency response of a filter is often of primary concern.*

> **proportional-integral-derivative controller**
> *A feedback controller that calculates error and applies correction based on proportional, integral, and derivative parameters.*

> **Fourier transform**
> *A transformation between time and frequency domains.*

> **floating-point**
> *A representation of a number with a changing number of digits after the decimal point according to range and precision.*

- **Basic Math Functions:** vector absolute value, vector addition, vector ***dot product***, vector multiplication, vector negate, vector offset, vector scale, vector shift, vector subtraction.
- **Controller Functions:** sine cosine, PID motor control, vector Clarke transform, vector inverse Clarke transform, vector Park transform, vector inverse Park transform.

- **Complex Math Functions:** *complex conjugate*, complex dot product, complex magnitude, complex magnitude squared, complex-by-complex multiplication, complex-by-real multiplication.
- **Fast Math Functions:** cosine, sine, square root.
- **Filtering Functions:** high precision Q1.31 biquad cascade filter, biquad cascade IIR filters using direct form I structure, biquad cascade IIR filters using direct form II transposed structure, convolution, partial convolution, correlation, finite impulse response (FIR) decimator, FIR filters, FIR lattice filters, FIR sparse filters, infinite impulse response (IIR) lattice filters, least mean square (LMS) filters, normalized LMS filters, FIR interpolator.
- **Interpolation Functions:** linear interpolation, bilinear interpolation.
- **Matrix Functions:** matrix addition, complex matrix multiplication, matrix initialization, matrix inverse, matrix multiplication, matrix scale, matrix subtraction, matrix transpose.
- **Statistics Functions:** maximum, mean, minimum, power, root mean square (RMS), standard deviation, variance.
- **Support Functions:** vector copy, vector fill, convert 32-bit floating-point value, convert 16-bit integer value, convert 32-bit integer value, convert 8-bit integer value.
- **Transform Functions:** complex FFT functions, radix-8 complex FFT functions, DCT type IV functions, real FFT functions, complex FFT tables.

> **dot product**
> *A scalar or inner product of equal length of sequence.*

> **complex conjugate**
> *A pair of complex numbers with identical real parts and opposite sign imaginary parts.*

A.7.3 Using the CMSIS-DSP Library with Keil μVision

To create a Keil μVision project using the CMSIS-DSP library, you must perform some configuration steps in addition to those described in Section A.3. First, open the **Manage Run-Time Environment** window. The basic configuration should be as in Figure A.3. In addition, select the **DSP** check-box under **CMSIS**.

There are some pre-processor macros used in CMSIS-DSP library functions. You should choose and define these according to hardware specifications. We summarize these macros below.

- **__FPU_PRESENT:** This macro must be defined if the target device has a *floating-point unit*.
- **Arm_MATH_CMx:** This macro must be defined according to the Cortex-M version of the processor. Define `Arm_MATH_CM4` for Cortex-M4 devices, `Arm_MATH_CM3` for Cortex-M3 devices, `Arm_MATH_CM0` for Cortex-M0 devices, and `Arm_MATH_-CM0PLUS` for Cortex-M0+ devices.
- **UNALIGNED_SUPPORT_DISABLE:** This macro must be defined if the silicon does not support unaligned memory access.
- **Arm_MATH_BIG_ENDIAN:** This macro must be defined for processors using the big-endian format.
- **Arm_MATH_MATRIX_CHECK:** This macro must be defined if the user wants to check the input and output sizes of matrices.
- **Arm_MATH_ROUNDING:** This macro must be defined if the user wants rounding to be used on support functions.

> **floating-point unit**
> *A processor structure that manipulates floating numbers more quickly than basic microprocessor circuitry does.*

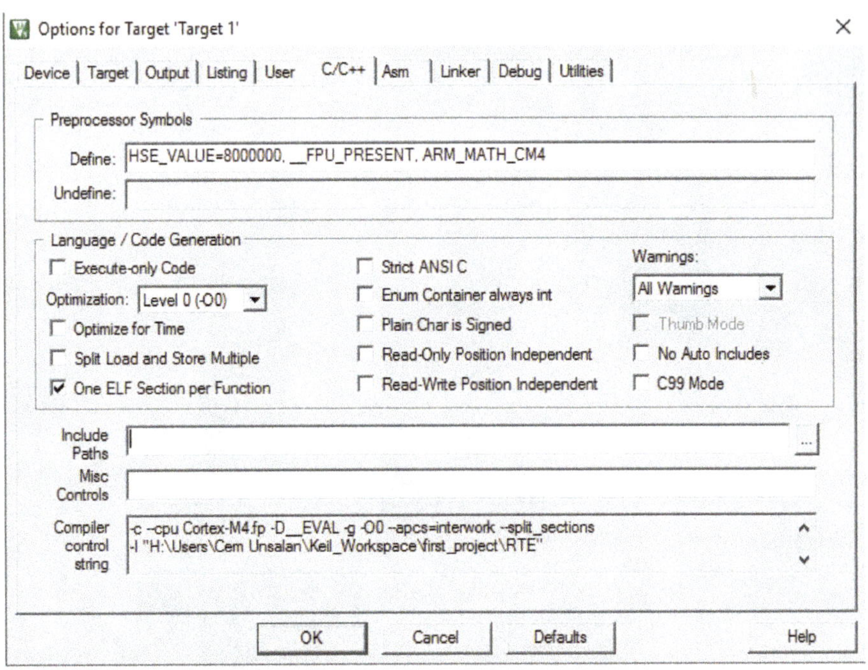

Figure A.13 CMSIS-DSP library definitions in the Options for Target window.

For the STM32F4 Discovery kit, open the **Options for Target** window. Then, go to the **C/C++** tab. In this tab, add constants `__FPU_PRESENT` and `Arm_MATH_CM4` to definitions as in Figure A.13. Finally, add the `\#include` "`arm_math.h`" line to the beginning of the `main.c` file. If complex FFT functions will be used, add the `\#include` "`arm_const_structs.h`" line to your code as well.

List of Symbols

n	Discrete-time signal time index
t	Continuous-time signal time index
$\delta[n]$	Discrete-time unit pulse signal
$\delta(t)$	Continuous-time unit pulse signal
$u[n]$	Discrete-time unit step signal
$u(t)$	Continuous-time unit step signal
$r[n]$	Discrete-time unit ramp signal
$e[n]$	Discrete-time exponential signal
$s[n]$	Discrete-time sinusoidal signal
ϵ	An arbitrarily small quantity
ω	Angular frequency for discrete-time signal in radians
Ω	Angular frequency for continuous-time signal in radians/second
π	Pi number
PI	Pi number constant
e	Euler's number, which denotes the natural logarithm base, 2.71828...
j	Square root of -1
$x[n]$	Discrete-time input signal
$y[n]$	Discrete-time output signal
$H\{\}$	Discrete-time system
$h[n]$	Impulse response of a discrete-time system
β	Discrete-time system feedforward constant coefficient
α	Discrete-time system feedback constant coefficient
γ	Discrete-time system coefficient
$*$	Convolution sum operation

$\circledast$	Circular convolution sum operation
$X(z)$	The Z-transform of a discrete-time input signal $x[n]$
$Y(z)$	The Z-transform of a discrete-time output signal $x[n]$
$H(z)$	Transfer function of a discrete-time system
z	Complex variable represented in polar form as $z = re^{j\theta}$. r is the radius term. θ is the angle term having a value between $[0, 2\pi]$ radians.
z_i	Zeros of a system
p_i	Poles of a system
$X(j\Omega)$	Fourier transform (FT) of a continuous-time signal $x(t)$
$X(e^{jw})$	Discrete-time Fourier transform (DTFT) of a discrete-time input signal $x[n]$
$Y(e^{jw})$	Discrete-time Fourier transform (DTFT) of a discrete-time output signal $y[n]$
$H(e^{jw})$	Frequency response of a discrete-time system
$X[k]$	Discrete Fourier transform (DFT) of a discrete-time input signal $x[n]$
$\widetilde{X}[k]$	Discrete-time Fourier series expansion (DTFSE) coefficients of a discrete-time input signal $x[n]$
$v[n]$	Discrete-time window function
$X_{ST}[k, n]$	Short-time Fourier transform (STFT) of a discrete-time input signal $x[n]$
T_S	Sampling period in seconds
f_S	Sampling frequency in Hertz
Ω_S	Sampling frequency in radians/second
Ω_M	Maximum angular frequency for continuous-time signal in radians/second
V_{in}	Analog voltage value of input signal
ω_S	Stop band frequency of a filter in radians
ω_S	Pass band frequency of a filter in radians
Ω_C	Cut-off frequency in radians/second
J	Cost function of an adaptive filter
μ	Step size of an adaptive filter
∂	Derivative operation

$d[n]$	Discrete-time desired output signal
$P[n]$	Normalization coefficient
τ	Convergence time of an adaptive filter
S	Sign bit of a number in float point representation
E	Exponent value of a number in float point representation
F	Fractional part of a number in float point representation
UQ	Fixed-point unsigned number representation
$T_{sampling}$	Sampling period in seconds

G

Glossary

accelerometer A device that measures change in velocity over time.

advanced high-performance bus An on-chip bus specification to connect and manage high clock frequency system modules in embedded systems.

advanced peripheral bus An on-chip bus specification with reduced power and interface complexity to connect and manage high clock frequency system modules in embedded systems.

aliasing The distortion of a sampled signal resulting from sampling a signal at a frequency that is too low.

amplitude The maximum deviation of a quantity from a reference level.

analog signal A continuous-time signal.

analog to digital converter A device that samples and quantizes an analog input signal to form a corresponding/representative digital signal.

API (application programming interface) A set of subroutine definitions, protocols, and tools for building software and applications.

array A data structure comprising a number of variables of the same type.

bandlimited signal A signal with power spectral density that is bound within a limited range of frequencies.

bit reversal Reverse the order of the bits in a digital word.

buffer A circuit in which data are stored while it is being processed or transferred.

capacitor A passive electrical component that is designed to accumulate a charge and store electrical energy.

center frequency The measure of a central frequency, between the upper and lower cut-off frequencies.

chirp signal A chirp is a signal in which the frequency increases (up-chirp) or decreases (down-chirp) over time.

codec (coder-decoder) A device or computer program that allows the encoding or decoding of a digital data stream or signal.

compiler A software program that translates codes written in a particular programming language into machine language.

complex conjugate A pair of complex numbers with identical real parts and opposite sign imaginary parts.

complex plane A geometrical representation of complex numbers established using a real axis and a perpendicular imaginary axis.

controller area network A vehicle bus standard designed to allow microcontrollers and devices to communicate with each other, in applications, without a host computer.

convergence The property of approaching a limit value.

counter A device, module, or variable that stores the number of times a particular event or process has occurred.

crystal resonator An oscillator circuit that uses the mechanical resonance of a vibrating crystal of piezoelectric material.

cut-off frequency The frequency in which the output power of the system decreases to a given proportion of the power in the passband.

data buffer A region of physical memory storage used to temporarily store data while they are being moved from one place to another.

Data Watchpoint and Trace Trace history of executed instructions and data for Arm Cortex-M devices.

debug The process of detecting and solving errors in software programs.

debugger A software program that allows users to detect errors in other software programs.

decibel A logarithmic unit of the ratio between the power of two signals.

decode The second step of a CPU operation cycle, in which the instruction inside the instruction register is decoded.

difference equation A difference equation characterizes a discrete-time system in terms of its input and output sequences.

digital signal A discrete-time signal comprising a sequence of numerical values from a finite, discrete set in a digital computer or digital hardware.

digital to analog converter A device that converts a digital value to its corresponding analog value (e.g., voltage).

direct memory access A mechanism whereby data may be transferred from one memory location to another (including memory-mapped peripheral interfaces) without loading, or independently of, the CPU.

DMIPS (Dhrystone Million Instructions per Second) A measure of a computer's processor speed.

dot product A scalar or inner product of equal length of sequence.

double precision A computer number format that occupies 8 bytes (64 bits) in computer memory and represents a wide, dynamic range of values by using a floating point.

driver A program that controls the interface between a computer and a peripheral device.

Euler's formula Also called Euler's expansion formula. $e^{jx} = cos(x) + jsin(x)$, where e is the base of the natural logarithm and j is the imaginary unit.

execute The last step of a CPU operation cycle in which the CPU carries out the decoded information.

exponential A function that raises a given constant to the power of its argument.

feedforward A non-recursive path from the input to the output of a system.

fetch The first step of a CPU operation cycle. The next instruction is fetched from the memory address that is currently stored in the program counter (PC) and stored in the instruction register (IR).

filter A system. In the context of signal processing, it is conventional to refer to systems as filters and the frequency response of a filter is often of primary concern.

First In First Out FIFO is a method for organizing and manipulating a data buffer, where the oldest (first) entry, or "head" of the queue, is processed first.

float The keyword for a floating-point data type with single precision.

floating input An input pin with no signal source or termination connected.

floating-point A representation of a number with a changing number of digits after the decimal point according to range and precision.

floating-point unit A processor structure that manipulates floating numbers more quickly than basic microprocessor circuitry does.

floor function A function that maps a real number to the largest previous or the smallest following integer.

hardware abstraction layer A low level software program that allows a high level program to interact with a hardware device at a general or abstract level, rather than at a detailed hardware level.

header file A file that contains function and macro declarations shared among source files.

highpass filter A filter that attenuates the low frequency components of a signal.

impulse train An input signal that consists of an infinite series of impulse units equally separated in time.

input/output A circuit in an embedded system that connects the system to the external world.

integrated development environment A software application that combines the tools for software development.

inter-IC sound A serial communication bus interface designed to connect digital audio devices.

interrupt A signal to the processor emitted by hardware or software to indicate an event that needs immediate attention.

interrupt flag A register bit used for indicating related interrupt status.

Fourier transform A transformation between time and frequency domains.

Laplace transform A transformation method used to map a continuous function of time to a continuous function of complex frequency.

local variable A variable that is accessible only from the function or block in which it is declared.

lowpass filter A filter that attenuates the high frequency components of a signal.

magnitude response The measure of the output magnitude of a system or device in response to an input stimulus.

memory protection unit The hardware in a computer that controls memory access rights.

MEMS (Micro-Electro-Mechanical Systems) Microscopic mechanical or electro-mechanical devices.

multimedia card A memory card standard used for solid-state storage.

multiply and accumulate unit A microprocessor circuit that carries a multiplication operation followed by accumulation.

natural frequency The frequency at which a system tends to oscillate on its own.

nested vectored interrupt controller (NVIC) ARM-based interrupt handler hardware.

normalized angular frequency f/f_s, where f is the angular frequency (rad/sec), which is the phase change of a sinusoidal signal per unit of time, and f_s is the sampling frequency (samples/sec). It has the unit of rad/sample.

on-the-go A USB specification that allows a USB device to act as a host, allowing other USB devices to connect to themselves. Also called USB on-the-go.

open-drain An output pin driven by a transistor, that pulls the pin to only one voltage.

oscillator A circuit that produces a periodic, oscillating electronic signal, often a sine wave or a square wave.

oscilloscope A laboratory instrument commonly used to display and analyze the waveform of electronic signals.

ping–pong buffer A double buffering technique in which one buffer is being read while the other continues to receive data.

pointer A data type in a programming language that contains the address information of another value.

port An interface through which data can be passed into or out of a CPU or peripheral.

printed circuit board A board made of fiberglass, composite epoxy, or other laminate material and used for connecting different electrical components via etched or printed pathways.

Program Counter A CPU register that contains the memory location of the instruction being executed at the present time.

proportional–integral–derivative controller A feedback controller that calculates error and applies correction based on proportional, integral, and derivative parameters.

pull-down resistor A pull-down resistor (to a negative power supply voltage) ensures that a signal conductor adopts a valid (electrical) logic level (low) in the absence of any other connection.

pull-up resistor A pull-up resistor (to a positive power supply voltage) ensures that a signal conductor adopts a valid (electrical) logic level (high) in the absence of any other connection.

pulse width modulation A modulation technique that generates variable-width pulses to represent the amplitude of an analog input signal.

quantization The process of mapping sampled analog data into non-overlapping discrete subranges.

Reduced Instruction Set Computing (RISC) A microprocessor design strategy based on the simplified instruction set.

real-time operating system An operating system designed for real-time systems.

register A memory location tightly coupled to the CPU. Typically a CPU will have a small number of registers and these will be accessible using specific instructions.

RC oscillator An oscillator circuit that produces a periodic signal with its frequency determined by resistance and capacitance.

resistor A passive electrical component designed to implement electrical resistance as a circuit element.

sampling The process of obtaining values at specific time intervals.

sampling frequency The reciprocal of the time delay between successive samples in a discrete-time signal.

secure digital input output A circuit that allows the sending of data to external devices using Secure Digital (SD) specification.

serial peripheral interface A serial communication bus used to send data, with high speed, between microcontrollers and small peripherals.

sidelobes In a frequency graph, the lobes that represent secondary frequencies.

signal generator A laboratory instrument commonly used to generate repeating or non-repeating electronic signals in either the analog or the digital domain.

single precision A computer number format that occupies 4 bytes (32 bits) in computer memory and represents a wide dynamic range of values by using a floating point.

spectrogram The visual representation of the spectrum of frequencies in a signal.

stack A data structure in which items are removed in the reverse order from that in which they are added, so that the most recently added item is the first one removed.

static random access memory Static, as opposed to dynamic, RAM retains its data for as long as its power supply is maintained.

step invariance A transformation method in which both continuous and discrete-time filters have the same step response at the sampling instants.

twiddle factor Trigonometric constant coefficients multiplied by data in the course of FFT algorithm.

universal asynchronous receiver/transmitter A serial communication bus commonly used to send data, asynchronously, between microcontrollers and small peripherals.

universal synchronous/asynchronous receiver/transmitter A serial communication bus commonly used to send data, both synchronously and asynchronously, between micro-controllers and small peripherals.

voltage offset A DC voltage value added to an AC waveform.

wake-up interrupt controller A peripheral that can detect an interrupt and wake a processor from deep sleep mode.

watchdog A timer in an embedded system that is used to detect and recover from malfunctions.

XML (Extensible Markup Language) A software and hardware independent file format for storing and transporting data.

References

Arm, *Arm University Program RTOS Education Kit*, available via www.arm.com/education.

Boroujeny, B.F. (2013) *Adaptive Filters: Theory and Applications*, Wiley, Second edn.

Diniz, P.S.R. (2013) *Adaptive Filtering: Algorithms and Practical Implementation*, Springer, Fourth edn.

Haykin, S. (2014) *Adaptive Filter Theory*, Pearson, Fifth edn.

Kehtarnavaz, N. (2004) *Real-Time Digital Signal Processing Based on the TMS320C600*, Elsevier.

Kester, W. (2004) *The Data Conversion Handbook*, Elsevier.

Kuo, S.M., Lee, B.H., and Tian, W. (2013) *Real-Time Digital Signal Processing: Fundamentals, Implementations and Applications*, Wiley, Third edn.

Manolakis, D.G. and Ingle, V.K. (2011) *Applied Digital Signal Processing*, Cambridge University Press.

McClellan, J.H., Schafer, R.W., and Yoder, M.A. (2003) *Signal Processing First*, Prentice-Hall.

Mitra, S.K. (2010) *Digital Signal Processing*, McGraw-Hill, Fourth edn.

Nilsson, J.W. and Riedel, S.A. (2014) *Electric Circuits*, Pearson, Tenth edn.

Ogata, K. (1987) *Discrete-Time Control Systems*, Prentice Hall.

Oppenheim, A.W. and Schafer, R.W. (2009) *Discrete-Time Signal Processing*, Prentice Hall, Third edn.

Oppenheim, A.W., Willsky, A.S., and Nawab, S.H. (1997) *Signals and Systems*, Prentice Hall, Second edn.

Orfanidis, S. (1995) *Introduction to Signal Processing*, Prentice-Hall.

Poularikas, A.D. (2014) *Adaptive Filtering: Fundamentals of Least Mean Squares with MATLAB*, CRC Press.

Proakis, J.G. and Manolakis, D.K. (1995) *Digital Signal Processing: Principles, Algorithms and Applications*, Prentice Hall, Third edn.

Sayed, A.H. (2008) *Adaptive Filters*, Wiley.

Singh, A. and Srinivasan, S. (2003) *Digital Signal Processing Implementations: Using DSP Microprocessors*, Cengage Learning.

Smith, S.W. (1997) *The Scientist and Engineer's Guide to Digital Signal Processing*, California Technical Pub.

Ünsalan, C. and Gürhan, H.D. (2013) *Programmable Microcontrollers with Applications: MSP430 LaunchPad with CCS and Grace*, McGraw-Hill.

Welch, T.B., Wright, C.H.G., and Morrow, M.G. (2011) *Real-Time Digital Signal Processing from MATLAB to C with the TMS320C6x DSPs*, CRC Press, Second edn.

Widrow, B. and Stearns, S.D. (1985) *Adaptive Signal Processing*, Prentice-Hall.

Index

CPSIA information can be obtained
at www.ICGtesting.com
Printed in the USA
BVHW010227301019
562246BV00064B/1073/P

9 781911 531166